T0326349

Abiotic and Biotic Stresses in Soybean Production

Abiotic and Biotic Stresses in Soybean Production

Soybean Production
Volume 1

Edited by

Dr. Mohammad Miransari
**Department of Book & Article,
AbtinBerkeh Ltd. Company, Isfahan, Iran**

AMSTERDAM • BOSTON • HEIDELBERG • LONDON
NEW YORK • OXFORD • PARIS • SAN DIEGO
SAN FRANCISCO • SINGAPORE • SYDNEY • TOKYO
Academic Press is an Imprint of Elsevier

Academic Press is an imprint of Elsevier
125, London Wall, EC2Y 5AS, UK
525 B Street, Suite 1800, San Diego, CA 92101-4495, USA
225 Wyman Street, Waltham, MA 02451, USA
The Boulevard, Langford Lane, Kidlington, Oxford OX5 1GB, UK

Notices
Knowledge and best practice in this field are constantly changing. As new research and experience broaden our understanding, changes in research methods, professional practices, or medical treatment may become necessary.

Practitioners and researchers must always rely on their own experience and knowledge in evaluating and using any information, methods, compounds, or experiments described herein. In using such information or methods they should be mindful of their own safety and the safety of others, including parties for whom they have a professional responsibility.

To the fullest extent of the law, neither the Publisher nor the authors, contributors, or editors, assume any liability for any injury and/or damage to persons or property as a matter of products liability, negligence or otherwise, or from any use or operation of any methods, products, instructions, or ideas contained in the material herein.

British Library Cataloguing-in-Publication Data
A catalogue record for this book is available from the British Library

Library of Congress Cataloging-in-Publication Data
A catalog record for this book is available from the Library of Congress

ISBN: 978-0-12-801536-0

For information on all Academic Press publications
visit our website at http://store.elsevier.com/

  Working together
to grow libraries in
developing countries

www.elsevier.com • www.bookaid.org

Publisher: Nikki Levy
Acquisition Editor: Nancy Maragioglio
Editorial Project Manager: Billie Jean Fernandez
Production Project Manager: Nicky Carter
Designer: Maria Ines Cruz

Typeset by Thomson Digital
Printed and bound in the USA

Dedication

This book is dedicated to my parents,
my wife, and my two children,
who have supported me all the time.

Contents

List of contributors

Naveen Kumar Arora Department of Environmental Microbiology, Babasaheb Bhimrao Ambedkar University, Lucknow, Uttar Pradesh, India

Vineetha Mariam Cherian Department of Biological Sciences, Faculty of Science, Kuwait University, Safat, Kuwait

Narjes H. Dashti Department of Biological Sciences, Faculty of Science, Kuwait University, Safat, Kuwait

Masayuki Fujita Laboratory of Plant Stress Responses, Department of Applied Biological Science, Faculty of Agriculture, Kagawa University, Miki-cho, Kita-gun, Kagawa, Japan

Shivani Garg ICAR-Directorate of Soybean Research, Khandwa Road, Indore, Madhya Pradesh, India

Priyanka Gupta Stress Physiology and Molecular Biology Laboratory, School of Life Sciences, Jawaharlal Nehru University, New Delhi, India

Mirza Hasanuzzaman Department of Agronomy, Faculty of Agriculture, Sher-e-Bangla Agricultural University, Dhaka, Bangladesh

Hon-Ming Lam Centre for Soybean Research of the Partner State Key Laboratory of Agrobiotechnology and School of Life Sciences, The Chinese University of Hong Kong, Shatin, Hong Kong

Man-Wah Li Centre for Soybean Research of the Partner State Key Laboratory of Agrobiotechnology and School of Life Sciences, The Chinese University of Hong Kong, Shatin, Hong Kong

Mohammad Miransari Department of Book & Article, AbtinBerkeh Ltd. Company, Isfahan, Iran

Nacira Muñoz Centre for Soybean Research of the Partner State Key Laboratory of Agrobiotechnology and School of Life Sciences, The Chinese University of Hong Kong, Shatin, Hong Kong; Instituto de Fisiología y Recursos Genéticos Vegetales, Centro de Investigaciones Agropecuarias – INTA, Córdoba, Argentina; Cátedra de Fisiología Vegetal, FCEFyN – UNC, Córdoba, Argentina

Kamrun Nahar Laboratory of Plant Stress Responses, Department of Applied Biological Science, Faculty of Agriculture, Kagawa University, Miki-cho, Kita-gun, Kagawa, Japan; Department of Agricultural Botany, Faculty of Agriculture, Sher-e-Bangla Agricultural University, Dhaka, Bangladesh

Sai-Ming Ngai Centre for Soybean Research of the Partner State Key Laboratory of Agrobiotechnology and School of Life Sciences, The Chinese University of Hong Kong, Shatin, Hong Kong

Marcela Claudia Pagano Federal University of Minas Gerais, Belo Horizonte, Brazil

Ashwani Pareek Stress Physiology and Molecular Biology Laboratory, School of Life Sciences, Jawaharlal Nehru University, New Delhi, India

Mahmood-ur-Rahman Department of Bioinformatics and Biotechnology, Government College University, Faisalabad, Pakistan

Mehboob-ur-Rahman Plant Genomics and Molecular Breeding Laboratory, Agricultural Biotechnology Division, National Institute of Biotechnology and Genetic Engineering, Faisalabad, Pakistan

Gyanesh K. Satpute ICAR-Directorate of Soybean Research, Khandwa Road, Indore, Madhya Pradesh, India

Tayyaba Shaheen Department of Bioinformatics and Biotechnology, Government College University, Faisalabad, Pakistan

Muhammad Shahid Riaz Department of Bioinformatics and Biotechnology, Government College University, Faisalabad, Pakistan

Mahaveer P. Sharma ICAR-Directorate of Soybean Research, Khandwa Road, Indore, Madhya Pradesh, India

Manoj K. Sharma School of Biotechnology, Jawaharlal Nehru University, New Delhi, India

Rita Sharma Stress Physiology and Molecular Biology Laboratory, School of Life Sciences, Jawaharlal Nehru University, New Delhi, India

Sneh L. Singla-Pareek Plant Molecular Biology Group, International Centre for Genetic Engineering and Biotechnology, New Delhi, India

Donald L. Smith James McGill Professor Department of Plant Science, McGill University, Montréal, Canada

Sakshi Tewari Department of Environmental Microbiology, Babasaheb Bhimrao Ambedkar University, Lucknow, Uttar Pradesh, India

Yusuf Zafar Minister Technical, Permanent Mission of Pakistan to IAEA, Vienna, Austria

Foreword

The important role of academicians and researchers is to feed the world's increasing publication. However, such contributions must be directed using suitable and useful resources. My decision to write this book as well as other books and contributions, including research articles, has been mainly due to the duties I feel toward the people of the world. Hence, I have tried to prepare references that can be of use at different levels of science. I have spent a significant part of my research life working on the important legume crop soybean (*Glycine max* (L.) Merr.) at McGill University, Canada. Before that the other important research I conducted, with the help of my supervisor, Professor A.F. Mackenzie, was related to the dynamics of nitrogen in the soil and in the plants including wheat (*Triticum aestivum* L.) and corn (*Zea mays* L.) across the great province of Quebec. These experiments resulted in a large set of data, with some interesting and applicable results. Such experiments were also greatly useful for my experiments on the responses of soybean under stress. I conducted some useful, great, and interesting researches with the help of my supervisor, Professor Donald Smith, on the new techniques and strategies that can be used for soybean production under stress, both under field and greenhouse conditions. When I came to Iran, I continued my research on stress, however, for wheat and corn plants, at Tarbiat Modares University, with the help of my supervisors, Dr H. Bahrami and Professor M.J. Malakouti, and my great friend Dr. F. Rejali from Soil and Water Research Institute, Karaj, Iran, using some new, great, and applicable techniques and strategies. Such efforts have so far resulted in 60 international articles, 18 authored and edited textbooks, and 38 book chapters, published by some of the most prestigious world publishers including Elsevier and Academic Press. I hope that this contribution can be used by academicians and researchers all across the globe. I would be happy to have your comments and opinions about this volume.

Mohammad Miransari
AbtinBerkeh Ltd. Company, Isfahan, Iran

Preface

The word "stress" refers to a deviation from natural conditions. A significant part of the world is subjected to stresses such as salinity, drought, suboptimal root zone temperature, heavy metals, etc. The important role of researchers and academicians is to find techniques, methods, and strategies that can alleviate such adverse effects on the growth of plants. Soybean [*Glycine max* (L.) Merr.] is an important legume crop feeding a large number of people as a source of protein and oil. Soybean and its symbiotic bacteria, *Bradyrhizobium japonicum*, are not tolerant under stress. However, it is possible to use some techniques, methods, and strategies that may result in the enhanced tolerance of soybean and *B. japonicum* under stress. Some of the most recent and related details have been presented in this volume.

In Chapter 1, Pagano and Miransari have presented the details related to the importance of soybean production worldwide, under different conditions including stress, with reference to the world's great nations of soybean production including Brazil, USA, China, India, and Argentina.

In Chapter 2, Pareek et al. discuss cross talk which may exist between biotic and abiotic stresses. Such a contribution is with reference to molecular mechanisms that can result in soybean tolerance under stress. They have highlighted genes that enhance plant resistance under more than one type of stress.

In Chapter 3, Miransari has discussed details which may improve soybean response and growth under biotic and abiotic stresses. The strategies, methods, and techniques that can enhance the tolerance of soybean and its symbiotic bacteria, *B. japonicum*, under stress have been presented, reviewed, and analyzed.

Chapter 4 is about evaluating soybean response under biotic and abiotic stresses using the proteomic technique. The authors have suggested that due to the complexity of soybean response under stress, the use of "omics" can be useful and applicable. Accordingly, the latest soybean response under stress by using proteomic has been presented.

In Chapter 5, details related to the use of soybean inoculums affecting soybean production worldwide has been presented by Miransari. The efficient production of soybean inoculums, using the symbiotic *B. japonicum*, has been discussed. With respect to the importance of soybean worldwide, the use of inocula is now widespread, especially in developed nations.

In Chapter 6, Arora et al. have presented the use of plant growth promoting rhizobacteria (PGPR) for soybean production under stress, with special reference to the nutritional values of soybean. They have mentioned that because the production of tolerant species of soybean is not easy and the use of chemical fertilization can adversely effect the properties of soil, the use of PGPR may be among the most suitable methods for enhancing soybean growth and yield under stress.

In Chapter 7, the adverse effects of salinity on the growth and yield of soybean under salinity stress have been discussed by Miransari. The use of different techniques and methods that can increase soybean growth and yield under stress have been presented, among which the use of the signal molecule, genistein, for the alleviation of the stress may be the most interesting.

In Chapter 8, the effects of drought stress on soybean response have been presented by Rahman et al. They have suggested that producing and using the tolerant genotypes can be an advantageous method under drought stress with respect to different morphological and physiological properties of soybean. Accordingly, the related genetic techniques have been reviewed and analyzed.

In Chapter 9, Miransari has discussed a collection of details on the alleviation of heavy metal stress adversely affecting soybean growth. Soybean and its symbiotic *B. japonicum* are not tolerant to heavy metal stress, however, it may be possible to produce soybean and rhizobium species that are more tolerant under heavy metal stress. As a result, planting soybean in contaminated areas may be more likely, although the quality of soybean seeds must be precisely examined.

Among the most important stresses affecting soybean growth and yield in the cold areas of the world is suboptimal root zone temperature. In Chapter 10, Dashti et al. have presented the results of their own and those of other researchers on the effects of such a stress on soybean growth and yield production. They have accordingly suggested, tested, and proved that the use of signal molecule, genistein, is among the most useful methods for the alleviation of the stress.

In Chapter 11, Miransari has presented the interactions between nitrogen fertilization and the process of symbiotic N fixation by soybean and *B. japonicum*. It has been indicated that although N fertilization may be essential for soybean production, at extra amounts, the process of biological N fixation will be adversely affected. The most important details, especially the indication of the optimum rates of N fertilization in combination with biological N fixation, have been discussed.

In Chapter 12, Fujita et al. have presented the results of their own and those of other researchers on the stress of heat and the related mechanisms that can alleviate the stress on soybean growth and production. This is because a large part of the world is subjected to the stress of heat.

In Chapter 13, Miransari has discussed all the previously presented chapters. It discusses the most usable strategies, challenges, and future perspectives for soybean production under stress. Such details can be of great importance as they pave the way for the more efficient production of soybean under stress.

Mohammad Miransari
AbtinBerkeh Ltd. Company, Isfahan, Iran

Acknowledgments

I would like to appreciate all the authors for their contributions and wish them all the best for their future research and academic activities. My sincere appreciation and acknowledgments are also conveyed to the great Elsevier team of editorial and production including Ms Nancy Maragioglio, the acquisition editor, Ms Billie Jean Fernandez, the editorial project manager, Ms Nicky Carter, the production project manager, Ms. Maria Ines Cruz, the designer, and Thomson Digital, the typesetter, for being so helpful and friendly during writing, preparing, and producing this project.

Mohammad Miransari
AbtinBerkeh Ltd. Company, Isfahan, Iran

The importance of soybean production worldwide

1

Marcela Claudia Pagano*, Mohammad Miransari**
*Federal University of Minas Gerais, Belo Horizonte, Brazil; **Department of Book
& Article, AbtinBerkeh Ltd. Company, Isfahan, Iran

Introduction

Interest in the impact of agriculture on soil structure or changing soil species in-
habitants has increased (Pagano et al., 2011; Wall and Nielsen, 2012). For example,
soybean is one of the major crops planted worldwide affecting different aspects of
the ecosystem. Among the most important components of the ecosystem are soil
microbes. Accordingly, with respect to the high cultivation of soybean crop worldwide, some of the most important parameters related to the production of soybean
are presented among which the soil biota including rhizobia and mycorrhizal fungi
are of great significance.

Because soybean is among the most important agricultural crop worldwide, more
research is being done to find details related to the production of soybean under different conditions including stress. Data indicating the rate of soybean production in
different parts of the world can be used to improve production of soybean and alleviate factors including stresses, which adversely affect soybean yield. The role of soil
microbes is especially important affecting the production of soybean. Just a few countries such as the USA, Brazil, Argentina, China, and India dominate the production of
soybean worldwide.

In particular, it is supposed that the soil biotas do not affect the agro-ecosystem
function or the services provided by them (Wall and Nielsen, 2012). Among the most
cultivated crops (maize, rice, wheat), soybean (*Glycine max* (L.) Merr.) is the only
leguminous species that can be associated with rhizobia and arbuscular mycorrhizal
(AM) fungi, with potential to be further exploited.

Pagano and Covacevich (2011) reviewed the current information on the benefit
of AM fungi in agro-ecosystems, mentioning that the increasing recognition of the
impacts of agricultural intensification and use of agrochemicals adversely affect soil
quality, modifying the number, diversity, and activity of the soil microbiota, including the populations of symbiotic fungi. Thus, improved research aimed at crop yield
enhancement and sustainability is essential and must be achieved.

Mutualistic associations such as AM fungi have important potentials for soybean
production (Pagano, 2012; Simard and Austin, 2010). There is a growing use of
beneficial rhizospheric microorganisms as biofertilizers in agriculture and there is
a need to better understand the effects of multiple inocula on soybean growth and
physiology.

Abiotic and Biotic Stresses in Soybean Production. http://dx.doi.org/10.1016/B978-0-12-801536-0.00001-3

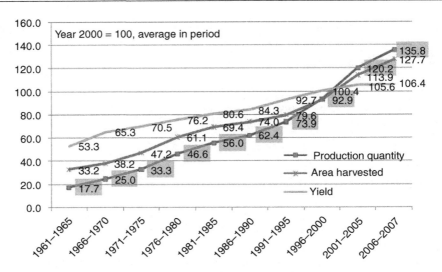

Figure 1.1 World soybean production quantity, area harvested, and yield: 1961–2007.
Source: FAOSTAT: http://faostat3.fao.org/home/E (Masuda and Goldsmith, 2009).

This chapter explores the current available information relevant to soybean production worldwide (Figure 1.1). Better knowledge of the wide variation in cultivation practices is important for understanding the ecology of soybean crops and for
management purposes. How a better management may be related to soil conditions
and microbial inocula will also be discussed. Better knowledge of the wide variation
in plant interactions is important for understanding the ecology of this crop and for
management purposes.

World soybean production

Soybean (*G. max* (L.) Merr.) originated in China and is a major source of protein
for humans and as a high-quality animal feed (FAO, 2003). Moreover, the presence
of important food supplements in soybean, and growing consumption, has resulted
in higher demands for soybean production. Soybean was originally domesticated in
China, with about 23,000 cultivars in Asia, and was introduced into the USA and
Brazil (López-López et al., 2010). For a brief history of world soybean dissemination,
see Rodríguez-Navarro et al. (2011). The term soybean possibly refers to the bean
from which soy sauce was manufactured.

Soybean constitutes one of the largest sources of vegetable oil and of animal
protein feed in the world (Sugiyama et al., 2015). It has the highest protein content
(40–42%) of all other food crops and is second only to groundnut with respect to the oil
content (18–22%) among food legumes (Robert, 1986). Moreover, soybean is used
for aquaculture and biofuel, as well as a protein source for the human diet (Masuda

and Goldsmith, 2009). Moreover, obesity and muscle fatigue can be prevented by soy protein (Agyei et al., 2015).

The USA produced more than 50% of the world soybean yield until the 1980s. However, nowadays Brazil and Argentina are also among the top world nations of producing soybean, following the USA. The major producers of soybean in the world include the USA, Brazil, Argentina, China, and India with more than 92% of the world's soybean production. It has also been produced in Africa since the twentieth century (Rodríguez-Navarro et al., 2011).

Brazil, as one of the tropical food giants, is now among the traditional "big five" grain exporters (America, Canada, Australia, Argentina, and the European Union). Since 1990, a third of world soybean exports has been accomplished by Brazil, second only to America which produces a quarter of the world's total soybean using just 6% of the country's arable land (The Economist, 2010).

Using fertilizer or biofertilizers, large amounts of nitrogen (N) are essential for the production of large soybean yields. The process of biological nitrogen (N_2) fixation by symbiotic soil bacteria, mainly *Bradyrhizobium*, is a less expensive source of N for soybean related to the use of chemical N fertilization. However, different factors determine the efficiency of the biological N fixation (related to the plant, rhizobia, symbiosis, and environmental stresses) (Rodríguez-Navarro et al., 2011). Accordingly, and with respect to the importance of soybean as a strategic crop, several governmental companies, universities centers, and individuals are researching different aspects of soybean production worldwide. Nowadays, more efficient inoculants are being used by farmers, which is the result of recent advances on soybean research, with a high value for the environment and sustainability of agro-ecosystems (Miransari, 2011a, b).

The recognition of soil microbes as important components of soil biodiversity is not largely integrated in strategies to conserve and manage these microorganisms. The financial value of soybean N fixation in Africa was evaluated by Chianu et al. (2010), indicating a higher rate of benefits for smallholder farmers. Accordingly, it was shown by the authors that the N fixing attribute of soybean in Africa is of high financial value.

They especially indicated the promiscuous varieties and recommended options, which may increase the chances of smallholder farmers to benefit from the process of N biological fixation. This is especially the case under those conditions where the quantities of inorganic fertilizers for increased soy productivity are inadequate. They mentioned that the promiscuous soybean varieties are not planted by the 19 African countries that produce soybean. However, the financial benefits from the process of N_2 fixation by promiscuous soybean can suitably illustrate how soil microbial biodiversity can sustain human welfare.

There are plenty of situations to further indicate the benefits of biological N fixation as, interestingly, some inoculated cultivars did not produce a higher yield than the uninoculated promiscuous varieties with the highest rate of production. Accordingly, the authors indicated that plant response to inoculation is complex and their recommendation was related to the selection and breeding of promiscuous soybean varieties in the case of Africa. Legume crops including soybean are able to nodulate with a wide variety of rhizobial strains in the soils being referred to as promiscuous (Mpepereki et al., 2000). Usually some uninoculated promiscuous varieties are able to produce similar yield levels, related to the promiscuous varieties, which are inoculated efficiently with rhizobia (Chianu et al., 2010).

It may not be a priority at this stage to focus on the development and production of inocula, due to the uncertainties resulted by different responses in many places and difficulties related to the production and conservation of inocula. In the future, greater profit may be obtained by the development and use of inocula, similarly to the production in countries such as Brazil, where a yield of 3 t ha^{-1} is relatively common, compared with an average yield of 1.1 t ha^{-1} for Africa (Chianu et al., 2010).

The importance of soybean production

In the future, a global crop demand is unavoidable, as the human population is steadily increasing (Tilman et al., 2011). In addition to population growth, agricultural production has not kept pace with estimated demand. Ray et al. (2013) compiled information on long-term production for maize, rice, wheat, and soybean, representing two-thirds of the total agricultural calorie demand. Those authors projected crop yields to 2050 indicating an increase of 1.3% year^{-1} for soybean, which is not at the level essential for providing people with their food by 2050. Soybean is among the 16 major crops (barley, cassava, groundnut, maize, millet, potato, oil palm, rapeseed, rice, rye, sorghum, soybean, sugar beet, sugarcane, sunflower, and wheat) cultivated worldwide (Foley et al., 2011). Thus, it is crucial that policy makers and land managers improve soybean research (see Masuda and Goldsmith, 2009).

Soybean is one of the major crops in five countries of South America, producing about 63% of the total cropped area (reviewed by Wingeyer et al., 2015). Its expansion resulted in a decrease in the cultivated area of other crops and native vegetation, and increased soybean production at an annual rate of ~6%. The main reason for the increase of soybean yield was a higher production area, related to the lower increase of grain yield (reviewed by Wingeyer et al., 2015).

The cultivation of soybean after maize in Canada is common; however, due to the presence of greater amounts of maize residues, its plantation under no-tillage, which may decrease its production. Such adverse effects are by influencing soil nitrogen and soybean nodulation, soybean emergence, growth, and development, as well as by impacting soil physical properties such as moisture and temperature (Vanhie et al., 2015).

Soybean is also cultivated as an important summer crop in Japan in rotation with winter wheat or as an upland crop fallow (Higo et al., 2013). Increasing the potential yield of soybean, especially with respect to the climate and genetic potential of crop requires more investigation, as well as taking into account the following: (1) maximum yield of a crop cultivar produced under certain environmental conditions; (2) adequate amounts of nutrients and water; and (3) controlling pests and diseases (reviewed by Salvagiotti et al., 2008).

Production and supply, stock levels, and soybean prices have changed along with the high demand of soybean by the population (MAPA, 2015; Masuda and

Goldsmith, 2009). Since 2005, the production of soybean in USA has been at its highest rate (89,507 million tons), over 33,640 million hectares (USDA, 2014).

Masuda and Goldsmith (2009) analyzed the production of soybean worldwide, as well as the area harvested and the related yield. The yearly rate of increase of soybean was at 4.6% from 1961 to 2007, with the average yearly production of 217.6 million tons in 2005–2007. They estimated that the yearly production of soybean will be at 2.2% and approach a yearly production of 371.3 million tons by 2030.

They accordingly indicated the following as the major factors affecting soybean production globally: (1) limitation of cultivable lands, and (2) the need for investment by the public, private concerns, and farmers to increase soybean yield. The substitution for other crops (cotton and sunflower), pasture, and native vegetation increased the cultivated soybean field areas and production by 36%. They mentioned that there has been a shift in the production area from the USA and Asia (China and India) to the USA and South America, including Argentina and Brazil.

Due to technological advances, 49% of grain production in Brazil is related to soybean. It is, especially, cultivated in the midwest and south regions of the country. The research and advances by the Brazilian Agricultural Research Corporation (Embrapa), in partnership with farmers, industry, and private research centers, has made the cultivation of soybean likely in the Cerrado grasslands. Such progress has also resulted in an increase yield production per hectare, competing with the major world production rates. However, the cultivation of soybean is conducted by the use of sustainable agricultural practices such as the use of no-tillage and integrated crop–livestock system (MAPA, 2015).

A single-gene transformation results in the production of genetically modified crops, such as a herbicide resistant crop (e.g., Roundup Ready soybeans) (Sobolevsky et al., 2005). Currently, soybean and corn, followed by canola and cotton, are the main transgenic crops, cultivated in the United States and some other countries (Argentina, Canada, and China). The production of genetically modified crops such as Roundup Ready soybeans is large in Argentina, which is the second biggest transgenic area worldwide; however, the effects of these biotechnologies have still to be further investigated (Qaim and Traxler, 2005).

Microbial associations

Rhizobia

Rhizobia are nitrogen-fixing bacteria classified and characterized by different systems. Beijerinck was able to isolate and cultivate a microorganism, named *Bacillus radiocicola*, from the nodules of legumes in 1888. However, Frank (1889) renamed it *Rhizobium leguminosarum* (Fred et al., 1932), which was retained in *Bergey's Manual of Determinative Bacteriology* (Holt et al., 1994).

Rhizobia are characterized on the basis of their growth rate on certain substrates, as fast and slow growers (Löhis and Hansen, 1921). Mean generation time of the slow and fast growing bacteria is greater and less than 6 h in selective broth media, respectively (Elkan, 1992). Until now, about 750 genera of legumes, containing 16,000–19,000 species, have been recognized; however, only a few have been examined (Allen and Allen, 1981).

The first accepted change in the rhizobial nomenclature was the establishment of *Bradyrhizobium* (Jordan, 1982). The strain of *Bradyrhizobium*, which is able to nodulate soybean, is characterized as *Bradyrhizobium japonicum*, the first recognized group of *Bradyrhizobium* strains (Young and Haukka, 1996). *Bradyrhizobium elkanii* (Kuykendall et al., 1992), possessing some specific phenotypic and genetic characters, indicates a number of species within the soybean nodulating bradyrhizobia.

Bradyrhizobium liaoningense are also among the other extra slow growing soybean rhizobia with the ability of forming a coherent DNA–DNA hybridization group (Xu et al., 1995). Moreover, some *Bradyrhizobium* strains, known as *Bradyrhizobium* sp., are not able to nodulate soybeans (Young, 1991). The current characterization of rhizobia is on the basis of gene sequencing for the 16S or small subunit of ribosomal RNA (Jarvis et al., 1997).

Four recognized species of *Bradyrhizobium* include: *B. japonicum* (Jordan, 1982); *B. elkanii* (Kuykendall et al., 1992); *B. liaoningense* (Xu et al., 1995), and *Bradyrhizobium* sp. (Young, 1991). As suggested by Young (1991), the *Bradyrhizobium* genus will not have new species allocated; however, the host name will be mentioned in parentheses.

The specific and compatible rhizobia nodulating soybean is *B. japonicum* (Cooper, 2007; Long, 1989; Rolfe, 1988). Soybean association with rhizobia, including *B. japonicum* and *B. elkanii*, provide about 50–60% of soybean nitrogen requirement supplied by the bacteria in nodules (Salvagiotti et al., 2008). Rhizobia are the bacteria, which include *Rhizobium, Bradyrhizobium, Sinorhizobium*, etc., surviving and reproducing in the soil, and fixing atmospheric N inside the nodules produced in the roots of their specific legume (reviewed by Denison and Kiers, 2004).

Laranjo et al. (2014) reviewed the rhizobial symbioses with the emphasis on mesorhizobia as legume inoculants. They have presented brief details of rhizobia including their rhizobial genomes, taxonomic diversity, and nodulation and nitrogen fixation genes. According to the above-mentioned details the term "rhizobia" includes the genus *Rhizobium, Bradyrhizobium, Sinorhizobium*, and *Mesorhizobium*. Moreover, rhizobia include Alphaproteobacteria (Rhizobiales) though some isolates of wild legumes belong to the class of Betaproteobacteria (Laranjo et al., 2014). Some research has also indicated that legumes are able to be nodulated once or several times during evolution (Sprent, 2007).

For a review on developments to improve symbiotic nitrogen fixation and productivity of grain legumes see Dwivedi et al. (2015). The main function of nodules on soybean roots is to fix the atmospheric N by the process of symbiotic nitrogen fixation, supplying nitrogen for plant growth and seed production. Sugiyama et al. (2015) reported changes in the rhizospheric bacteria and especially *Bradyrhizobium* during soybean growth, suggesting that the symbiosis of host plant with rhizobia may be selective.

In the last few years approximately 13,247 peer reviewed journal papers on soybean production have been produced, of which 731 focused on soil management (Table 1.1). Among the studies on soybean interactions with microorganisms (Table 1.1), research

Table 1.1 Journal articles dealing with soybean production worldwide

Keywords	Number of journal articles
Soybean production	13,247
Soybean production + soil	2,300
Soybean production + plant management	501
Soybean production + soil management	731
Soybean production + rhizobia	231
Soybean production + symbioses	147
Soybean production + arbuscular mycorrhizas	39

Database survey conducted on May 2015 (SCOPUS).

on rhizobia predominated (circa 231 reports existing for rhizobia in soybean) over mycorrhizal research (39 reports). Among an increasing number of reviews published on N_2 fixation in legumes, soybean in particular accounts for 20 documents in the SCOPUS database (Miransari et al., 2013; Rao, 2014; Uchida and Akiyama, 2013); however, other reports (Hungria et al., 2005a, b) are also available.

In a review paper, Salvagiotti et al. (2008) analyzed 637 data sets derived from 108 field studies in 17 countries published from 1966 to 2006 including nitrogen fixation and N fertilization in soybean. For a 1 kg increase in N accumulation in above-ground biomass, they found a mean linear increase of 0.013 Mg soybean seed yield. Their meta-analysis indicated that 50–60% of soybean N demand is by the process of biological N_2 fixation; however, the rate of N fixation decreases with increasing N fertilizer.

Moreover, the N that is harvested by soybean grain must be supplied by both the process of N fixation and chemical N fertilization. It was not possible to estimate the actual contribution of below-ground N and its variation, and more research work must be conducted to determine such details. In conclusion, those authors mentioned that the yield response of soybean to N fertilizer is determined by yield production, environment, and abiotic/biotic stresses, which decrease crop growth and the associated N demand. With respect to such constraints, the development of rhizobia, which are able to fix N_2 under stress, is essential for providing the host plant with N (Alves et al., 2003; Hungria and Vargas, 2000).

It has been shown that the efficiency of the symbiotic process depends on many factors including the host plant, bacteria, the process of symbiosis, and the environment. Among the most important constraints affecting plant growth and the process of fixation are the soils, which are not highly fertile, resulting in the limited availability of macro- and micronutrients (Campo et al., 2009).

In the absence of growth constraints and in the presence of soybean genotypes with a high rate of yield at levels above 4.5 Mg ha^{-1}, the response of well-nodulated soybean crops to N fertilization is likely. The deep use of (slow-release) fertilizer underneath the nodulation zone, or using N chemical fertilization during the reproductive stages in high-yielding environments, can significantly improve soybean yield (Salvagiotti et al., 2008).

Diaz et al. (2009) investigated the soybean response to inoculation and N application following long-term grass pasture due to conversion of pastures into soybean fields. They observed that inoculation of soybean host plant with rhizobia increased soybean grain yield, plant dry matter, N concentration, N accumulation, and grain N, although the quality of seed remained constant. In contrast, although the N fertilizer increased plant dry matter, it did not increase grain yield, with or without inoculation. Moreover, no increases in plant N or improved seed quality were detected. They accordingly suggested inoculating soybean seed when planted after long-term grass pasture, without chemical N fertilizer.

Cases of legume introduction in places where rhizobia were not present to nodulate the introduced crop indicate the essentiality of research work to determine rhizobial evolution. One of the most remarkable cases is the introduction of soybean in Brazil (Barcellos et al., 2007). The implications of biserrula, nodulated by *Mesorhizobium ciceri* (typically known for nodulating chickpea), introduced in Australia, is the other example of naturally occurring rhizobia, which are able to evolve and acquire, by the process of lateral gene transfer, genes essential for the inoculation of the introduced legume. A 5-year period was essential for the detection of rhizobia able to nodulate biserrula in Australian soils, different from the original inoculant (Nandasena et al., 2007).

Mycorrhizas

With regard to mycorrhizas, Miranda (2008) compiled information on AM fungi in crops from Cerrado, the Brazilian savannah. In line with earlier studies, she showed that soybean can be inoculated by four species of AM fungi including *Glomus etunicatum*, *Entrophospora colombiana*, *Acaulospora scrobiculata*, and *Gigaspora gigantea* in pots with autoclaved native soil, fertilized with P_2O_5 and lime.

She showed that *G. etunicatum* was the most efficient inoculum followed by *E. colombiana*, increasing soybean production by four times relative to the uninoculated control. She also found that the plant production in the inoculated pastures (*Andropogon guayanus* and *Stylosanthes guianensis*) was more responsive to fungal inoculation. Usually, soybean crop is inoculated with a lesser rate of mycorrhizal colonization than for maize. Hence, the crop rotation can benefit soybean plant with a higher rate of AM fungi in the first year of soybean–maize rotation (Miranda et al., 2005).

Perez-Brandan et al. (2014) reported changes in soil microbial diversity in soybean fields. They analyzed soybean monoculture, soybean–maize rotation, and native vegetation. According to their research, a higher rate of carbon in microbial biomass and a higher rate of glomalin-related soil protein was found under the rotation system than in monoculture. Such results indicate that agricultural intensification can deteriorate soil biological, chemical, and physical properties. They also indicate that functional diversity was less in monocultures than in rotation and native vegetation.

It is known that the intensive use of land can affect biodiversity and as a result the changes in the composition or species diversity of aboveground communities can affect soil communities (Suleiman et al., 2013). Research on glomalin or glomalin-related soil protein is increasing in agro-ecosystems (Curaqueo et al., 2010;

Redmile-Gordon et al., 2014). It is believed that AM hyphae decompose and liberate glomalin residue in the soil (Treseder and Turner, 2007). The protein is extracted from the soil by autoclaving in citrate solutions and their easy evaluation and little soil demand (generally 1 g) support their assessment. Curaqueo et al. (2010) found higher values related to the mycorrhizal hyphae, glomalin-related soil protein, and water stable aggregates under no-tillage relative to the conventional tillage in a rotation experiment with wheat in Chile.

Junior et al. (2013) tested the nodulation and mycorrhization of transgenic soybean under greenhouse conditions after using glyphosate. They determined an increased number of nodules using Roundup until 15 days after the application. However, after that period, the inoculated control presented more nodules. They observed no influence of glyphosate in the root colonization by AM fungi.

It is known that *Bradyrhizobium* strains may not be tolerant to the presence of glyphosate application, thus decreasing the host plant nodulation (Bohm and Rombaldi, 2010; Reis et al., 2010). The process of biological N fixation contributes to the high production of soybean and this technology uses the selected strains of *B. japonicum* and *B. elkanii* as inocula (Hungria et al., 2005a, b).

Malty et al. (2006) tested Roundup on three strains of *Bradyrhizobium* and on three AM fungi (*Glomus etunicatum*, *Gigaspora margarita*, and *Scutellospora heterogama*) in culture and in soil. The growth of *Bradyrhizobium* spp. and AM fungi in culture medium decreased at concentrations greater than the optimum concentration. Germination and growth of AM fungal spores was affected more in the Gigasporaceae representatives than in *Glomus*. In conclusion the results indicated that soil application of herbicide up to a rate equivalent to 10 L ha^{-1} did not affect nodulation and mycorrhization of soybean.

It is known that ecosystem services are affected by soil properties, soil conditions (e.g., moisture, temperature), and the biological processes within soil, as well as management, which will select the strongest and more efficient organisms in the ecosystem (Dominati et al., 2010; Pagano, 2013). However, better soil management will depend on regional understanding and cooperation between researchers, policy makers, and the community.

It is also known that the number, diversity, and activity of both free and symbiotic fungi are modified by crops (Kahiluoto et al., 2009; Nyfeler et al., 2011; Pagano et al., 2011). That is why the ecosystem services of soils need to be given greater recognition as the impact of agriculture on soil structure or changing the inhabitant species of soil is crucial. Efforts to restore ecosystem services need to take into account sustainable rural incomes and community participation.

Among soil ecosystem services, AM fungi protect soil structure and plant roots against disease or drought (Simard and Austin, 2010). Mycorrhizal fungi can significantly affect plant growth (Smith and Read, 2008) as different AM fungal communities are present in different land use systems (Sene et al., 2012; Stürmer and Siqueira, 2011). In mycorrhizal fungal association, 20% of the host plant photosynthetic C can be moved to the fungus (Lerat et al., 2003).

Cotton et al. (2015) analyzed the AM fungal communities in the roots of soybean in fields exposed to higher levels of O_3 and CO_2 (as predicted for 2050). On increasing the rate of CO_2 exposure, there were only differences created in the community composition of AM fungi (increased ratio of Glomeraceae to Gigasporaceae). Due to its importance as a major crop in many parts of the world and in internationally competitive agriculture, more research on soybean management can contribute to a higher yield of soybean worldwide.

Interestingly, Juge et al. (2012) tested the inoculation of three microorganisms (*Azospirillum*, *B. japonicum*, and *Glomus irregulare*) in soybean production. They found different effects of the tested microbes on shoot biomass; however, they mentioned that the fact that coinoculation effects on nodulation are strain dependent and must be considered. Higo et al. (2013) analyzed the diversity and vertical distribution of AM fungi under two soybean rotational systems in Japan. They found the effect of crop rotation on AM fungal communities with specific AM fungi associated with soybean. In Argentina, Grümberg et al. (2015) showed the significant role of AM fungi in alleviating drought effects on soybean. They also pointed out differences between mixtures of AM fungi isolates and single strain inocula, proposing an effective selection of AM fungi for soybean.

The importance of better soybean management

Plant and soil management

Interest in soil management and sustainable production has increased worldwide. Moreover, tillage practices with higher efficiency have contributed to the success of cropland yields, although there has been a recent expansion of monocropping soybean production. Such a method in current agricultural practices can lead to a decrease in soil quality even though the no-tillage practices may improve such effects (Wingeyer et al., 2015).

The removal of pasture from crop cultivations, together with the increased frequency of soybean cultivation, and the conversion of native vegetation into farmland constitute may adversely affect soil quality. In this regard, Vanhie et al. (2015) reviewed the potential strategies to address using the high levels of maize residues for soybean production under no-till. It is a common practice to encourage environmental benefits such as reduced soil erosion, fuel usage, and carbon emissions (Seta et al., 1993; Yiridoe et al., 2000).

Plant residues from the previous crop can suppress the activity of pathogens by enhancing the general microbial activity. Although the debris increases the microbial activity, it can also enhance the activity of pathogens by preventing a decrease in the inoculum density, such as *Macrophomina phaseolina*, which causes charcoal rot in soybean (Baird et al., 2003).

High maize yield significantly increases the amount of residues, which is also a result of changes in the cropping system influencing maize residue decomposition. Such aspects can also decrease the non-till practices affecting no-till and conventional tillage of soybean. Vanhie et al. (2015) suggested the following strategies to manage high quantities of maize residues when planting soybean: (1) removal of residue; (2) proper handling of residue distribution and orientation; (3) instead of no-tillage practices, minimum tillage be used; (4) application of nitrogen; and (5) use of more efficient planting equipment. However, additional research is required to fully work out these options.

Wingeyer et al. (2015) investigated the details related to soil conditions and research affecting soil degradation in South America (Argentina, southern Brazil, Bolivia, Paraguay, and Uruguay). They suggested that such degradation must be controlled by conducting research to prevent the negative effects on soil quality. They indicated that the properties of different regional soils must be evaluated and accordingly scientific recommendations be presented so that the environmental degradation be prevented.

Similarly, because climate change can adversely affect agricultural intensification, more attention is given to the ecologically sustainable use of land (Borie et al., 2006). It is also essential to use more efficient agricultural practices so that the plants can be acclimated to the climate and the increasing population can have their needs (for food, fodder, and fuel) met. The properties of soil have also to be restored and soil fertility enhanced by using proper agricultural practices (reviewed by Lal, 2009). Although increasing the rate of atmospheric CO_2 exposure can promote plant growth and soil C input, it may also affect the decomposition of soil microbes (Covacevich and Berbara, 2010; van Groenigen et al., 2014).

The following indicates how soil may be used in a sustainable manner: (1) suitable method of seed cultivation and planting; (2) rotation of crop; (3) enhancing soil fertility; (4) using proper agricultural practices including mulch residue, no-tillage, and cover crops; (5) appropriate use of nutrients; (6) applying biochar; and (7) use of crops which have been genetically modified and improved (reviewed by Lal, 2009).

Organic cover and the rate of water available in soils and plants are among the imporant parameters that have been better recongized and determine the response of surface and subsurface soil to global change (Torn et al., 2015). Studies on soil aggregate stability and soil resistance are increasingly being investigated (Lal, 2010; Powlson et al., 2011).

Castro and Crusciol (2013) determined the response of soybean under the no-tillage system in Brazil. For the adjustment of soil pH they used limestone or slag (silicates of calcium and magnesium). The chemical attributes of soil were evaluated 6, 12, and 18 months after the application of the chemical compounds and it was indicated that slag is an efficient and effective source for the adjustment of soil acidity. This was because slag increased the grain yield of soybean in the treated plots, related to the control treatment without the use of chemical compounds.

Soybean rhizobial inoculants

With regard to rhizobial inoculants, some reports showed frequent contamination of inocula (Gemell et al., 2005; Herrmann et al., 2013). The inocula of mycorrhizal fungi generally contain a few viable propagules, decreasing the rate of colonization (Faye et al., 2013).

Hungria et al. (2005a, b) evaluated the details related to the inoculant production and application. They indicated that the process of biological nitrogen fixation is an important process preventing degradation and hence the adverse effects of soil on crop yield worldwide, mainly in tropical regions such as Brazil and Africa. Benefits could be enhanced by using efficient and competitive species of rhizobia which are of high quality and available in adequate quantities in the soils under cultivation for legumes including soybean.

Hungria and colleagues briefly mention the long period of rhizobial inoculant production and their related products, which are of poorer quality. Moreover, they explain that for the production of successful inoculants, a selected strain, which has been used for a long time under specific environmental conditions, must be used in the presence of persistent species. Soybean was brought to and cultivated in Brazil in 1882 with cultivation of large rate of fields using bradyrhizobial inoculants from the USA. The important point was the successful use of a bradyrhizobial strain that was tolerant to natural acid soils. Next, the host plant demand for nitrogen increased due to the higher mean production of soybean yields (2765 kg ha^{-1} in 2003).

Inoculated bradyrhizobia are found in most soybean fields; however, strains have to be selected that are more efficient and competitive and are able to provide soybean with most of its essential N. Thus, adapted strains, which can result in the production of higher grain yield, have been selected. Under the conditions of producing commercial soybean, four strains, which are of high capacity for nitrogen fixation, have been selected and used commonly; however, the selection strategies continue to help farm owners. Among the selected strains, the variant strains of SEMIA 566 and CB 1809 resulted in the highest yield of soybean cultivar BR 133 as well as a higher rate of nodulation in fields from south Brazil. The N fixation potential of both strains was not significantly different from that of the fertilized control (200 kg N ha^{-1}) (Hungria et al., 2005a, b).

The number of rhizobia decreases with time as affected by the environmental conditions, properties of soil, and the bacterial strain. However, some research has shown 5–15 years' persistence of inoculant in the soil. Using new and more efficient strains in place of established *Bradyrhizobium* inoculants is difficult and must be done on the basis of frequent reinoculation (Hungria et al., 2005a, b). For example, the supplanting of the CPAC 15 strain must involve reinoculation yearly, which may result in higher costs.

For instance, using molecular methods the strains with a higher persistency and related factors, which may affect such persistency, must be indicated. The rhizobia might be incompatible with the use of micronutrients, seed-applied pesticides, or the size of small seeds, limiting the number of bacteria. The most effective method of using inoculants is to place them near the seed/seedling; hence, they can be inoculated directly in the soil furrow as liquid, peat, or granules (and not mixed with fertilizers) (Hungria et al., 2005a, b).

However, the inoculation of soil may be more costly as a higher rate of inocula is essential. For example, in Brazil inoculating seeds with broth inoculants in the furrow or 2.5 cm underneath the seed may result in a higher rate of soybean nodulation (Hungria et al., 2005a, b). Among the most important parameters affecting the efficiency of inoculant industry is rhizobial biodiversity, which can result in the selection of a higher number of suitable strains. However, the selection of strains with high temperature resistance for soybean in Iran was based only on 56 strains (Rahmani et al., 2009).

The interactions between soybean and soil microbes and the related details of the agronomical and most relevant genetic aspects of soybean rhizobia symbiosis have been reviewed by Rodríguez-Navarro et al. (2011). However, they accordingly mentioned that more details are essential on a molecular basis, indicating the specificity of cultivar–strain and occupancy of nodules by rhizobia competitors. Thus to produce more efficient commercial inoculants and develop symbiotic interactions for the other important agricultural crops, such constraints must be resolved.

Inoculation of soybean, under field conditions, has been efficiently used in the USA, Brazil, and Argentina. However, the high populations of indigenous soil rhizobia can adversely affect the successful use of inoculants in certain areas. More than 105 soybean rhizobia per gram of soil have been found in most Chinese soils, which interfere with the occupancy of nodules by the inoculant (Rodríguez-Navarro et al., 2011).

Due to the need for the sustainable use of agricultural practices, more attention has been given to the process of biological nitrogen fixation and rhizobial symbioses. Several studies have revealed the diversity of *Rhizobium* species with respect to their genetic and phenotypic properties, which can be used for the study of the evolutionary associations among the specific species (reviewed by Laranjo et al., 2014). Advances in molecular genetics of rhizobia have contributed to a better understanding of such plant symbionts. A number of research projects have been conducted on the use of mesorhizobia isolated from the nodules of chickpea (*Cicer arietinum*), which is among the most important legumes and is nodulated by *Mesorhizobium* species (Laranjo et al., 2004, 2008, 2012).

The rapid evolution of mesorhizobia has been shown in a review by Laranjo et al. (2014). They mentioned that the first genetic transfer of *B. elkanii* and *Sinorhizobium fredii* by a strain of *B. japonicum* in symbiosis with soybean was done using the technique of lateral transfer of chromosomal symbiosis islands in the field. However, in *Mesorhizobium* strains, the symbiosis genes are rarely found in plasmids, as they are commonly located in chromosomal symbiosis islands (reviewed by Laranjo et al., 2014).

Bai et al. (2003) isolated three strains of *Bacillus* from inside the nodules of vigorous soybean grown under field conditions and tested their coinoculation with *B. japonicum* on the growth of soybean. The coinoculation of *Bacillus* strains with *B. japonicum* increased the number and weight of soybean nodules, as well as the shoot and root weight, total nitrogen, total biomass, and grain yield (Bai et al., 2003). They recommended a selected strain (*B. thuringiensis* NEB17) for use as plant growth promoting rhizobacteria (PGPR) in soybean production systems under the conditions of suboptimal root zone temperatures. Under the circumstance that the growing season is not long, soybean growth and N fixation are negatively affected; however, PGPR can alleviate such effects on the growth of soybean plants.

In Pakistan, the average yield of soybean is low relative to the other top producing nations. Low soil fertility, as a result of intensive cropping, and cultivating in a small area are important constraints of soybean cultivation. Hence, research on the process of biological fixation is providing more options to increase soybean yields under different conditions including stress. Using N-fixing and P-solubilizing bacteria, such as *Pseudomonas*, resulted in a higher soybean yield relative to the single use of P_2O_5 fertilization (Afzal et al., 2010). In *Pseudomonas sensu stricto*, several species with the ability of solubilizing phosphate *in vitro* (reviewed by Peix et al., 2003) from the ribosomal RNA group I (Palleroni, 1992) are included.

In their experiments, Afzal et al. (2010) indicated that soybean coinoculation with *Pseudomonas* strain 54RB (P-solubilization index of 4.1) resulted in a higher production rate of auxins and gibberellins than with *B. japonicum* strain TAL 377. The strain with the highest rate of phytohormones also increased soybean growth and yield the most (Afzal et al., 2010). The dual inoculation of *Bradyrhizobium–Pseudomonas* (biological fertilization) with the addition of triple superphosphate (P_2O_5) (chemical fertilization) was the most effective treatment on soybean growth, yield (12 and 38% increase compared with the single use of P_2O_5), and yield components.

Although previous experiments have indicated that coinoculation with microbial species may synergistically and antagonistically affect the responses occurring, it has been indicated that the number of soil microbes may increase in the presence of P_2O_5, confirming that P_2O_5 assists the growth and colonization of *Bradyrhizobium*. As an energy source (adenosine triphosphate) affecting the reduction of N_2 to NH_3, phosphorus deficiency can negatively affect the process of photosynthesis, symbiotic N_2 fixation, nodule development, and root growth (Pereira and Bliss, 1989; Vadez et al., 1996).

Accordingly, Afzal et al. (2010) recommended the consortium use of beneficial microbial for important crops in Pakistan as well as the use of this technology by farmers as indicated by extension workers. Additionally, few rhizobial strains may be tolerant under different stresses such as salinity, drought, acidity, and heavy metal, and constitutes unique PGPR (reviewed by Deshwal et al., 2013).

It is also known that AM fungi can significantly enhance phosphorus use efficiency and N accumulation in the host plant. As a result mycorrhizal fungi affect the association between phosphorus P utilization efficiency and symbiotic N fixation (Tang et al., 2001) and the mechanisms, which may affect phosphorus P uptake and utilization by the host plant (Bucher et al., 2001; Jia et al., 2004). It is also indicated that plant phosphorus P utilization determines shoot growth and N_2 fixation (Rodino et al., 2009).

Soybean can establish tripartite symbiotic associations with rhizobia and AM fungi (Lisette et al., 2003); however, few results are available on their effects on plant growth, or their association with root architecture as well as with N and P availability. Xie et al. (1995) reported that soybean coinoculation with *B. japonicum* 61-A-101 and mycorrhizal fungi resulted in a more efficient colonization by *Glomus mosseae*, and increased N and P uptake by the host plant.

Similarly, it was indicated that coinoculation with rhizobia and AM fungi can favor the growth and yield of faba bean (Li et al., 2004). However, none of such favorable effects of coinoculation were reported in green gram (Saxena et al., 1997) and pea (Blilou et al., 1999). With regard to the synergistic association between rhizobia and mycorrhiza, Tajini et al. (2012) tested the benefits of coinoculation with those microbes on common bean as a practical method for agricultural development in marginal lands with P deficiency. As the dual symbiosis with rhizobia and mycorrhizal fungi, under P deficient conditions, may improve symbiotic N fixation in leguminous plants, they suggested this practice is environmentally and economically recommendable.

Wang et al. (2011) investigated the effects of coinoculation with mycorrhizal fungi and rhizobia on soybean growth with respect to the properties of root architecture and availability of N and P in a field experiment. According to their results root architecture and mycorrhizal fungal colonization were positively correlated. They indicated that a soybean genotype with a deep root network had greater mycorrhizal fungal colonization at low P, and more efficient nodulation under high P concentration than the shallow root genotype.

They also found that the synergistic association between rhizobia and AM fungi is dependent on N and P status affecting soybean growth. Such a coinoculation also increased soybean growth under low P and/or low N levels (increased shoot dry weight, along with plant N and P content). Root architecture determined the effects of coinoculation as the genotype with deep roots benefited more from coinoculation than the genotype with shallow roots. Thus, their results clarified some previous unknowns about such a tripartite association when nutrients are limited, indicating a theoretical aspect for planting soybean with coinoculants under field conditions.

It is known that inoculation with efficient rhizobia at ordinary rates cannot greatly increase the seed yield of soybean because the presence of less efficient native rhizobia restricts the occupation ratio of soybean nodules by inoculated rhizobial strains (Kvien et al., 1981; Weaver and Frederick, 1974). Accordingly, a higher rate of nodule occupancy by rhizobial inoculation increases soybean yield. Such an approach may be achieved by, for example, improving the method of inoculation and using the more efficient techniques (Takahashi et al., 1996).

For the screening and production of efficient and competitive strains, a high number of useful strains were isolated from recombinant and mutagenized rhizobia (Maier and Graham, 1990; Williams and Phillips, 1983). Yamakawa and Saeki (2013) compiled the inoculation methods, using effective *Bradyrhizobium* strains, to increase the yield of soybean in the south-west area of Japan.

Some rhizobial strains are capable of solubilizing nonsolubilizing P in the soil, which increases the rate of plant growth. Halder and Chakrabarty (1993) indicated that inorganic phosphate can be solubilized by a large number of *Bradyrhizobium* strains. Moreover, Chabot et al. (1996) found that the solubilization of phosphate by strains of *Rhizobium leguminosarum* bv. *phaseoli* was the most important mechanism enhancing the growth of maize and lettuce.

Antoun et al. (1998) also determined the solubilization of phosphate by *Bradyrhizobium* sp. (*Lupinus*). Similarly, Dashti et al. (1997) indicated phosphate solubilization and the subsequent acceleration of nodulation by PGPR increased N fixation activity in soybean under suboptimal root zone heat. It is known that a successful colonization of the legume rhizosphere is achieved if the *Bradyrhizobium* inocula constitute a large component in relation to the indigenous microorganisms for organic compounds excreted by the root (Van der Merwe et al., 1974). The high rate of indigenous bacteria as well as their rapid or slow grow influences the colonization of the host plant by rhizobial inocula.

In previous research, Anderson (1957) found a reduced number of nodules formed by *R. leguminosarum* biovar *trifolii* in *Trifolium repens* L., which was prevented by several bacteria without the ability to produce antibiotics. Similarly, Plazinski and Rolfe (1985) reported that the strains of *Azospirillium*, which did not produce antibiotics, decreased the nodulation of *Trifolium subterrancum* and *T. repens*. The bacterial strains producing antibiotics adversely affect the activity of many of the indigenous bacteria and, due to the reduced competition, proliferate and nodulate more extensively in the rhizosphere (Li and Alexander, 1988)

Mycorrhizal fungal inoculants of soybean

More research is available on the association of soybean with rhizobia than with mycorrhizal fungal association. Soybean is generally responsive to inoculation with *Glomus*; however, soybean response to inoculation with other genera such as *Gigaspora* has not been great (Nogueira and Cardoso, 2000). This is because the fungus absorbs P, resulting in the inhibition of plant P transporters. An increased concentration of micronutrients in plant tissues can be related to colonization by AM fungi. For example, colonized soybean can have a higher rate of zinc uptake than in plants fertilized with P (Cardoso, 1985).

In their greenhouse experiment, Nogueira and Cardoso (2000) tested the effects of two different mycorrhizal species fungi (*G. margarita* and *Glomus intraradices*), on the production of soybean under increasing P levels (0, 25, 50, 100, and 200 mg kg^{-1}). Increasing rates of P decreased root colonization as well as the total and active external mycelium in both fungal species. Due to a faster growth, *G. intraradices* inoculated soybean roots more efficiently and produced more active external mycelium than *G. margarita*. Soybean inoculation with *G. intraradices* and production of the active external mycelium increased with time and decreased with increasing P rates.

Other researchers, such as Minhoni et al. (1993), also investigated soybean inoculation by mycorrhizal fungi. They found that under increasing levels of P fertilization, root colonization by *Glomus macrocarpum* slowly decreased. Previously, Faquin (1988) indicated that there were no differences at 90 mg kg^{-1} of P and Siqueira et al. (1984) also detected an inverse association between P availability and root colonization in a sandy soil.

Moreover, inoculation with AM fungi can alleviate the negative effects of drought on soybean growth and prevent the premature senescence of nodules brought about by stress (Porcel et al., 2003). If mycorrhizal fungi and rhizobia are properly combined, such a combination enhances plant growth and resistance to pathogens (Aysan and Demir, 2009) and improves nodulation and nitrogen fixation (Barea et al., 2002). In the future, the role of mycorrhizal fungi inocula will be better illustrated in sustainable agriculture.

Other soybean inoculants

Rhizospheric ecology is of major interest for agronomists because it constitutes a combination of different interactions among the microorganisms and the environment surrounding roots affecting plant growth (Glick, 2012). Development of new technologies, which help in the understanding of such effects and hence benefit microorganisms, is essential. Legumes are greatly responsive to their rhizospheric microbes, especially rhizobia (Glick, 2012). The microbiota in the legume rhizosphere can be beneficial to plant growth and yield production by enhancing the processes of recycling, mineralization, and nutrient uptake. Moreover, microbes can increase plant growth by the production of plant growth regulating substances such as phytohormones, vitamins, and amino acids (Raaijmakers et al., 2008).

Estimation of crop loss by pathogens is not well documented; however, it may range from 7% to 15%, affecting major world crops (wheat, maize, soybean, potato, and rice) due to fungi and bacteria (Oerke, 2005). The endophytic, symbiotic, or free-living association of soil bacteria, PGPR, results in the promotion of plant growth by enhancing the acquisition of plant nutrients or influencing the intensity of plant hormone, or by alleviating the adverse effects of pathogens (Glick, 2012).

Rhizobial bacteria, as the symbionts of legumes, are able to fix atmospheric N_2 by the process of biological nitrogen fixation, providing one of the major macronutrients to the host plant. The process of biological N fixation is very promising because the production of nitrate fertilizers is expensive and is not recommended environmentally (high amounts of nonrenewable fossil energy are essential for the production of chemical fertilization resulting in the release of greenhouse gases) (reviewed by Laranjo et al., 2014). However, rhizobia may also be used as nonsymbiotic PGPR for the production of nonlegume crops (rice or wheat), which are of economic significance. Such details indicate the importance of rhizobia and the high rate of research on the use of rhizobia as models of mutualistic associations benefiting sustainable agriculture (reviewed by Laranjo et al., 2014).

Interestingly, using soil bacteria, which are nonrhizobia, for the inoculation of plants has also recently become the center of attention. For example, it has been indicated that *Azospirillum* is able to increase plant growth and seed yields under different conditions including stress by the use of different mechanisms, including the production of plant hormones and the increasing phosphate uptake by plant roots. *Azospirillum* coinoculated with rhizobia can enhance nodulation and N fixation in the host plant (Rodríguez-Navarro et al., 2011).

Interaction with biochar

With respect to the use of biochar as a source of soil amendment for the production of soybean, there is an increasing rate of research in this area. Biochar can increase CEC and base saturation nine fold over that in control soils, and significantly increases available K, Ca, Mg, total N, and P (Glaser et al., 2002). Iijima et al. (2015) recommended using nodule bacteria with biochar in the subsoil employing a technique called crack fertilization to enhance soybean yield. However, there are also potentially important differences that necessitate testing biochar for the negative effects it may have. Using biochar affected soybean production by increasing: (1) pH; (2) soil C; and (3) the surface area of subnanopores; and (4) by decreasing soil bulk density compared with the control (Mukherjee et al., 2014).

In Thailand, for example, the effects of quail litter biochar at 0, 24.6, 49.2, 73.8, 98.4, and 123 g per pot were tested on soybean growth. Biochar enhanced soil fertility and increased soybean production an optimum level of 98.4 g per pot mixture. Increasing the amount of biochar increased the nutrient contents in the soil; however, there were no effects on quantities of heavy metal residues in the leaves and seeds. Levels higher than 98.4 g per pot mixture were not recommended for plant growth because of the attendant alkalinity, affecting soil pH (Suppadit et al., 2012). Lastly, there are no adequate data available from which to draw conclusions about the usefulness of biochar. Complementary details can be obtained from other reviews (Biederman and Harpole, 2013; Lehmann et al., 2011).

Conclusions

In the introduction to this chapter, the interest in the impact of agriculture on soil structure or in changing the makeup of soil species was mentioned. It is because such effects can influence the production of soybean worldwide. The importance of soybean production was presented with respect to the parameters involved, which may affect such production. Among such parameters the most important is the use of rhizobial inocula. The use of mycorrhizal fungi is also of significance, which must be investigated further. Preserving agro-ecosystem services can be decisive to buffer the negative effects of global change. Throughout the chapter, the biotic associations of soybean have been pointed out. Mycorrhizal fungi and rhizobia have greater potential as biofertilizers, but further studies are required to understand the full role of soil microbes in association with soybean. Literature has proposed that few rhizobia can survive under unfavorable conditions in soil. Bioinoculants of rhizobial strains efficiently improve soybean growth and productivity. Finally, this chapter showed that soybean management can play a crucial role in the future of soybean production but more research is needed.

Acknowledgments

Marcela Pagano is grateful to FAPEMIG, to CAPES for postdoctoral scholarships, CAPES-PNPD Process No. 23038.007147/2011-60 and MCTI/CNPQ No. 14/2012 (Universal Grant) from Conselho Nacional de Desenvolvimento Científico e Tecnológico (CNPq). We would also like to thank Dr Tadayoshi Masuda, from Kinki University, Japan, for providing us with Figure 1.1.

References

Afzal, A., Bano, A., Fatima, M., 2010. Higher soybean yield by inoculation with N-fixing and P-solubilizing bacteria. Agron. Sustain. Dev. 30, 487–495.

Agyei, D., Potumarthi, R., Danquah, M.K., 2015. Food-derived multifunctional bioactive proteins and peptides: applications and recent advances. In: Gupta, V.K., Tuohy, M.G., O'Donovan, A., Lohani, M. (Eds.), Biotechnology of Bioactive Compounds: Sources and Applications. Wiley Blackwell, Chichester, pp. 507–524.

Allen, O.N., Allen, E.K., 1981. The Leguminosae. University of Wisconsin Press, Madison, WI.

Alves, B.J.R., Boddey, R.M., Urquiaga, S., 2003. The success of BNF in soybean in Brazil. Plant Soil 252, 1–9.

Anderson, K.J., 1957. The effect of soil microorganisms on the plant-rhizobia association. Phyton 8, 59–73.

Antoun, H., Beauchamp, C.J., Goussard, N., Chabot, R., Lalande, R., 1998. Potential of *Rhizobium* and *Bradyrhizobium* species as plant growth promoting rhizobacteria on non-legumes: effect on radishes (*Raphanus sativus* L.). Plant Soil 204, 57–67.

Aysan, E., Demir, S., 2009. Using arbuscular mycorrhizal fungi and *Rhizobium leguminosarum*, Biovar Phaseoli against *Sclerotinia sclerotiorum* (Lib.) de bary in the common bean *Phaseolus vulgaris* L. Plant Pathol. J. 8, 74–78.

Bai, Y., Zhou, X., Smith, D.L., 2003. Enhanced soybean plant growth resulting from coinoculation of *Bacillus* strains with *Bradyrhizobium japonicum*. Crop Sci. 43, 1774–1781.

Baird, R.E., Watson, C.E., Scruggs, M., 2003. Relative longevity of *Macrophomina phaseolina* and associated mycobiota on residual soybean roots in soil. Plant Dis. 87, 563–566.

Barcellos, F.G., Menna, P., da Silva Batista, J.S., Hungria, M., 2007. Evidence of horizontal transfer of symbiotic genes from a *Bradyrhizobium japonicum* inoculant strain to indigenous diazotrophs *Sinorhizobium* (Ensifer) *fredii* and *Bradyrhizobium elkanii* in a Brazilian Savannah soil. Appl. Environ. Microbiol. 73, 2635–2643.

Barea, J.M., Azcon, R., Azcon-Aguilar, C., 2002. Mycorrhizosphere interactions to improve plant fitness and soil quality. Antonie van Leeuwenhoek 81, 343–351.

Biederman, L., Harpole, W.S., 2013. Biochar and its effects on plant productivity and nutrient cycling: a meta-analysis. Glob. Change Biol. Bioenergy 5, 202–214.

Blilou, J., Ocampo, J., Garcia-Garrido, J., 1999. Resistance of pea roots to endomycorrhiza fungus or *Rhizobium* correlates with enhanced levels of endogenous salicylic acid. J. Exp. Bot. 50, 1663–1668.

Bohm, G.B., Rombaldi, C.V., 2010. Efeito da transformação genética e da aplicação do glifosato na microbiota do solo, fixação biológica de nitrogênio, qualidade e segurança de grãos de soja geneticamente modificada. Ciência Rural 40, 213–221.

Borie, F., Rubio, R., Rouanet, J.L., Morales, A., Borie, G., Rojas, C., 2006. Effects of tillage systems on soil characteristics, glomalin and mycorrhizal propagules in a Chilean Ultisol. Soil Till. Res. 88, 253–261.

Bucher, M., Rausch, C., Daram, P., 2001. Molecular and biochemical mechanisms of phosphorus uptake into plants. J. Plant Nutr. Soil Sci. 164, 209–221.

Campo, R.J., Silva Araujo, R., Hungria, M., 2009. Molybdenum-enriched soybean seeds enhance N accumulation, seed yield, and seed protein content in Brazil. Field Crops Res. 110, 219–224.

Cardoso, E.J.B.N., 1985. Efeito de micorriza vesículo-arbuscular e fosfato-de-rocha na simbiose soja-*Rhizobium*. Rev. Brasileira Ciência Solo 9, 125–130.

Castro, G.S.A., Crusciol, C.A.C., 2013. Yield and mineral nutrition of soybean, maize, and Congo signal grass as affected by limestone and slag. Pesqui. Agropecu. Bras. 48, 673–681.

Chabot, R., Antoun, H., Cescas, M.P., 1996. Growth promotion of maize and lettuce by phosphate-solubilizing *Rhizobium leguminosarum* biovar. phaseoli. Plant Soil 184, 311–321.

Chianu, J.N., Huising, J., Danso, S., Okoth, P., Chianu, J.N., Sanginga, N., 2010. Financial value of nitrogen fixation in soybean in Africa: increasing benefits for smallholder farmers. J. Life Sci. 4, 50–59.

Cooper, J.E., 2007. Early interactions between legumes and rhizobia: disclosing complexity in a molecular dialogue. J. Appl. Microbiol. 103, 1355–1365.

Cotton, T.E., Fitter, A.H., Miller, R.M., Dumbrell, A.J., Helgason, T., 2015. Fungi in the future: interannual variation and effects of atmospheric change on arbuscular mycorrhizal fungal communities. New Phytol. 205, 1598–1607.

Covacevich, F., Berbara, R.L.L., 2010. The impact of climate changes on belowground: how the CO_2 increment affects arbuscular mycorrhiza? In: Thangadurai, E., Hijri, M., Busso, C.A. (Eds.), Perspectives in Mycorrhizal Research. Bioscience Publishers, Enfield, NH, pp. 203–211.

Curaqueo, G., Acevedo, E., Cornejo, P., Seguel, A., Rubio, R., Borie, F., 2010. Tillage effect on soil organic matter, mycorrhizal hyphae and aggregates in a Mediterranean agroecosystem. Rev. Ciencia Suelo Nutr. Veg. 10, 12–21.

Dashti, N., Zhang, F., Hynes, R., Smith, D.L., 1997. Application of plant growth-promoting rhizobacteria to soybean [*Glycine max* (L.) Merr.] increases protein and dry matter yield under shortseason conditions. Plant Soil 188, 33–41.

Denison, R.F., Kiers, E.T., 2004. Why are most rhizobia beneficial to their plant hosts, rather than parasitic? Microbes Infect. 6, 1235–1239.

Deshwal, V.K., Singh, S.B., Kumar, P., Chubey, A., 2013. Rhizobia unique plant growth promoting rhizobacteria: a review. Int. J. Life Sci. 2, 74–86.

Diaz, D.A.R., Pedersen, P., Sawyer, J.E., 2009. Soybean response to inoculation and nitrogen application following long-term grass pasture. Crop Sci. 49, 1058–1062.

Dominati, E., Patterson, M., Mackay, A., 2010. A framework for classifying and quantifying the natural capital and ecosystem services of soils. Ecol. Econ. 69, 1858–1868.

Dwivedi, S.L., Kanwar, L., Sahrawat, Hari, Upadhyaya, D., Mengoni, A., Galardini, M., Bazzicalupo, M., Biondi, E.G., Hungria, M., Kaschuk, G., Blair, M.W., Ortiz, R., 2015. Advances in host plant and rhizobium genomics to enhance symbiotic nitrogen fixation in grain legumes. Adv. Agron. 129, 1–116.

Elkan, G.H., 1992. Taxonomy of the rhizobia. Can. J. Microbiol. 38, 446–450.

FAO (Food and Agriculture Organization), 2003. http://apps.fao.org.

Faquin, V., Cinética da absorção de fosfato, nutrição mineral, crescimento e produção da soja sob influência de micorriza vesículo-arbuscular (MVA). Piracicaba, Escola Superior de Agricultura "Luiz de Queiroz", Universidade de São Paulo, 1988. 136 pp.

Faye, A., Dalpé, Y., Ndung'u-Magiroi, K., Jefwa, J., Ndoye, I., Diouf, M., Lesueur, D., 2013. Evaluation of commercial arbuscular mycorrhizal inoculants. Can. J. Plant Sci. 93, 1201–1208.

Foley, J.A., Ramankutty, N., Brauman, K.A., Cassidy, E.S., Gerber, J.S., Johnston, M., Mueller, N.D., O'Connell, C., Ray, D.K., West, P.C., Balzer, C., Bennett, E.M., Carpenter, S.R., Hill, J., Monfreda, C., Polasky, S., Rockström, J., Sheehan, J., Siebert, S., Tilman, D., Zaks, D.P.M., 2011. Solutions for a cultivated planet. Nature 478, 337–342.

Frank, B., 1889. Uber die pilzsymbiose der leguminosen. Berichte Deutschen Botanischen Gesellschaft 7, 332–346.

Fred, E.B., Baldwin, I.L., McCoy, E., 1932. Root nodule bacteria and leguminous plants. University of Wisconsin Studies in Science No. 5. University of Wisconsin, Madison.

Gemell, L.G., Hartley, E.J., Herridge, D.F., 2005. Point-of-sale evaluation of preinoculated and custom-inoculated pasture legume seed. Anim. Prod. Sci. 45, 161–169.

Glaser, B., Lehmann, J., Zech, W., 2002. Ameliorating physical and chemical properties of highly weathered soils in the tropics with charcoal – a review. Biol. Fertil. Soils 35, 219–230.

Glick, B.R., 2012. Plant growth-promoting bacteria: mechanisms and applications. Scientifica 2012, 1–15.

Grümberg, B.C., Urcelay, C., Shroeder, M.A., Vargas-Gil, S., Luna, C.M., 2015. The role of inoculum identity in drought stress mitigation by arbuscular mycorrhizal fungi in soybean. Biol. Fertil. Soils 51, 1–10.

Halder, A., Chakrabarty, P.K., 1993. Solubilization of inorganic phosphate by *Rhizobium*. Folia Microbiol. 38, 325–330.

Herrmann, L., Atieno, M., Brau, L., Lesueur, D., 2013. Microbial quality of commercial inoculants to increase BNF and nutrient use efficiency. In: de Bruijn, F.J. (Ed.), Molecular Microbial Ecology of the Rhizosphere. Wiley-Blackwell, Hoboken, NJ.

Higo, M., Isobe, K., Yamaguchi, M., Drijber, R.A., Jeske, E.S., Ishii, R., 2013. Diversity and vertical distribution of indigenous arbuscular mycorrhizal fungi under two soybean rotational systems. Biol. Fertil. Soils 49, 1085–1096.

Holt, J.G., Krieg, N.R., Sneath, P.H.A., Staley, J.T., Williams, S.T., 1994. Bergey's Manual of Determinative Bacteriology. Williams and Wilkins Press, Baltimore, MD.

Hungria, M., Vargas, M., 2000. Environmental factors affecting N_2 fixation in grain legumes in the tropics, with an emphasis on Brazil. Field Crops Res. 65, 151–164.

Hungria, M., Loureiro, M.F., Mendes, I.C., Campo, R.J., Graham, P.H., 2005a. Inoculant preparation, production and application. In: Werner, D., Newton, W.E. (Eds.), Nitrogen Fixation in Agriculture, Forestry, Ecology, and the Environment. Kluwer, Dordrecht, pp. 223–253.

Hungria, M., Franchini, J.C., Campo, R.J., Graham, P.H., 2005b. The importance of nitrogen fixation to soybean cropping in South America. In: Werner, D., Newton, W.E. (Org.), Nitrogen Fixation in Agriculture: Forestry Ecology and Environment. Kluwer Academic Publishers, Dordrecht, pp. 25–42.

Iijima, M., Yamane, K., Izumi, Y., Daimon, H., Motonaga, T., 2015. Continuous application of biochar inoculated with root nodule bacteria to subsoil enhances yield of soybean by the nodulation control using crack fertilization technique. Plant Prod. Sci. 18, 197–208.

Jarvis, B.D.W., van Berkum, P., Chen, W.X., Nour, S.M., Fernandez, M.P., Cleyet-Marel, J.C., Gillis, M., 1997. Transfer of *Rhizobium loti, Rhizobium huakuii, Rhizobium ciceri, Rhizobium mediterraneum* and *Rhizobium tianshanense* to *Mesorhizobium* gen. nov. Int. J. Syst. Bacteriol. 47, 895–898.

Jia, Y., Gray, V.M., Straker, C.J., 2004. The influence of rhizobium and arbuscular mycorrhizal fungi on nitrogen and phosphorus accumulation by *Vicia faba*. Ann. Bot. 94, 251–258.

Jordan, D.C., 1982. Transfer of *Rhizobium japonicum* Buchanan 1980 to *Bradyrhizobium* gen. nov., a genus of slow growing root nodule bacteria from leguminous plants. Int. J. Syst. Bacteriol. 32, 136–139.

Juge, C., Prévost, D., Bertrand, A., Bipfubusa, M., Chalifour, F.-P., 2012. Growth and biochemical responses of soybean to double and triple microbial associations with *Bradyrhizobium*, *Azospirillum* and arbuscular mycorrhizae. Appl. Soil Ecol. 61, 147–157.

Junior, A.F.C., Reis, M.R., Santos, G.R., Erasmo, E.A., Chagas, L.F.B., 2013. Nodulation and mycorrhization of transgenic soybean after glyphosate application. Semina Ciências Agrárias 34, 3675–3682.

Kahiluoto, H., Ketoja, E., Vestberg, M., 2009. Contribution of arbuscular mycorrhiza to soil quality in contrasting cropping systems. Agric. Ecosyst. Environ. 134, 36–45.

Kuykendall, L.D., Saxena, B., Devine, T.E., Udell, S.E., 1992. Genetic diversity in *Bradyrhizobium japonicum* Jordan 1982 and a proposal for *Bradyrhizobium elkani* sp. nov. Can. J. Microbiol. 38, 501–505.

Kvien, C.S., Ham, G.E., Lambert, J.W., 1981. Recovery of introduced *Rhizobium japonicum* strains by soybean genotypes. Agron. J. 73, 900–905.

Lal, R., 2009. Soils and food sufficiency. A review. Agron. Sustain. Dev. 29, 113–133.

Lal, R., 2010. Enhancing eco-efficiency in agro-ecosystems through soil C sequestration. Crop Sci. 50, S120–S131.

Laranjo, M., Machado, J., Young, J.P.W., Oliveira, S., 2004. High diversity of chickpea *Mesorhizobium* species isolated in a Portuguese agricultural region. FEMS Microbiol. Ecol. 48, 101–107.

Laranjo, M., Alexandre, A., Rivas, R., Velázquez, E., Young, J.P.W., Oliveira, S., 2008. Chickpea rhizobia symbiosis genes are highly conserved across multiple *Mesorhizobium* species. FEMS Microbiol. Ecol. 66, 391–400.

Laranjo, M., Young, J.P.W., Oliveira, S., 2012. Multilocus sequence analysis reveals multiple symbiovars within *Mesorhizobium* species. System. Appl. Microbiol. 35, 359–367.

Laranjo, M., Alexandre, A., Oliveira, S., 2014. Legume growth-promoting rhizobia: an overview on the *Mesorhizobium* genus. Microbiol. Res. 169, 2–17.

Lehmann, J., Rillig, M.C., Thies, J., Masiello, C.A., Hockaday, W.C., Crowley, D., 2011. Biochar effects on soil biota: a review. Soil Biol. Biochem. 43, 1812–1836.

Lerat, S., Lapointe, L., Gutjahr, S., Piché, Y., Vierheilig, H., 2003. Carbon partitioning in a split-root system of arbuscular mycorrhizal plants is fungal and plant species dependent. New Phytol. 157, 589–595.

Li, D.M., Alexander, M., 1988. Co-inoculation with antibiotic producing bacteria to increase colonization and nodulation by rhizobia. Plant Soil 108, 211–219.

Li, S.M., Li, L., Zhang, F.S., 2004. Enhancing phosphorus and nitrogen uptake of faba bean by inoculating arbuscular mycorrhizal fungus and *Rhizobium leguminosarum*. J. China Agric. Univ. 9, 11–15.

Lisette, J., Xavier, C., Germida, J.J., 2003. Selective interactions between arbuscular mycorrhizal fungi and *Rhizobium leguminosarum* bv. *viceae* enhance pea yield and nutrition. Biol. Fertil. Soils 37, 261–267.

Löhis, F., Hansen, R., 1921. Nodulating bacteria of leguminous plant. J. Agric. Res. 20, 543–556.

Long, S.R., 1989. Rhizobium-legume nodulation: life together in the underground. Cell 56, 203–214.

López-López, A., Rosenblueth, M., Martínez, J., Martínez-Romero, E., 2010. Rhizobial symbioses in tropical legumes and non-legumes. Soil Biology and Agriculture in the Tropics. Springer, Berlin, Heidelberg, pp. 163–184.

Maier, J., Graham, L., 1990. Mutant strains of *Bradyrhizobium japonicum* with increased symbiotic N_2 fixation rates and altered Mo metabolism properties. Appl. Environ. Microbiol. 56, 2341–2346.

Malty, J.D.S., Siqueira, J.O., Moreira, F.M.S., 2006. Effects of glyphosate on soybean symbiotic microorganisms, in culture media and in greenhouse. Pesqui. Agropecu. Bras. 41, 285–291.

MAPA, 2015. Ministério da Agricultura, Pecuária e Abastecimento, Esplanada dos Ministérios – Bloco D – Brasília/DF.

Masuda, T., Goldsmith, P.D., 2009. World soybean production: area harvested, yield, and long-term projections. Int. Food Agribus. Manag. Rev. 12, 143–161.

Minhoni, M.T.A., Cardoso, E.J.B.N., Eira, A.F., 1993. Efeitos da adição de fosfato de rocha, bagaço de cana-de-açúcar, fosfato solúvel e fungo micorrízico no crescimento e na absorção de nutrientes por plantas de soja. Rev. Brasileira Ciência Solo 17, 173–178.

Miranda, J.C.C., 2008. Cerrado, Micorriza Arbuscular Ocorrência e Manejo. Embrapa, Planaltina 169 pp.

Miranda, J.C.C., Vilela, L., Miranda, L.N., 2005. Dynamics and contribution of arbuscular mycorrhiza in culture systems with crop rotation. Pesqui. Agropecu. Bras. 40, 1005–1014.

Miransari, M., 2011a. Interactions between arbuscular mycorrhizal fungi and soil bacteria. Review article. Appl. Microbiol. Biotechnol. 89, 917–930.

Miransari, M., 2011b. Soil microbes and plant fertilization. Review article. Appl. Microbiol. Biotechnol. 92, 875–885.

Miransari, M., Riahi, H., Eftekhar, F., Minaie, A., Smith, D.L., 2013. Improving soybean (*Glycine max* L.) N_2 fixation under stress. J. Plant Growth Regul. 32, 909–921.

Mpepereki, S., Javaheri, F., Davis, P., Giller, K.E., 2000. Soybeans and sustainable agriculture: "Promiscuous" soybeans in southern Africa. Field Crops Res. 65, 137–149.

Mukherjee, A., Lal, R., Zimmerman, A.R., 2014. Effects of biochar and other amendments on the physical properties and greenhouse gas emissions of an artificially degraded soil. Sci. Total Environ. 487, 26–36.

Nandasena, K.G., O'Hara, G.W., Tiwari, R.P., Sezmis, E., Howieson, J.G., 2007. In situ lateral transfer of symbiosis islands results in rapid evolution of diverse competitive strains of mesorhizobia suboptimal in symbiotic nitrogen fixation on the pasture legume Biserrula pelecinus L. Environ. Microbiol. 9, 2496–2511.

Nogueira, M.A., Cardoso, E.J.B.N., 2000. External mycelium production by arbuscular mycorrhizal fungi and growth of soybean fertilized with phosphorus. Rev. Brasileira Ciência Solo 24, 329–338.

Nyfeler, D., Huguenin-Elie, O., Suter, M., Frossard, E., Lüscher, A., 2011. Grass-legume mixtures can yield more nitrogen than legume pure stands due to mutual stimulation of nitrogen uptake from symbiotic and non-symbiotic sources. Agric. Ecosyst. Environ. 140, 155–163.

Oerke, E.C., 2005. Crop losses to pests. J. Agric. Sci. 144, 31–43.

Pagano, M.C. (Ed.), 2012. Mycorrhiza: Occurrence in Natural and Restored Environments. Nova Science Publishers, New York.

Pagano, M.C., 2013. Plant and soil biota: crucial for mitigating climate change. In: Interactions of Forests, Climate, Water Resources, and Humans in a Changing Environment. Br. J. Environ. Climate Change SCIENCEDOMAIN Int. 3 (2), 188–196.

Pagano, M.C., Covacevich, F., 2011. Arbuscular mycorrhizas in agroecosystems. In: Fulton, S.M. (Ed.), Mycorrhizal Fungi: Soil, Agriculture and Environmental Implications. Nova Science Publishers, New York, pp. 35–65.

Pagano, M.C., Schalamuk, S., Cabello, M.N., 2011. Arbuscular mycorrhizal parameters and indicators of soil health and functioning: applications for agricultural and agroforestal systems. In: Miransari, M. (Ed.), Soil Microbes and Environmental Health. Nova Science Publishers, New York, pp. 267–276.

Palleroni, N.J., 1992. Present situation of the taxonomy of aerobic pseudomonads. In: Galli, E., Silver, S., Witholt, B. (Eds.), Pseudomonas: Molecular Biology and Biotechnology. American Society for Microbiology, Washington, DC, pp. 105–115.

Peix, A., Rivas, R., Mateos, P.F., Martínez-Molina, E., Rodríguez-Barrueco, C., Velazquez, E., 2003. Pseudomonas rhizosphaerae sp. nov., a novel species that actively solubilizes phosphate in vitro. Int. J. Syst. Evol. Microbiol. 53, 2067–2072.

Pereira, P.A.A., Bliss, F.A., 1989. Selection of common bean (Phaseolus vulgaris L.) for N_2 fixation at different levels of available phosphorus under field and environmentally-controlled conditions. Plant Soil 115, 75–82.

Perez-Brandan, C., Huidobro, J., Grumberg, B., Scandiani, M.M., Luque, A.G., Meriles, J.M., Vargas-Gil, S., 2014. Soybean fungal soil-borne diseases: a parameter for measuring the effect of agricultural intensification on soil health. Can. J. Microbiol. 60, 73–84.

Plazinski, J., Rolfe, B.G., 1985. Influence of Azospirillum strains on the nodulation of clovers by Rhizobium strains. Appl. Environ. Microbiol. 49, 984–989.

Porcel, R., Barea, J.M., Ruiz-Lozano, J.M., 2003. Antioxidant activities in mycorrhizal soybean plants under drought stress and their possible relationship to the process of nodule senescence. New Phytol. 157, 135–143.

Powlson, D.S., Gregory, P.J., Whalley, W.R., Quinton, J.N., Hopkins, D.W., Whitmore, A.P., Hirsch, P.R., Goulding, K.W.T., 2011. Soil management in relation to sustainable agriculture and ecosystem services. Food Policy 36, S72–S87.

Qaim, M., Traxler, G., 2005. Roundup Ready soybeans in Argentina: farm level and aggregate welfare effects. Agric. Econ. 32, 73–86.

Raaijmakers, J.M., Timothy, C., Paulitz, T., Steinberg, C., Alabouvette, C., Moënne-Loccoz, Y., 2008. The rhizosphere: a playground and battlefield for soil borne pathogens and beneficial microorganisms. Plant Soil 321, 341–361.

Rahmani, H.A., Saleh-Rastin, N., Khavazi, K., Asgharzadeh, A., Fewer, D., Kiani, S., Lindstrom, K., 2009. Selection of thermotolerant bradyrhizobial strains for nodulation of soybean (*Glycine max* L.) in semi-arid regions of Iran. World J. Microbiol. Biotechnol. 25, 591–600.

Rao, D.L.N., 2014. Recent advances in biological nitrogen fixation in agricultural systems. Proc. Ind. Natl. Sci. Acad. 80, 359–378.

Ray, D.K., Mueller, N.D., West, P.C., Foley, J.A., 2013. Yield trends are insufficient to double global crop production by 2050. PLoS One 8, 6.

Redmile-Gordon, M.A., Brookes, P.C., Evershed, R.P., Goulding, K.W.T., Hirsch, P.R., 2014. Measuring the soil-microbial interface: extraction of extracellular polymeric substances (EPS) from soil biofilms. Soil Biol. Biochem. 72, 163–171.

Reis, M.R., Silva, A.A., Pereira, J.L., Freitas, M.A.M., Costa, M.D., Silva, M.C.S., Santos, E.A., França, A.C., Ferreira, G.L., 2010. Impacto do glyphosate associado com endossulfan e tebuconazole sobre microrganismos endossimbiontes da soja. Planta Daninha Viçosa 28, 113–121.

Robert, J.W., 1986. The Soybean Solution: Meeting World Food Needs. NIT-College of Agriculture, University of Illinois at Urbana, Champaign, USA, 1 Bulletin, pp. 4–27.

Rodino, A.P., Metrae, R., Guglielmi, S., Drevon, J.J., 2009. Variation among common-bean accessions (*Phaseolus vulgaris* L.) from the Iberian Peninsula for N_2-dependent growth and phosphorus requirement. Symbiosis 47, 161–174.

Rodríguez-Navarro, D.N., Oliver, I.M., Contreras, M.A., Ruiz-Sainz, J.E., 2011. Soybean interactions with soil microbes, agronomical and molecular aspects. Agron. Sustain. Dev. 31, 173–190.

Rolfe, B.G., 1988. Flavones and isoflavones as inducing substances of legume nodulation. Biofactors 1, 3–10.

Salvagiotti, F., Cassman, K.G., Specht, J.E., Walters, D.T., Weiss, A., Dobermann, A., 2008. Nitrogen uptake, fixation and response to fertilizer N in soybeans: a review. Field Crops Res. 108, 1–13.

Saxena, A.K., Rathi, S.K., Tilak, K.V.B.R., 1997. Differential effect of various endomycorrhizal fungi on nodulating ability of green gram by *Bradyrhizobium* sp. (*Vigna*) strain S24. Biol. Fertil. Soils 24, 175–178.

Sene, G., Samba-Mbaye, R., Thiao, M., Khasa, D., Kane, A., et al., 2012. The abundance and diversity of legume-nodulating rhizobia and arbuscular mycorrhizal fungal communities in soil samples from deforested and man-made forest systems in a semiarid Sahel region in Senegal. Eur. J. Soil Biol. 52, 30–40.

Seta, A., Blevins, R., Frye, W., Barfield, B., 1993. Reducing soil erosion and agricultural chemical losses with conservation tillage. J. Environ. Qual. 22, 661–665.

Simard, S., Austin, M.E., 2010. The role of mycorrhizas in forest soil stability with climate change. In: Simard, S., Austin, M.E. (Eds.), Climate Change and Variability. INTECH Open Access Publisher, pp. 275–302.

Siqueira, J.O., Hubbell, D.H., Valle, R.R., 1984. Effects of P on formation of the VAM symbiosis. Pesqui. Agropecu. Bras. 19, 1465–1474.

Smith, S.E., Read, D.J., 2008. Mycorrhizal Symbiosis. Elsevier, New York.

Sobolevsky, A., Moschini, G., Lapan, H., 2005. Genetically modified crops and product differentiation: trade and welfare effects in the soybean complex. Am. J. Agric. Econ. 87, 621–644.

Sprent, J.I., 2007. Evolving ideas of legume evolution and diversity: a taxonomic perspective on the occurrence of nodulation. New Phytol. 174, 11–25.

Stürmer, S.L., Siqueira, J.O., 2011. Species richness and spore abundance of arbuscular mycorrhizal fungi across distinct land uses in Western Brazilian Amazon. Mycorrhiza 21, 255–267.

Sugiyama, A., Ueda, Y., Takase, H., Yazaki, K., 2015. Do soybeans select specific species of *Bradyrhizobium* during growth? Commun. Integr. Biol. 8, e992734.

Suleiman, A.K.A., Manoeli, L., Boldo, J.T., Pereira, M.G., Roesch, L.F.W., 2013. Shifts in soil bacterial community after eight years of land-use change. Syst. Appl. Microbiol. 36, 137–144.

Suppadit, T., Phumkokrak, N., Poungsuk, P., 2012. The effect of using quail litter biochar on soybean (*Glycine max* [L.] Merr.). Chilean J. Agric. Res. 72, 244–251.

Tajini, F., Trabelsi, M., Drevon, J.J., 2012. Combined inoculation with *Glomus intraradices* and *Rhizobium tropici* CIAT899 increases phosphorus use efficiency for symbiotic nitrogen fixation in common bean (*Phaseolus vulgaris* L.). Saudi J. Biol. Sci. 19, 157–163.

Takahashi, T., Ito, A., Suzuki, H., 1996. Interaction between effective bacteria and host plant. In: Ishizuka, J. (Ed.), Studies on Nodule Formation and Nitrogen Fixation in Legume Crops. Ministry of Agriculture, Forestry and Fisheries, Tokyo, pp. 92–107.

Tang, C., Hinsinger, P., Jaillard, B., Rengel, Z., Drevon, J., 2001. Effect of phosphorus deficiency on the growth, symbiotic N_2 fixation and proton release by two bean (*Phaseolus vulgaris*) genotypes. Agronomie 21, 683–689.

The Economist, 2010. The miracle of the cerrado: Brazil has revolutionised its own farms. Can it do the same for others? The Economist, Aug. 26, 2010, CREMAQ, Piauí.

Tilman, D., Balzer, C., Hill, J., Befort, B.L., 2011. Global food demand and the sustainable intensification of agriculture. Proc. Natl. Acad. Sci. USA 108, 20260–20264.

Torn, M.S., Chabbi, A., Crill, P., Hanson, P.J., Janssens, I.A., Luo, Y., Pries, C.H., Rumpel, C., Schmidt, M.W.I., Six, J., Schrumpf, M., Zhu, B., 2015. A call for international soil experiment networks for studying, predicting, and managing global change impacts. Soil Discuss. 2, 133–151.

Treseder, K.K., Turner, K.M., 2007. Glomalin in ecosystems. Soil Sci. Soc. Am. J. 71, 1257–1266.

Uchida, Y., Akiyama, H., 2013. Mitigation of post harvest nitrous oxide emissions from soybean ecosystems: a review. Soil Sci. Plant Nutr. 59, 477–487.

Vadez, V., Rodier, F., Payre, H., Drevon, J.J., 1996. Nodule permeability to O_2 and nitrogenase-linked respiration in bean genotypes varying in the tolerance of N_2 fixation to P deficiency. Plant Physiol. Biochem. 34, 871–878.

Van der Merwe, S.P., Strijdom, B.W., Uys, C.J., 1974. Groundnut response to seed inoculation under extensive agricultural practices in South African soils. Phytophylactics 6, 295–302.

van Groenigen, K.J., Qi, X., Osenberg, C.W., Luo, Y., Hungate, B.A., 2014. Faster decomposition under increased atmospheric CO_2 limits soil carbon storage. Science 344, 508–509.

Vanhie, M., Deen, W., Lauzon, J.D., Hooker, D.C., 2015. Effect of increasing levels of maize (*Zea mays* L.) residue on no-till soybean (*Glycine max* Merr.) in Northern production regions: a review. Soil Tillage Res. 150, 201–210.

Wall, D.H., Nielsen, U.N., 2012. Biodiversity and ecosystem services: is it the same below ground? Nat. Educ. Knowledge 3, 8.

Wang, X., Pan, Q., Chen, F., Yan, X., Liao, H., 2011. Effects of co-inoculation with arbuscular mycorrhizal fungi and rhizobia on soybean growth as related to root architecture and availability of N and P. Mycorrhiza 21, 173–181.

Weaver, R.W., Frederick, L.R., 1974. Effect of inoculum rate on competitive nodulation of *Glycine max* L. Merrill. II. Field studies. Agron. J. 66, 233–236.

Williams, L.E., Phillips, D.A., 1983. Increased soybean productivity with a *Rhizobium japonicum* mutant. Crop Sci. 23, 246–250.

Wingeyer, A.B., Amado, T.J.C., Pérez-Bidegain, M., Studdert, G.A., Varela, C.H.P., Garcia, F.O., Karlen, D.L., 2015. Soil quality impacts of current South American agricultural practices. Sustainability 7, 2212–2242.

Xie, Z., Staehelin, C., Vierheili, H., Wiemken, A., Jabbouri, S., Broughton, W.J., Vogeli-Lange, R., Boller, T., Xie, Z.P., 1995. Rhizobial nodulation factors stimulate mycorrhizal colonization of undulating and non-nodulating soybeans. Plant Physiol. 108, 1519–1525.

Xu, L.M., Ge, C., Cui, Z., Li, J., Fan, H., 1995. *Bradyrhizobium liaoningensis* sp. nov. isolated from the root nodules of soybean. Int. J. Syst. Bacteriol. 45, 706–711.

Yamakawa, T., Saeki, Y., 2013. Inoculation methods of *Bradyrhizobium japonicum* on soybean in South-West area of Japan. A Comprehensive Survey of International Soybean Research – Genetics, Physiology, Agronomy and Nitrogen Relationships. INTECH Open Access Publisher, pp. 83–114.

Yiridoe, E., Vyn, T., Weersink, A., Hooker, D., Swanton, C., 2000. Farm-level profitability analysis of alternative tillage systems on clay soils. Can. J. Plant Sci. 80, 65–73.

Young, J.P.W., 1991. Phylogenetic classification of nitrogen-fixing organisms. In: Stacey, G., Burris, R.H., Evans, H.J. (Eds.), Biological Nitrogen Fixation. Chapman and Hall, New York, pp. 43–86.

Young, J.A.W., Haukka, K., 1996. Diversity and phylogeny of rhizobia. New Phytol. 133, 87–94.

Signaling cross talk between biotic and abiotic stress responses in soybean

2

Priyanka Gupta*, Rita Sharma*, Manoj K. Sharma**,
Mahaveer P. Sharma†, Gyanesh K. Satpute†, Shivani Garg†,
Sneh L. Singla-Pareek‡, Ashwani Pareek*
*Stress Physiology and Molecular Biology Laboratory, School of Life Sciences, Jawaharlal Nehru University, New Delhi, India; **School of Biotechnology, Jawaharlal Nehru University, New Delhi, India; †ICAR-Directorate of Soybean Research, Khandwa Road, Indore, Madhya Pradesh, India; ‡Plant Molecular Biology Group, International Centre for Genetic Engineering and Biotechnology, New Delhi, India

Introduction

The cultivated soybean (*Glycine max* (L.) Merr.) is a dominant oilseed crop in the world. It is not only a rich source of proteins and isoflavones, but also exerts a positive impact on the environment by establishing a symbiotic relationship with nitrogen-fixing bacteria (Singh and Hymowitz, 1999). According to the United States Soybean Export Council (USSEC, 2008), soybean is estimated to contribute about 56% of oilseed and 30% of total vegetable oil production worldwide. With an annual production of 261 million tons and average productivity of 2533 kg ha^{-1}, soybean is globally cultivated on 103 million ha of land (FAOSTAT, 2013).

The top five soybean producing countries in the world, the United States, Brazil, Argentina, China, and India, together account for more than 90% of the world's soybean production (Leff et al., 2004). However, declining global acreage, unpredictable environmental conditions, and severe disease epidemics are a major threat to soybean production in India and worldwide. The major diseases affecting soybean production include soybean cyst nematode, root knot nematode, reniform nematode, bacterial blight, soybean rust, frogeye leaf spot, collar rot, bud blight, yellow mosaic virus, and charcoal rot (Haile et al., 1998; Wrather et al., 1997; Yang et al., 1999; Yorinori et al., 2005).

Several abiotic stresses including drought, flooding, salinity, heat, and cold adversely affect productivity in soybean (Tran and Mochida, 2010; Valliyodan and Nguyen, 2008). To cope with these challenges, significant efforts have been made by researchers toward elucidating stress signaling pathways in response to individual stresses. However, under natural growth conditions, plants are often exposed to more than one stress, simultaneously or in a sequential manner.

Recent evidence shows that these stresses have synergistic or antagonistic effects on each other due to extensive cross talk in underlying signaling cascades (Atkinson and Urwin, 2012; Sharma et al., 2013). Elucidating this cross talk is essential

Abiotic and Biotic Stresses in Soybean Production. http://dx.doi.org/10.1016/B978-0-12-801536-0.00002-5

to engineer plants with tolerance to multiple stresses. In this chapter, we discuss advances made in soybean in this area and highlight key transcription factors (TFs) and signaling components regulating multiple stress tolerance demonstrated using transgenic approaches (Table 2.1). These components of stress response machinery are not only important candidates for further investigation, but also provide clues about the mechanism of stress cross talk in soybean. Additionally, the potential of microRNAs (miRNAs) and arbuscular mycorrhizal (AM) fungi in regulating stress cross talk and tolerance to multiple stresses has been discussed.

Transcription factors as key mediators of stress cross talk

TFs play an essential role in the stress response by directly regulating the expression of several stress response-related genes. In addition to their transcriptional regulation capacity, TFs play a key role in integrating diverse signaling pathways in response to biotic and abiotic stresses. Several genes belonging to APETALA2/ethylene response factor (AP2/ERF), MYB, WRKY, plant homeodomain (PHD), NAC, basic leucine zipper (bZIP), and GT transcription factor (GT-1 binding site related element binding factors) families have been shown to play key roles in orchestrating stress-regulatory cross talk in model systems, rice and *Arabidopsis* (Cao et al., 2006; Chen et al., 2010; Lindemose et al., 2013; Liu et al., 2012; Mengiste et al., 2003; Nakashima et al., 2007; Vannini et al., 2006; Xiao et al., 2013; Zhu et al., 2010). In the sections below, we summarize the emerging evidence of the role of TFs belonging to these families in regulating stress cross talk in soybean.

AP2/ERF family of transcription factors

AP2/ERF TFs, characterized by a AP2/ERF DNA-binding domain, play a very important role in stress response and tolerance (Liu et al., 1998; Morran et al., 2011; Oh et al., 2005). Several members of this family have already been shown to regulate stress cross talk in model plant systems, rice, and *Arabidopsis* (Cao et al., 2006; Zhu et al., 2010). The AP2/ERF family is further divided into five subfamilies, including AP2, ERF, drought responsive element binding proteins (DREB), related to ABI3/VP1 (RAV), and Soloist based on the sequence structure and number of AP2/ERF domains in the encoded protein (Nakano et al., 2006).

More than 380 genes, with complete AP2/ERF domain, have been identified in soybean (Mochida et al., 2009; Wang et al., 2010). Expression analysis revealed that nine genes including *GmERF039*, *GmERF056*, *GmERF057*, *GmERF061*, *GmERF069*, *GmERF079*, *GmERF081*, *GmERF089*, and *GmERF098* are induced by hormone treatments and biotic/abiotic stresses (Zhang et al., 2008b). In another study, Zhai et al., 2013 showed that one of the members of the ERF gene family, *GmERF7*, is induced by multiple abiotic stresses including drought, salt, ethylene (ET), abscisic acid (ABA), and methyl jasmonate (MeJA).

These results suggest the involvement of AP2/ERF family genes in cross talk between biotic and abiotic stress signaling pathways. So far, six AP2/ERF family genes have been functionally characterized for their involvement in tolerance to multiple

Table 2.1 Genes involved in biotic and abiotic stress cross talk in soybean

S. No.	Gene names	Annotations	Functions	References
Genes conferring tolerance to both biotic and abiotic stresses				
1	*GmERF3*	TF	Salinity and drought tolerance and resistance to *Ralstonia solanacearum, Alternaria alternata*, and TMV	Zhang et al. (2009a)
2	*GmERF057*	TF	Salinity tolerance and resistance to *R. solanacearum*	Zhang et al. (2008b)
3	*GmCAM4*	Calmodulin	Salinity tolerance and resistance to *Phytophthora sojae, Alternaria tenuissima*, and *Phomopsis longicolla*	Rao et al. (2014)
Genes conferring tolerance to multiple abiotic stresses				
1	*GmDREB2*	TF	Salinity and drought tolerance	Chen et al. (2007)
2	*GmDREB3*	TF	Salt, cold, and drought tolerance	Chen et al. (2009)
3	*GmERF4*	TF	Salinity and drought tolerance	Zhang et al. (2010)
4	*GmERF089*	TF	Salinity and drought tolerance	Zhang et al. (2008b)
5	*GmNAC20*	TF	Salinity and cold tolerance	Hao et al. (2011)
6	*GmbZIP44*	TF	Salinity and freezing tolerance	Liao et al. (2008b)
7	*GmbZIP62*	TF	Salinity and freezing tolerance	Liao et al. (2008b)
8	*GmbZIP1*	TF	Salinity, drought, and freezing tolerance	Gao et al. (2011)
9	*GmWRKY54*	TF	Salinity, drought, and cold tolerance	Zhou et al. (2008)
10	*GmGT-2A*	TF	Salinity, drought, and cold tolerance	Xie et al. (2009)
11	*GmGT-2B*	TF	Salinity, drought, and cold tolerance	Xie et al. (2009)
12	*GmPHD2*	TF	Salinity and oxidative-stress tolerance	Wei et al. (2009)
13	*GmMYB76*	TF	Salinity and cold tolerance	Liao et al. (2008b)
14	*GmMYB177*	TF	Salinity and cold tolerance	Liao et al. (2008b)
15	*GmMYBJ1*	TF	Oxidative, drought, and cold tolerance	Su et al. (2014)
16	*GmUBC2*	Ubiquitin conjugating enzyme	Salinity and drought tolerance	Zhou et al. (2010)
Genes conferring tolerance to multiple biotic stresses				
	GmMPK4	MAPK	Negatively regulates resistance to SMV and *P. manschurica*	Liu et al. (2011)

abiotic stresses using transgenic approaches. These include two DREB (*GmDREB2* and *GmDREB3*) and four ERF subfamily genes (*GmERF3*, *GmERF4*, *GmERF057*, and *GmERF089*). The detailed information on these genes is summarized below.

Heterologous expression of GmDREB2 and GmDREB3 confers tolerance to multiple abiotic stresses

DREBs are an important class of proteins that bind to dehydration responsive elements (DREs) in the regulatory regions of stress-responsive genes (Lata and Prasad, 2011; Zhang et al., 2009b). They have been categorized into six subclasses named A1–A6 (Sakuma et al., 2002; Sharoni et al., 2011). Chen et al., 2007 showed that transcripts of an A5 subclass gene, *GmDREB2*, accumulate in response to high salt, drought, cold, and ABA treatment. When overexpressed in *Arabidopsis* under the control of constitutive CaMV35S promoter or stress-inducible responsive to dehydration 29A (RD29A) promoter, the transgenic plants could survive up to 200 mM salt concentration and drought stress (Chen et al., 2007).

The transgenic plants showed enhanced expression of stress-inducible marker genes *RD29A* and *cold-regulated 15a* (*COR15a*) genes. Both *RD29A* and *COR15a* genes contain DREs in their regulatory regions and play a role in the stress response in *Arabidopsis,* downstream of *AtDREB1A* (Chen et al., 2007). Furthermore, overexpression of *GmDREB2* in tobacco plants resulted in accumulation of high levels of an osmolyte, proline, which can help the plant survive under abiotic stress conditions (Chen et al., 2007).

Another DREB gene of soybean, *GmDREB3*, is induced in response to low temperature stress. Heterologous expression of this gene in *Arabidopsis* confers enhanced tolerance to multiple abiotic stresses including drought, salt, and cold (Chen et al., 2009). Transgenic tobacco plants overexpressing *GmDREB3* are tolerant to drought and high salt stress (Chen et al., 2009). Comparative osmolyte profiling of wild type (WT) and transgenic plants revealed higher accumulation of free proline in the transgenic plants after sixteen days of drought treatment. The chlorophyll content of the transgenic *Arabidopsis* and tobacco plants was also significantly higher than that of WT plants.

These findings clearly indicate that DREB genes have a conserved role in regulating abiotic stress tolerance. Higher accumulation of osmoprotectants in transgenic plants suggests that better water retention capability of the transgenic plants helped them to cope with abiotic stresses (Chen et al., 2009).

GmERF3 acts at the interface of biotic and abiotic stresses

GmERF3 belongs to a novel class of ERFs which bind to both GCC box and DRE/CRT-core motifs with different binding affinities (Zhang et al., 2009a). Expression of *GmERF3* is induced by plant hormones, including salicylic acid (SA), jasmonic acid (JA), ET and ABA, and soybean mosaic virus (SMV), as well as abiotic stresses (high salinity and drought), indicating that it might have a role in connecting biotic and abiotic stress signaling pathways (Zhang et al., 2009a).

Ectopic expression of *GmERF3* in transgenic tobacco plants conferred enhanced resistance against bacterial pathogen *Ralstonia solanacearum,* fungal pathogen *Alternaria alternata,* and tobacco mosaic virus (TMV). The transgenic plants were also tolerant to salinity and dehydration stresses. Although the mechanism of its function is not clear yet, data presented suggest that *GmERF3* is one of the very few positive transcriptional regulators, whose overexpression can lead to resistance in all three types of pathogens (bacterial, fungal, and viral) and tolerance to abiotic stresses. Such genes have great potential for crop improvement.

Constitutive expression of GmERF4 in Arabidopsis leads to enhanced tolerance to salt and drought stress

Another ERF family gene, *GmERF4,* which contains ERF-associated amphiphilic repression (EAR) motif, essential for transcriptional repression, was identified and characterized by Zhang et al. (2010). Expression of this gene was upregulated in response to both biotic and abiotic stresses including salt, drought, cold, and SMV inoculation and phytohormones including JA, ET, SA, and ABA (Zhang et al., 2010). When overexpressed in tobacco, although no distinct phenotypic difference was observed between transgenic and WT plants in response to the bacterial pathogen *Ralstonia dolaanacearum,* the expression of several defense-related genes including pathogenesis-related 1 (*PR1*), *PR2, PR4, osmotin,* and *systemic acquired resistance 8.2 (SAR8.2)* were downregulated (Zhang et al., 2010). Conversely, transgenic plants were better adapted to salt and drought stress compared to WT tobacco plants. These findings suggest that *GmERF4* may be acting as a negative regulator of defense pathways but positively regulates tolerance to abiotic stresses.

Overexpression of GmERF057 and GmERF089 confers enhanced tolerance to abiotic and/or biotic stresses in tobacco

Zhang et al. (2008b) created individual overexpression of a B2 group gene, *GmERF057,* and the B5 group gene *GmERF089* in tobacco plants using CaMV35S constitutive promoter. Ectopic expression of *GmERF057* in tobacco not only led to increased tolerance against salinity stress but also showed enhanced resistance against bacterial pathogen *R. solanacearum* (Zhang et al. 2008b). Overexpression of *GmERF089* in tobacco plants conferred tolerance against drought and salt stress in a developmental stage-specific manner (Zhang et al., 2008b). These results point to the distinct functions of the different ERF family genes.

NAC family of transcription factors

The NAC family of TFs are plant-specific. These TFs are characterized by a DNA binding NAC domain at the N-terminus and a transcriptional activation domain at the C-terminus. They play a significant role in biotic and abiotic stress responses (Nuruzzaman et al., 2013; Puranik et al., 2012). The NAC domain was initially discovered in three different genes, namely *NAM* (no apical meristem) of petunia, *ATAF1/2*

(*Arabidopsis* transcription activating factors), and *CUC1/2* (cup-shaped cotyledon) of *Arabidopsis*. Hence, it was abbreviated with the initial letters of these genes as NAC.

Genome-wide analysis of soybean identified 152 NAC family genes (Le et al., 2011; Mochida et al., 2009; Pinheiro et al., 2009). Quantitative real time polymerase chain reaction analysis revealed that *GmNAC3* and *GmNAC4* are induced by osmotic stress, ABA, JA, and salinity (Pinheiro et al., 2009). On the other hand, *GmNAC5* is not induced by ABA but upregulated in response to wounding, salinity, cold, and JA treatment (Jin et al., 2013). These results suggest that NAC TFs are involved in abiotic stress response in either an ABA-dependent (*GmNAC3* and *GmNAC4*) or -independent (*GmNAC5*) manner. Drought and salt stress induced expression of *GmNAC002*, *GmNAC010*, *GmNAC012*, *GmNAC013*, *GmNAC015*, and *GmNAC028* in an ABA-independent manner (Tran et al., 2009). Furthermore, *GmNAC002*, *GmNAC013*, *GmNAC015*, and *GmNAC028* are also induced in response to freezing stress.

GmNAC11 *and* GmNAC20 *regulate stress tolerance through activation of the DREB/CBF–COR pathway*

Recent work by Hao et al. (2011) revealed that high salt concentration, drought, and 1-naphthylacetic acid treatments induced the expression of both *GmNAC11* and *GmNAC20*. Also, ABA and cold stress upregulated the expression of *GmNAC11* and *GmNAC20*, respectively (Hao et al., 2011). To further characterize the gene functions *in planta,* both genes were independently overexpressed in *Arabidopsis*. Although the electrolyte leakage was significantly lower in both transgenic plants (Hao et al., 2011), those overexpressing *GmNAC20* showed enhanced tolerance to salinity and cold stress, whereas those overexpressing *GmNAC11* displayed tolerance to salinity stress only.

Furthermore, expressions of several stress response-related genes including *DREB1A/CBF3* (dehydration-responsive element binding factor 1A/C domain binding factor 3), *KIN2* (kinase 2), *COR15A*, and *RD29A/COR78* were significantly induced in transgenic plants overexpressing *GmNAC20* in response to cold stress. On the other hand, expression of *DREB1C/CBF2*, which is a negative regulator of *DREB1A*, was downregulated. *GmNAC20* activates *DREB1A/CBF3* by direct regulation as well as by suppressing *DREB1C/CBF2*, which is a negative regulator of *DREB1A*. These findings suggest *GmNAC20*-mediated tolerance to cold stress is conferred through activation of the DREB/CBF–COR pathway.

Expression of several abiotic stress marker genes including *DREB1A*, *ERD11* (early response to dehydration 11), *COR15A*, *ERF5*, *RAB18* (responsive to ABA 18), and *KAT2* (K^+ transporter of *Arabidopsis thaliana*) genes was also higher in *GmNAC11* transgenic plants. Authors have shown that both *GmNAC20* and *GmNAC11* modulate expression of *DREB1A* by directly binding to its promoter region (Hao et al., 2011). Further characterization of these genes in soybean will provide insights into their precise functions.

bZIP family of transcription factors

bZIP TFs are characterized by the presence of the bZIP domain. The bZIP domain has a basic region for binding to DNA in a sequence-specific manner and a leucine zipper for dimerization (Jakoby et al., 2002). Most of the bZIP TFs bind to the ABA-responsive elements in the promoter regions of downstream genes and are involved in ABA-dependent stress responses (Liao et al., 2008a, c). These TFs have demonstrated roles in biotic and abiotic stresses (Alves et al., 2013; Liu et al., 2014).

In the soybean genome, about 150 bZIP TFs have been annotated (Mochida et al., 2009). Expression analysis of 131 GmbZIP family genes revealed differential accumulation of eleven genes in response to salt, cold, and drought stress, whereas 24 genes responded to at least two stresses (Liao et al., 2008c). In another study, expression of four bZIP gene family members, *GmbZIP44*, *GmbZIP46*, *GmbZIP62*, and *GmbZIP78*, were analyzed under salinity, cold, drought, and ABA (Liao et al., 2008a). Except for *GmbZIP46*, whose expression was downregulated under drought and salt stress, all three genes were induced in response to all the stresses indicating their involvement in the abiotic stress response (Liao et al., 2008c).

Ectopic expression of GmbZIP44, GmbZIP62, and GmbZIP78 *confer salt and freezing tolerance in* Arabidopsis

Ectopic expression of *GmbZIP44*, *GmbZIP62*, and *GmbZIP78* in *Arabidopsis* results in reduced sensitivity toward ABA compared with WT plants at the germination stage. The degree of insensitivity varied among different transgenic plants, with *GmbZIP62* overexpressing plants being less insensitive. Since ABA insensitive mutants show a high rate of transpiration because of the reduced stomatal closure (Assmann et al., 2000), water loss in transgenic plants overexpressing *GmbZIP44*, *GmbZIP62*, and *GmbZIP78* is much higher than in WT plants (Liao et al., 2008c).

Transgenic seeds overexpressing each of these genes were more sensitive toward osmotic stress as the germination efficiency of transgenic seeds on 200 and 400 mM mannitol was quite low as compared with WT plants. But *GmbZIP44* and *GmbZIP62* conferred enhanced tolerance to salinity stress. Germination percentage of the transgenic seeds of *GmbZIP44* and *GmbZIP62* genes was comparatively higher under salinity stress. At the same time, transgenic plants overexpressing *GmbZIP44* and *GmbZIP62* were more tolerant to freezing stress, while *GmbZIP78* conferred partial tolerance to cold stress. All three transgenic plants accumulated high levels of proline compared with WT plants.

Further expression of downstream genes involved in ABA signaling and other stress-related genes was also altered in the transgenic plants. The significantly upregulated genes included *ABI1* (ABA insensitive 1), *ABI2*, *ERF5*, *KIN1*, *COR15A*, and *COR78*. Expression of a key proline biosynthetic enzyme encoding gene, *P5CS1* (delta 1-pyrroline-5-carboxylate synthetase 1), was also higher in *GmbZIP62* and *GmbZIP78* transgenic plants. On the other hand, expression of *ABF3* (ABA responsive element-binding factor 3) and *ABF4* was repressed in the *GmbZIP62* and *GmbZIP78* overexpressing plants but did not change in *GmbZIP44* overexpressing transgenic plants.

On the contrary, *DREB2A* and *COR47* were repressed in all plants. ABA signaling plays an important role in abiotic stresses. Insensitivity of the three bZIP family genes of soybean toward ABA indicate that these genes may be working as negative regulators of ABA by activating the repressors *ABI1* and *ABI2,* and repressing the positive regulators of ABA signaling *ABF3* and *ABF4* (Liao et al., 2008c).

Overexpression of GmbZIP1 enhances tolerance to abiotic stresses in transgenic Arabidopsis, tobacco, and wheat

GmbZIP1 belonging to the ABA-responsive element binding protein (AREB) subfamily of bZIP TFs is highly induced in response to ABA, drought, high salt, and low temperature stress in soybean (Gao et al., 2011). Overexpression of *GmbZIP1* in *Arabidopsis* and tobacco conferred tolerance against multiple abiotic stresses. Transgenic *Arabidopsis* and tobacco plants overexpressing *GmbZIP1* were more sensitive to ABA but tolerant to drought, freezing, and salt stresses.

Distinct phenotypic changes observed in the transgenic tobacco plants overexpressing *GmbZIP1* include quick stomatal opening under the influence of ABA. Also, the stomatal aperture in transgenic tobacco plants was narrower compared to that in WT plants, thereby reducing the water loss during dehydration stress. In addition, transgenic *Arabidopsis* plants accumulated high levels of osmoprotectants compared with WT plants. Expression of several genes regulating ABA signaling (*ABI1* and *ABI2*), guard cell ion channels (*KAT1* and *KAT2*), and stress response (*RD29B* and *RAB18*) was upregulated in the transgenic plants (Gao et al., 2011). Overexpression of *GmbZIP1* in Chinese wheat variety BS93 also led to enhanced drought tolerance indicating that *GmbZIP1* is an excellent genetic resource for engineering abiotic stress tolerance in crop plants (Gao et al., 2011).

WRKY family of transcription factors

WRKY TFs are characterized by the highly conserved WRKY domain at the N-terminus and a novel zinc finger-like motif (C_2–H_2 or C_2–H–C motif) at the C-terminus (Zhang and Wang, 2005). WRKY TFs are known to regulate stress response by binding to the *cis*-regulatory elements of both biotic and abiotic stress responsive genes (Luo et al., 2013; Pandey and Somssich, 2009). About 200 WRKY family of transcription factor genes have been annotated in soybean (Mochida et al., 2009; Wang et al., 2010). Zhou et al. (2008) reported differential accumulation of 64 WRKY genes under abiotic stresses. Out of these, seven are induced in response to salt, drought, and cold, whereas 14 genes are induced in response to both salt and drought stress.

Overexpression of GmWRKY54 confers enhanced tolerance to salt and drought stress in Arabidopsis

A group II WRKY family gene of soybean, *GmWRKY54*, is induced by both drought and salt stress (Zhou et al., 2008). Heterologous expression of *GmWRKY54* in *Arabidopsis* conferred enhanced tolerance to drought, salt, and cold stress. To understand the mechanism of tolerance, authors analyzed the expression of several stress-responsive marker genes and showed that *DREB2A, RD29B,* and *ERD10* are upregulated in transgenic plants, whereas expression of *STZ/Zat10* (salt tolerance zinc finger), which is a transcriptional repressor, is downregulated (Zhou et al., 2008). Based on these results, the authors hypothesized that *GmWRKY54* might confer abiotic stress tolerance by promoting the *DREB2A* pathway while suppressing the *STZ/Zat10* pathway. Further experiments would be required to validate their hypothesis.

GT factors

GT factors comprise a plant-specific family of TFs that specifically bind to GT elements in the promoter region of various genes (Zhou, 1999). The DNA binding domain of GT factors is characterized by a trihelix (helix–loop–helix–loop–helix) structure and they are therefore also known as trihelix TFs. Diverse functions governed by GT factors include flower morphogenesis (Brewer et al., 2004), seed maturation (Gao et al., 2009), and light-regulated responses (Chattopadhyay et al., 1998; Dehesh et al., 1990). Osorio et al. (2012) identified 63 putative trihelix/GT TFs in soybean. Expression analysis using SuperSAGE revealed that eleven of them are differentially regulated in response to biotic and abiotic stresses (Osorio et al., 2012).

GmGT-2A and GmGT-2B confer tolerance to multiple abiotic stresses

The functions of two soybean trihelix–GT factor genes, *GmGT-2A* and *GmGT-2B*, have been characterized in *Arabidopsis* (Xie et al., 2009). In soybean, both genes are upregulated in response to ABA, cold, drought, and salt treatments. Overexpression of each of these genes in *Arabidopsis* significantly improved plant tolerance to drought, cold, and salinity stress. Analysis of gene expression in transgenic plants revealed upregulation of several stress response-related genes (*AZF1*, *STZ*, *RHL41/ZAT12*, and *DREB2A*), ABA signaling components (*LTP3*, *LTP4*, *PAD3*, and *UGT71B6*), and several MYB family of TFs (Xie et al., 2009). Cumulative action of these genes is responsible for enhanced abiotic stress tolerance. Since the set of genes induced by both transgenic plants harboring *GmGT-2A* and *GmGT-2B* overlap, they function via a common pathway (Xie et al., 2009).

PHD finger transcription factors

PHD finger proteins are a group of proteins containing the Zn^{2+} motif. The PHD finger domain is mostly found in nuclear proteins involved in chromatin-mediated gene regulation. The N-terminal domain of these proteins specifically recognizes the "GTGGAG" sequence in the *cis*-element. Expression profiling of six ALFIN-type PHD finger genes including *GmPHD1*, *GmPHD2*, *GmPHD3*, *GmPHD4*, *GmPHD5*, and *GmPHD6* under high salt, drought, ABA, and cold stress revealed their differential accumulation (Wei et al., 2009). *GmPHD4* showed induction in response to all four stresses, whereas the other five genes were induced by at least two abiotic stresses.

Ectopic expression of GmPHD2 confers salt and oxidative stress tolerance in Arabidopsis

GmPHD2, induced by all four stresses including salt, drought, ABA, and cold stress in soybean, has been functionally characterized in *Arabidopsis* using transgenic approaches. Ectopic expression of *GmPHD2* in *Arabidopsis* conferred tolerance against both salt and oxidative stress (Wei et al., 2009). Transgenic plants grow better compared to WT plants at the seedling as well as the mature stage. Expression of *CBF2/DREB1C* (repressor of *CBF1/DREB1B* and *CBF3/DREB1A*), *STRS1/2* (stress response suppressor 1/2), and MAP3K was suppressed in transgenic plants, whereas genes encoding *WAK5* (wall-associated kinase 5), *TPP* (phosphate transporter), *ABI5* (ABA-insensitive 5), and peroxidases were upregulated.

Upregulation of peroxidases led to the hypothesis that transgenic plants confer abiotic stress tolerance by diminishing oxidative stress. With this hypothesis, germination efficiency of the transgenic and WT plants was checked in the presence of paraquat, a toxic herbicide that leads to reactive oxygen species generation, and transgenic plants were found to grow better than WT plants. At the same time, accumulation of H_2O_2 was also lower in transgenic plants. Analysis of membrane integrity during salt stress as a measure of electrolyte leakage shows that transgenic *Arabidopsis* plants maintain better membrane integrity compared with WT plants (Wei et al., 2009). Overall, *GmPHD2* may be working at the junction of the oxidative and salinity stress response pathways. Further investigations are required to reveal its precise mechanism of action.

MYB family of transcription factors

Plant MYB TFs, characterized by the presence of a R2/R3-MYB DNA binding domain, regulate numerous biochemical pathways in response to environmental stress conditions (Martin and Paz-Ares, 1997; Nagaoka and Takano, 2003). About 800 MYB family genes have been predicted in the soybean genome (Wang et al., 2010). Expression of 10 genes has been shown to be induced upon treatment with salt, drought, and cold stress (Liao et al., 2008b).

Overexpression of GmMYB76, GmMYB92, and GmMYB177 conferred enhanced tolerance to salinity and cold stress in Arabidopsis

Three GmMYB genes including *GmMYB76*, *GmMYB92*, and *GmMYB177* are upregulated in response to abiotic stresses. When overexpressed in *Arabidopsis*, transgenic plants overexpressing each of them could germinate under high salt concentration (600 mM NaCl). Transgenic plants overexpressing each of *GmMYB76* and *GmMYB177* were also tolerant to cold stress. Further, expression analysis of downstream genes showed that transcripts of several stress-associated genes including *RD29B*, *DREB2A*, *P5CS*, *RD17*, *ERD10*, and *COR78/RD29A* were higher in the *GmMYB76* overexpressing plants. Expression of *DREB2A*, *RD17*, and *P5CS* was upregulated in *GmMYB92* overexpressing plants. Similarly, expression of *rd29B*, *ABI2*, *DREB2A*, *RD17*, *P5CS*, *ERD10*, *COR6.6*, *ERD11*, and *COR78* was higher in *GmMYB177* overexpressing plants (Liao et al., 2008b). However, no change to the transcripts of *RAB18* and *ABI1*, positive regulators of ABA signaling, was observed in any of the three transgenic plants suggesting that they might be conferring tolerance via an ABA-independent pathway (Liao et al., 2008b).

Transgenic Arabidopsis plants overexpressing GmMYBJ1 are more tolerant to drought, cold, and oxidative stress

Su et al. (2014) identified a novel soybean gene *GmMYBJ1*, induced under drought, cold, salt, and ABA treatments. To check the *in planta* function of *GmMYBJ1*, transgenic *Arabidopsis* plants overexpressing *GmMYBJ1* were generated. Interestingly, transgenic plants were more tolerant to drought, cold, and oxidative stress compared to WT plants (Su et al., 2014). Malondialdehyde (MDA) is an index of lipid peroxidation in response to stress. Biochemical analysis revealed that transgenic plants accumulated lower MDA content in response to drought stress compared with WT plants. At the same time, the water retention capability of transgenic plants was better than that of WT plants.

Also, in response to cold stress, levels of soluble sugars were higher in transgenic plants than in WT plants. Analysis of the transcript abundance of the downstream genes showed upregulation of several dehydration and cold stress-associated genes including *AtRD29B*, *AtCOR47*, *AtCOR78*, *AtCOR15a*, and *AtP5CS*. However, *AtDRE-B2A* was downregulated in transgenic plants. The given data suggests that *GmMybJ1* probably functions by a similar mechanism to that of *AtMYB41* and *AtMYB102*, homologs of *GmMybJ1* in *Arabidopsis*, by modulating the DREB pathway and osmolyte biosynthesis, and regulating the expression of stress-responsive genes.

Role of phytohormones and signaling components in stress regulatory cross talk in soybean

Phytohormones play an essential role in plant development as well as biotic and abiotic stress tolerance (Catinot et al., 2008; Gray, 2004; Peleg and Blumwald, 2011). Studies focusing on perturbations in phytohormone flux and its impact on signaling in response to biotic and/or abiotic stresses have led to exciting new developments in deciphering stress cross talk. Below, we have summarized examples elaborating the role of phytohormones and signaling components in stress-regulatory cross talk in soybean.

Overexpression of GmCaM4 in soybean enhances resistance to pathogens and salinity stress

Calcium (Ca^{2+}) serves as a universal secondary messenger in various phases of plant development (Sanders et al., 1999). Harsh environmental conditions such as drought, salinity, cold, pathogens, etc. elevate the cytosolic Ca^{2+} and activate downstream signaling pathways (Dey et al., 2010; Kiegle et al., 2000). Calmodulins are calcium binding proteins, which sense cytosolic calcium ion levels. Rao et al. (2014) found that overexpression of *GmCaM4* (*G. max* gene coding for calmodulin) in soybean enhances the resistance response against oomycetes *Phytophthora sojae* and two necrotrophic fungal pathogens, *Alternaria tenuissima* and *Phomopsis longicolla*.

Transgenic plants exhibited enhanced accumulation of JA and induction of PR genes. Conversely, silencing of *GmCaM4* resulted in reduced expression of PR genes. The GmCaM4 protein physically interacts with MYB2, which has a demonstrated role in regulating salt and dehydration-responsive gene expression (Jae et al., 2005). Heterologous expression of *GmCaM4* in *Arabidopsis* has been previously reported to enhance the transcription of AtMYB2-regulated genes and confer salt tolerance (Jae et al., 2005). Therefore, *GmCAM4* might confer abiotic stress tolerance through MYB2 though other possible mechanisms are equally plausible.

Soybean homolog of AtMPK4 *confers enhanced resistance to smv and downy mildew*

Protein phosphorylation is one of the key posttranslational modifications governing various cellular processes across the different genera of living organisms. Mitogen-activated protein kinases (MAPKs) regulate phosphorylation events to transduce the signal across and within the cell (Nakagami et al., 2005). A genome-wide analysis of the MAPK gene family revealed 38 MAPK, 11 MAPKK, and 150 MAPKKK genes in soybean (Neupane et al., 2013). The role of these GmMAPKs has been recognized in both biotic and abiotic stresses.

Liu et al., 2011 identified and characterized *GmMPK4*, an ortholog of *Arabidopsis MPK4*. The role of *Arabidopsis MPK4* in the defense response is well demonstrated (Petersen et al., 2000). Transgenic soybean plants silenced for *GmMPK4* showed hyperactivation of the defense response correlated with enhanced resistance against viral pathogen SMV as well as oomycete pathogen *Peronospora manshurica* (downy mildew). An enhanced defense response was associated with hyper production and accumulation of SA and H_2O_2. The intracellular movement and replication machinery for the viral particle is impaired in the *GmMPK4* silenced plants. Based on the expression data, the authors suggested that reduced size of SMV–N–GUS foci in the *GmMPK4* silenced plants could be responsible for compromised intercellular and/or intracellular movement of virus as indicated by downregulation of genes involved in microtubule and actin formation (Liu et al., 2011). On the other hand, enhanced accumulation of lignin in cell walls could also be contributing to enhanced tolerance to downy mildew (Hamiduzzaman et al., 2005). Overall, the data support that *GmMPK4* has a conserved role as a negative regulator of the defense response in both *Arabidopsis* and soybean.

Ectopic expression of GmUBC2 *in* Arabidopsis *leads to enhanced tolerance against salt and drought stress*

Protein ubiquitination plays a very important role in various physiological processes in eukaryotes. Ubiquitination of target protein is carried out by group of enzymes including ubiquitin-activating enzyme (E1), ubiquitin-conjugating enzyme (E2), and ubiquitin ligase (E3). This enzyme complex maintains balance in the cellular proteome.

This enzyme complex has been shown to play a key role in the regulation of plant growth and development (Conti et al., 2014; Maier et al., 2013) but its role in stress responses still needs to be investigated.

Taking this further, Zhou et al. (2010) identified a ubiquitin-conjugating enzyme encoding gene, *GmUBC2*, in soybean. *GmUBC2* is upregulated in response to ABA, salt, and osmotic stress (Zhou et al., 2010). Heterologous expression of *GmUBC2* in *Arabidopsis* led to enhanced tolerance to both salt and drought stress. Further, expression analysis of genes involved in ion homeostasis, osmolyte synthesis, and oxidative stress revealed upregulation of the vacuolar antiporter *AtNHX1* (*A. thaliana* sodium/hydrogen antiporter 1), NO_3^-/H^+ exchanger *AtCLCa* (chloride channel a), *AtP5CS1*, and *AtCCS* (copper chaperone for superoxide dismutase). On the other hand, no change in the expression of plasma membrane antiporter, *SOS1* (salt overly sensitive 1), stress-responsive TF, *DREB2A*, and *RD29b* was observed (Zhou et al., 2010). Enhanced activity of *AtP5CS1*, *AtCCS*, and *AtNHX1* resulted in hyper accumulation of proline, high superoxide dismutase (SOD) activity, and more sodium compartmentalization in transgenic plants ultimately leading to tolerance against drought and salt stress.

Involvement of microRNAs in abiotic and biotic stress regulation in soybean

miRNAs are a family of small noncoding RNAs that negatively regulate the posttranscriptional gene expression in a sequence-specific manner (He and Hannon, 2004). In plants, miRNAs are known to regulate the diverse aspects of development, physiology, biotic, and abiotic stress responses. The first report on the identification of miRNAs involved in nodulation in soybean was delivered by Subramanian et al. (2008). Thereafter, several studies identified many more miRNAs in soybean and their role in physiological processes (Song et al., 2011; Wang et al., 2009; Zhang et al., 2008a) but knowledge about the role of soybean miRNA in environmental stresses remained unclear.

Work done by Kulcheski et al. (2011) identified novel miRNAs with possible involvement in water deficit and rust–stress responses. Taking advantage of the high-throughput sequencing, Kulcheski et al. (2011) constructed and analyzed four different small RNA libraries from roots of drought-sensitive and tolerant soybean varieties as well as from leaves of rust-susceptible and resistant soybean cultivars. Based on this study, they identified 24 families of novel miRNA. Expression analysis of selected candidates revealed their differential expression in response to biotic and abiotic stresses in soybean. Novel miRNA, MIR-Seq07, was highly expressed in both sensitive and tolerant soybean cultivars in response to drought. *In silico* analysis predicted a gene encoding the fructose biphosphate aldolase enzyme as a key downstream target of MIR-Seq07. This enzyme is an important constituent of both glycolytic and gluconeogenic pathways and the pentose phosphate cycle in plants.

Therefore, Kulcheski and co-workers hypothesized that higher expression of MIR-Seq07 in roots of both soybean varieties during dehydration stress may be related to degradation of aldolase enzyme, and, hence, to the overall slowdown of metabolic pathways during drought stress. Surprisingly, the same miRNA, MIR-Seq07, was downregulated in response to fungal infection in both the susceptible and resistant cultivars. Sequence analysis suggests Leucine-rich-repeats (LRR) genes implicated in defense response as potential targets of MIR-Seq07 (Kulcheski et al., 2011). Although the precise role of MIR-Seq07 in both drought tolerance and rust resistance needs further investigation, it is a very good candidate for investigating the miRNA-mediated regulation of stress cross talk in plants.

Mycorrhiza-mediated approaches for conferring biotic and abiotic stress tolerance in soybean

The role of microorganisms in plant growth promotion and disease control is well known (Bashan, 1998; Compant et al., 2005; Glick, 2012). However, the role of microbes in management of biotic and abiotic stresses is gaining in importance in relation to the alleviation of stresses in crop plants, thus opening a new and emerging application in agriculture. AM fungi, the most common mycorrhizal association with higher plants, account for up to 50% of the biomass of soil microbes (Cardoso and Kuyper, 2006; Olsson et al., 1999).

The ability of AM fungi to increase the host-plant uptake of relatively immobile nutrients, in particular phosphorus, and several micronutrients enhances the ability of plants to cope with stress conditions associated with nutrient and water deficiencies (Schreiner et al., 1997). AM fungi enhance tolerance to several biotic and abiotic stresses by the production of antioxidants, suppression of ET production, stabilization of soil structure, enhancement of osmolyte production, and improvement of ABA regulation (Alami et al., 2000; Aroca et al., 2008; Grover et al., 2011; Kohler et al., 2008; Saravanakumar and Samiyappan, 2007).

Employing these microbes provides an alternative strategy to genetic engineering and plant breeding technologies for increasing crop stress tolerance. In fact, a strong association between soybean roots and mycorrhiza has been reported (Khalil et al., 1994). Here, we have summarized the current understanding of the role of AM fungi in engineering multiple stress tolerance in soybean plants. Several studies have reported enhanced tolerance to biotic and abiotic stresses in soybean plants colonized with mycorrhiza. *Glomus* species increase host tolerance to plant nematodes such as *Heterodera glycines* (Tylka et al., 1991) and *Gigaspora margarita* (Carling et al., 1989) in soybean.

Lambais (2000) indicated that an elicitor produced from an extract of extraradical mycelium of *Glomus intraradices* was able to induce the synthesis of phytoalexin in soybean cotyledons (Morandi, 1996). Similarly, Dehne et al. (1978) demonstrated increased concentration of antifungal chitinase in AM roots. Morandi (1996) reported increased concentrations of phytoalexin-like isoflavonoid compounds in AM roots compared with those in nonmycorrhizal (nonAM) soybean. These compounds could account for the increased resistance to fungal and nematode root pathogens. Enhanced accumulation of coumestrol in soybean decreases the development of pathogenic nematodes more consistently than that of the fungi (Hussey and Roncondori, 1982).

There have also been reports on increased tolerance to abiotic stresses due to AM colonization. Foliar drought tolerance was increased in plants colonized by a semiarid mix of AM fungi relative to plants colonized by *G. intraradices* (Augé et al., 2003). AM fungi contribute to favorable soil conditions by forming water-stable aggregates that correlate positively with root and AM mycelium development irrespective of nitrogen source (Bethlenfalvay et al., 1999).

In the earliest work on the subject, Safir et al. (1971, 1972) concluded that AM symbiosis probably influences the water relations in soybean plants. In some instances, however, stomatal parameters are altered by AM symbiosis without altering leaf hydration (Allen and Boosalis, 1983; Osundina, 1995; Stahl and Smith, 1984). Perhaps the single most important indicator of a drought resistance strategy is the dehydration tolerance of a species, that is, tissue capacity for withstanding low water potential conditions.

Augé et al. (2001) demonstrated that AM symbiosis can alter the lethal leaf water potential (Ψ) of soybean. Soybean plants showed relatively higher shoot and seed dry weights in plants inoculated with *Glomus etunicatum*, which could be ascribed to their higher relative water content (RWC), suggesting that drought avoidance is the main mechanism of plant–microbe association in alleviation of water stress in soybean (Aliasgharzad et al., 2006). Safir et al. (1972) reported that the presence of mycorrhiza decreases root resistance to water flow in soybean, which could lead to improved water balance, high RWC, and turgor in leaves.

Schellenbaum et al. (1998) suggested that AM fungi are a strong competitor for root-allocated carbon under conditions limiting photosynthesis and that the fungal disaccharide trehalose is greatly increased in mycorrhizal plants during drought. Most organisms naturally accumulate trehalose under stress in a two-step process by the activity of the enzymes trehalose-6-phosphate synthase (TPS) and trehalose-6-phosphate phosphatase. Transgenic plants overexpressing TPS have shown enhanced drought tolerance in spite of minute accumulation of trehalose, in amounts too small to provide a protective function. However, overproduction of TPS in plants has also been associated with pleiotropic growth aberrations (Karim et al., 2007).

Like several other metabolizable sugars, feeding trehalose exogenously to higher plants altered the activity of different enzymes. Exogenous trehalose also enhanced sucrose synthase and alkaline invertase activities but concomitantly reduced acid invertase activity and sucrose levels in soybean roots (Schellenbaum et al., 1998). Tholkappian et al. (2001) reported that mycorrhizal soybean plants contained lower levels of proline accumulation than nonmycorrhizal plants. The nitrogen content and pod number per plant increased in mycorrhizal soybean as opposed to nonmycorrhizal plants.

Reduced oxidative damage to lipids and proteins in nodules of mycorrhizal plants under drought stress can alleviate drought-induced nodule senescence in legume plants (Porcel et al., 2003; Ruiz-Lozano et al., 2001). Moreover, AM symbiosis can increase the glutathione reductase activity considerably both in roots and nodules of soybean plants subjected to drought stress (Porcel et al., 2003). AM inoculation also enhanced proline content in these plants compared with uninoculated plants, which may help in osmotic adjustment of AM-mediated plants (Sharma et al., 2012).

Porcel and Ruiz-Lozano (2004) measured the activities of antioxidant enzymes for correlation with the oxidative damage to lipids. Results showed that there was no relationship between the antioxidant activities and the decrease in lipid peroxidation in roots and shoots of drought affected AM plants. In addition, only shoot SOD activities showed a significant interaction between mycorrhization and water regime, while no significant interaction was observed for the other activities. Porcel et al. (2003) obtained similar results with *Glomus mosseae* inoculation in soybean plants. They reported that leaf water potential was higher in stressed AM soybean than in corresponding nonAM plants.

Ghorbanli et al. (2004) reported the effect of NaCl and AM fungi on antioxidant enzymes in the shoots and roots of soybean. Activities of SOD, peroxidase (POD), and catalase were increased in the shoots of both mycorrhizal and nonAM plants grown under salinity. Mycorrhizal plants had greater SOD, POD, and ascorbate peroxidase activity under salinity stress. Bressano et al. (2010) reported no change or a significant increase in growth parameters in response to oxidative stress in AM soybean plants inoculated with *G. mossese* or *G. intraradices*. Since oxidative stress is common to several environmental stresses, oxidative stress regulation might be the key strategy adopted by AM to protect the plant from different biotic and abiotic stresses.

Conclusions and future perspectives

Biotic and abiotic stresses have always been a major threat to plant health and productivity. However, with the global climate change and environmental pollution, there is an increasing risk of exposure of the plants to more than one stress under field conditions. Several lines of evidence now indicate extensive cross talk between stress-signaling pathways (Atkinson and Urwin, 2012; Fujita et al., 2006; Sharma et al., 2013). Therefore, to predict the effect of one stress on the outcome of another, it is imperative to elucidate the cross talk in stress signaling pathways. As summarized in this chapter, transcriptome profiling in response to biotic and abiotic stresses in soybean identified several candidate genes involved in stress cross talk, paving the way for further characterization of gene functions in response to more than one type of stress. As also summarized, several of the candidate transcription factor genes, when overexpressed, confer tolerance to more than one stress (Figure 2.1). Induction of common stress responsive marker genes by different TFs as illustrated in Figure 2.2

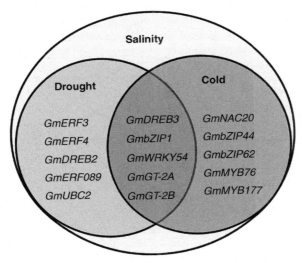

Figure 2.1 Venn diagram depicting genes conferring tolerance to a combination of salinity, drought, and cold stress in soybean. The genes have been classified based on the tolerance phenotype observed in transgenic plants.

might be one of the means through which different pathways converge to affect the stress response. However, detailed investigation is required to understand the precise mechanism of stress cross talk and devise strategies for engineering multistress tolerance in crop plants.

It would be interesting to determine if the known mediators of stress cross talk in model systems rice and *Arabidopsis* have conserved roles in soybean also. For instance, *NPR1* (nonexpressor of pathogenesis-related protein 1) of *Arabidopsis* is not only the master regulator of SA-mediated defense response but also affects tolerance to abiotic stresses (Srinivasan et al., 2009; Wu et al., 2012). Overexpression of rice *NH1* (homolog of NPR1) confers resistance against several fungal and bacterial pathogens in rice while rendering the plants more susceptible to abiotic stresses (Yuan et al., 2007). Sandhu et al. (2009) identified two functional homologs of *Arabidopsis NPR1*, *GmNPR1-1*, and *GmNPR1-2*, in soybean. Maldonado et al. (2014) reported that overexpression of *SNC1* (suppressor of NPR1, constitutive) and silencing of *SNI1* (suppressor of NPR1, inducible), known to regulate NPR1 activity, significantly inhibited maturation of soybean cyst nematodes. Further investigation is required to understand the mechanism of nematode resistance and check if soybean orthologs of *NPR1* have any role in other stress signaling pathways. Furthermore, due to difficulty in soybean transformation, most of the genes, to date, have been characterized in heterologous systems. Since there is a possibility for these genes to exhibit different functions in different species, these candidates would be worthy of further functional analysis in soybean.

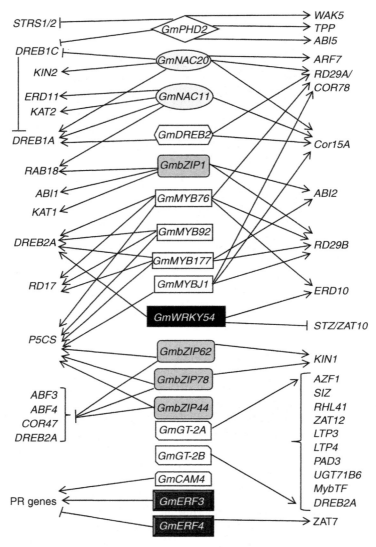

Figure 2.2 Schematic representation of signal cross talk observed in response to biotic and abiotic stresses in soybean. The TFs are shown in the center. Arrows represent upregulation and blunt-end lines represent downregulation of expression. Signals from most of the TFs converge to common stress-responsive genes demonstrated by upregulation of these genes in transgenic plants overexpressing candidate TFs. Different shapes and shades of gray have been used to differentiate genes belonging to different transcription factor families.

Acknowledgments

The authors would like to thank the Department of Biotechnology and Department of Science and Technology, Ministry of Science and Technology for the financial support to the lab. The Award of Research Fellowship from Council of Scientific and Industrial Research to PG, Ramalingaswami Fellowship to RS and MKS is gratefully acknowledged.

References

Alami, Y., Achouak, W., Marol, C., Heulinm, T., 2000. Rhizosphere soil aggregation and plant growth promotion of sunflowers by an exopolysaccharide-producing *Rhizobium* sp. strain isolated from sunflower roots. Appl. Environ. Microbiol. 66, 3393–3398.

Aliasgharzad, N., Neyshabouri, M.R., Salimi, G., 2006. Effects of arbuscular mycorrhizal fungi and *Bradyrhizobium japonicum* on drought stress of soybean. Biologia 19, S324–S326.

Allen, M.F., Boosalis, M.G., 1983. Effects of two species of VA mycorrhizal fungi on drought tolerance of winter wheat. New Phytol. 93, 67–76.

Alves, M.S., Dadalto, S.P., Gonçalves, A.B., De Souza, G.B., Barros, V.A., Fietto, L.G., 2013. Plant bZIP transcription factors responsive to pathogens: a review. Int. J. Mol. Sci. 14, 7815–7828.

Aroca, R., Vernieri, P., Ruiz-Lozano, J.M., 2008. Mycorrhizal and non-mycorrhizal *Lactuca sativa* plants exhibit contrasting responses to exogenous ABA during drought stress and recovery. J. Exp. Bot. 59, 2029–2041.

Assmann, S.M., Synder, J.A., Lee, Y.R.J., 2000. ABA-deficient (*aba1*) and ABA-insensitive (*abi1-1* and *abi2-1*) mutants of *Arabidopsis* have a wild-type stomatal response to humidity. Plant Cell Environ. 23, 387–395.

Atkinson, N.J., Urwin, P.E., 2012. The interaction of plant biotic and abiotic stresses: from genes to the field. J. Exp. Bot. 63, 3523–3543.

Augé, R.M., Kubikova, E., Moore, J.L., 2001. Foliar dehydration tolerance of mycorrhizal cowpea, soybean and bush bean. New Phytol. 151, 535–541.

Augé, R.M., Moore, J.L., Cho, K., Stutz, J.C., Sylvia, D.M., Al-Agely, A.K., Saxton, A.M., 2003. Relating foliar dehydration tolerance of mycorrhizal *Phaseolus vulgaris* to soil and root colonization by hyphae. J. Plant Physiol. 160, 1147–1156.

Bashan, Y., 1998. Inoculants of plant growth-promoting bacteria for use in agriculture. Biotechnol. Adv. 16, 729–770.

Bethlenfalvay, G.J., Cantrell, I.C., Mihara, K.L., Schreiner, R.P., 1999. Relationships between soil aggregation and mycorrhizae as influenced by soil biota and nitrogen nutrition. Biol. Fertil. Soils 28, 356–363.

Bressano, M., Curetti, M., Giachero, L., Vargas, G.S., Cabello, M., March, G., Ducasse, D.A., Luna, C.M., 2010. Mycorrhizal fungi symbiosis as a strategy against oxidative stress in soybean plants. J. Plant Physiol. 18, 1622–1626.

Brewer, P.B., Howles, P.A., Dorian, K., Griffith, M.E., Ishida, T., Kaplan-Levy, R.N., Kilinc, A., Smyth, D.R., 2004. *PETAL LOSS*, a trihelix transcription factor gene, regulates perianth architecture in the *Arabidopsis* flower. Development 131, 4035–4045.

Cao, Y., Song, F., Goodman, R.M., Zheng, Z., 2006. Molecular characterization of four rice genes encoding ethylene-responsive transcriptional factors and their expressions in response to biotic and abiotic stress. J. Plant Physiol. 163, 1167–1178.

Cardoso, I.M., Kuyper, T.W., 2006. Mycorrhizas and tropical soil fertility. Agric. Ecosyst. Environ. 116, 72–84.

Carling, D.E., Leiner, R.H., Westphale, P.C., 1989. Symptoms, signs and yield reduction associated with *Rhizoctonia* disease of potato induced by tuber-borne inoculum of *Rhizoctonia solani* AG-3. Am. Potato J. 66, 693–701.

Catinot, J., Buchala, A., Abou-Mansour, E., Métraux, J.-P., 2008. Salicylic acid production in response to biotic and abiotic stress depends on isochorismate in *Nicotiana benthamiana*. FEBS Lett. 582, 473–478.

Chattopadhyay, S., Puente, P., Deng, X.-W., Wei, N., 1998. Combinatorial interaction of light-responsive elements plays a critical role in determining the response characteristics of light-regulated promoters in *Arabidopsis*. Plant J. 15, 69–77.

Chen, M., Wang, Q.-Y., Cheng, X.-G., Xu, Z.-S., Li, L.-C., Ye, X.-G., Xia, L.-Q., Ma, Y.-Z., 2007. *GmDREB2*, a soybean DRE-binding transcription factor, conferred drought and high-salt tolerance in transgenic plants. Biochem. Biophys. Res. Commun. 353, 299–305.

Chen, M., Xu, Z., Xia, L., Li, L., Cheng, X., Dong, J., Wang, Q., Ma, Y., 2009. Cold-induced modulation and functional analyses of the DRE-binding transcription factor gene, *GmDREB3*, in soybean (*Glycine max* L.). J. Exp. Bot. 60, 121–135.

Chen, H., Lai, Z., Shi, J., Xiao, Y., Chen, Z., Xu, X., 2010. Roles of *Arabidopsis WRKY18*, *WRKY40* and *WRKY60* transcription factors in plant responses to abscisic acid and abiotic stress. BMC Plant Biol. 10, 281.

Compant, S., Duffy, B., Nowak, J., Cle, C., Barka, E.A., 2005. Use of plant growth-promoting bacteria for biocontrol of plant diseases: principles, mechanisms of action and future prospects. Appl. Environ. Microbiol. 71, 4951–4959.

Conti, L., Nelis, S., Zhang, C., Woodcock, A., Swarup, R., Galbiati, M., Tonelli, C., Napier, R., Hedden, P., Bennett, M., Sadanandom, A., 2014. Small ubiquitin-like modifier protein SUMO enables plants to control growth independently of the phytohormone gibberellin. Dev. Cell 28, 102–110.

Dehesh, K., Bruce, W.B., Quail, P.H., 1990. A *trans*-acting factor that binds to a GT-motif in a phytochrome gene promoter. Science 250, 1397–1399.

Dehne, H.W., Schonbeck, F., Baltruschat, H., 1978. The influence of endotrophic mycorrhiza on plant diseases 3. Chitinase-activity and the omithine-cycle. Z. Pßanzenkr Pflanzenschutz 85, 666–678.

Dey, S., Ghose, K., Basu, D., 2010. *Fusarium* elicitor-dependent calcium influx and associated ros generation in tomato is independent of cell death. Eur. J. Plant Pathol. 126, 217–228.

FAOSTAT, 2013. http://www.faostat.fao.org/site/339/default.aspx.

Fujita, M., Fujita, Y., Noutoshi, Y., Takahashi, F., Narusaka, Y., Yamaguchi-Shinozaki, K., Shinozaki, K., 2006. Crosstalk between abiotic and biotic stress responses: a current view from the points of convergence in the stress signaling networks. Curr. Opin. Plant Biol. 9, 436–442.

Gao, M.J., Lydiate, D.J., Li, X., Lui, H., Gjetvaj, B., Hegedus, D.D., Rozwadowski, K., 2009. Repression of seed maturation genes by a trihelix transcriptional repressor in *Arabidopsis* seedlings. Plant Cell 21, 54–71.

Gao, S.Q., Chen, M., Xu, Z.S., Zhao, C.P., Li, L., Xu, H., Tang, Y., Zhao, X., Ma, Y.Z., 2011. The soybean *GmbZIP1* transcription factor enhances multiple abiotic stress tolerances in transgenic plants. Plant Mol. Biol. 75, 537–553.

Ghorbanli, M., Ebrahimzadeh, H., Sharifi, M., 2004. Effects of NaCl and mycorrhizal fungi on antioxidative enzymes in soybean. Biol. Plant. 48, 575–581.

Glick, B.R., 2012. Plant growth-promoting bacteria: mechanisms and applications. Scientifica, 2012, 1–15.

Gray, W.M., 2004. Hormonal regulation of plant growth and development. PLoS Biol. 2, e311.

Grover, M., Ali, S.Z., Sandhya, V., Rasul, A., Venkateswarlu, B., 2011. Role of microorganisms in adaptation of agriculture crops to abiotic stresses. World J. Microbiol. Biotechnol. 27, 1231–1240.

Haile, F.J., Higley, L.G., Specht, J.E., 1998. Soybean cultivars and insect defoliation: yield loss and economic injury levels. Agron. J. 90, 344–352.

Hamiduzzaman, M.M., Jakab, G., Barnavon, L., Neuhaus, J.-M., Mauch-Mani, B., 2005. β-Aminobutyric acid-induced resistance against downy mildew in grapevine acts through the potentiation of callose formation and jasmonic acid signaling. Mol. Plant Microbe Interact. 18, 819–829.

Hao, Y.J., Wei, W., Song, Q.X., Chen, H.W., Zhang, Y.Q., Wang, F., Zou, H.F., Lei, G., Tian, A.G., Zhang, W.K., Ma, B., Zhang, J.S., Chen, S.Y., 2011. Soybean NAC transcription factors promote abiotic stress tolerance and lateral root formation in transgenic plants. Plant J. 68, 302–313.

He, L., Hannon, G.J., 2004. MicroRNAs: small RNAs with a big role in gene regulation. Nat. Rev. Genet. 5, 522–531.

Hussey, R.S., Roncondori, R.W., 1982. Vesicular-arbuscular mycorrhizae may limit nematode activity and improve plant growth. Plant Dis. 66, 9–14.

Jae, H.Y., Chan, Y.P., Jong, C.K., Won, D.H., Mi, S.C., Hyeong, C.P., Min, C.K., Byeong, C.M., Man, S.C., Yun, H.K., Ju, H.L., Ho, S.K., Sang, M.L., Hae, W.Y., Chae, O.L., Yun, D.J., Sang, Y.L., Woo, S.C., Moo, J.C., 2005. Direct interaction of a divergent CaM isoform and the transcription factor, *MYB2*, enhances salt tolerance in *Arabidopsis*. J. Biol. Chem. 280, 3697–3706.

Jakoby, M., Weisshaar, B., Dröge-Laser, W., Vicente-Carbajosa, J., Tiedemann, J., Kroj, T., Parcy, F., 2002. bZIP transcription factors in *Arabidopsis*. Trends Plant Sci. 7, 106–111.

Jin, H., Xu, G., Meng, Q., Huang, F., Yu, D., 2013. *GmNAC5*, a NAC transcription factor, is a transient response regulator induced by abiotic stress in soybean. Scientific World J., 2013, 768972.

Karim, S., Aronsson, H., Ericson, H., Pirhonen, M., Leyman, B., Welin, B., Mäntylä, E., Palva, E.T., Van Dijck, P., Holmström, K.O., 2007. Improved drought tolerance without undesired side effects in transgenic plants producing trehalose. Plant Mol. Biol. 64, 371–386.

Khalil, S., Loynachan, T.E., Tabatabai, M.A., 1994. Mycorrhizal dependency and nutrient uptake by improved and unimproved corn and soybean cultivars. Agron. J. 86, 949–958.

Kiegle, E., Moore, C.A., Haseloff, J., Tester, M.A., Knight, M.R., 2000. Cell-type-specific calcium responses to drought, salt and cold in the *Arabidopsis* root. Plant J. 23, 267–278.

Kohler, J., Hernández, J.A., Caravaca, F., Roldán, A., 2008. Plant growth-promoting rhizobacteria and arbuscular mycorrhizal fungi modify alleviation biochemical mechanisms in water-stressed plants. Funct. Plant Biol. 35 (2), 141–151.

Kulcheski, F.R., de Oliveira, L.F., Molina, L.G., Almerão, M.P., Rodrigues, F.A., Marcolino, J., Barbosa, J.F., Stolf-Moreira, R., Nepomuceno, A.L., Marcelino-Guimarães, F.C., Abdelnoor, R.V., Nascimento, L.C., Carazzolle, M.F., Pereira, G.A., Margis, R., 2011. Identification of novel soybean microRNAs involved in abiotic and biotic stresses. BMC Genomics 12, 307.

Lambais, M.R., 2000. Regulation of plant DR genes in arbuscular mycorrhizae. In: Podila, G.K., Douds, Jr., D.D. (Eds.), Current Advances in Mycorrhizae Research. APS Press, St. Paul, Minnesota, pp. 45–59.

Lata, C., Prasad, M., 2011. Role of DREBs in regulation of abiotic stress responses in plants. J. Exp. Bot. 62, 4731–4748.

Le, D.U., Nishiyama, R.I.E., Watanabe, Y.A., Mochida, K.E., Yamaguchi-Shinozaki, K.A., 2011. Genome-wide survey and expression analysis of the plant-specific NAC transcription factor family in soybean during development and dehydration stress. DNA Res. 18 (4), 1–14.

Leff, B., Ramankutty, N., Foley, J.A., 2004. Geographic distribution of major crops across the world. Global Biogeochem. Cycles 18, 1–27.

Liao, Y., Zhang, J.-S., Chen, S.Y., Zhang, W.K., 2008a. Role of soybean *GmbZIP132* under abscisic acid and salt stresses. J. Integr. Plant Biol. 50, 221–230.

Liao, Y., Zou, H.F., Wang, H.W., Zhang, W.K., Ma, B., Zhang, J.S., Chen, S.Y., 2008b. Soybean *GmMYB76*, *GmMYB92*, and *GmMYB177* genes confer stress tolerance in transgenic *Arabidopsis* plants. Cell Res. 18, 1047–1060.

Liao, Y., Zou, H.-F., Wei, W., Hao, Y.-J., Tian, A.-G., Huang, J., Liu, Y.-F., Zhang, J.-S., Chen, S.-Y., 2008c. Soybean *GmbZIP44*, *GmbZIP62* and *GmbZIP78* genes function as negative regulator of ABA signaling and confer salt and freezing tolerance in transgenic *Arabidopsis*. Planta 228, 225–240.

Lindemose, S., O'Shea, C., Jensen, M.K., Skriver, K., 2013. Structure, function and networks of transcription factors involved in abiotic stress responses. Int. J. Mol. Sci. 14, 5842–5878.

Liu, Q., Kasuga, M., Sakuma, Y., Abe, H., 1998. Two transcription factors, DREB1 and DREB2, with an EREBP/AP2 DNA binding domain separate two cellular signal transduction pathways in drought- and low-temperature-responsive gene expression, respectively, in *Arabidopsis*. Plant Cell 10, 1391–1406.

Liu, J.Z., Horstman, H.D., Braun, E., Graham, M.A., Zhang, C., Navarre, D., Qiu, W.L., Lee, Y., Nettleton, D., Hill, J.H., Whitham, S.A., 2011. Soybean homologs of *MPK4* negatively regulate defense responses and positively regulate growth and development. Plant Physiol. 157, 1363–1378.

Liu, C., Wu, Y., Wang, X., 2012. bZIP transcription factor *OsbZIP52/RISBZ5*: a potential negative regulator of cold and drought stress response in rice. Planta 235, 1157–1169.

Liu, C., Mao, B., Ou, S., Wang, W., Liu, L., Wu, Y., Chu, C., Wang, X., 2014. *OsbZIP71*, a bZIP transcription factor, confers salinity and drought tolerance in rice. Plant Mol. Biol. 84, 19–36.

Luo, X., Bai, X., Sun, X., Zhu, D., Liu, B., Ji, W., Cai, H., Cao, L., Wu, J., Hu, M., Liu, X., Tang, L., Zhu, Y., 2013. Expression of wild soybean WRKY20 in *Arabidopsis* enhances drought tolerance and regulates ABA signalling. J. Exp. Bot. 64, 2155–2169.

Maier, A., Schrader, A., Kokkelink, L., Falke, C., Welter, B., Iniesto, E., Rubio, V., Uhrig, J.F., Hülskamp, M., Hoecker, U., 2013. Light and the E3 ubiquitin ligase COP1/SPA control the protein stability of the MYB transcription factors *PAP1* and *PAP2* involved in anthocyanin accumulation in *Arabidopsis*. Plant J. 74, 638–651.

Maldonado, A., Youssef, R., McDonald, M., Brewer, E., Beard, H., Matthews, B., 2014. Modification of the expression of two NPR1 suppressors, SNC1 and SNI1, in soybean confers partial resistance to the soybean cyst nematode, *Heterodera glycines*. Funct. Plant Biol. 41, 714–726.

Martin, C., Paz-Ares, J., 1997. MYB transcription factors in plants. Trends Genet. 13, 67–73.

Mengiste, T., Chen, X., Salmeron, J., Dietrich, R., 2003. The *Botrytis susceptible1* gene encodes an R2R3 MYB transcription factor protein that is required for biotic and abiotic stress responses in *Arabidopsis*. Plant Cell 15, 2551–2565.

Mochida, K., Yoshida, T., Sakurai, T., Yamaguchi-Shinozaki, K., Shinozaki, K., Tran, L.S.P., 2009. *In silico* analysis of transcription factor repertoire and prediction of stress responsive transcription factors in soybean. DNA Res. 16, 353–369.

Morandi, D., 1996. Occurrence of phytoalexins and phenolic compounds in endomycorrhizal interactions, and their potential role in biological control. Plant Soil 185, 241–251.

Morran, S., Eini, O., Pyvovarenko, T., Parent, B., Singh, R., Ismagul, A., Eliby, S., Shirley, N., Langridge, P., Lopato, S., 2011. Improvement of stress tolerance of wheat and barley by modulation of expression of DREB/CBF factors. Plant Biotechnol. J. 9, 230–249.

Nagaoka, S., Takano, T., 2003. Salt tolerance-related protein STO binds to a Myb transcription factor homologue and confers salt tolerance in *Arabidopsis*. J. Exp. Bot. 54, 2231–2237.

Nakagami, H., Pitzschke, A., Hirt, H., 2005. Emerging MAP kinase pathways in plant stress signalling. Trends Plant Sci. 10, 339–346.

Nakano, T., Suzuki, K., Fujimura, T., Shinshi, H., 2006. Genome-wide analysis of the ERF gene family in *Arabidopsis* and rice. Plant Physiol. 140, 411–432.

Nakashima, K., Tran, L.S.P., Van Nguyen, D., Fujita, M., Maruyama, K., Todaka, D., Ito, Y., Hayashi, N., Shinozaki, K., Yamaguchi-Shinozaki, K., 2007. Functional analysis of a NAC-type transcription factor *OsNAC6* involved in abiotic and biotic stress-responsive gene expression in rice. Plant J. 51, 617–630.

Neupane, A., Nepal, M.P., Piya, S., Subramanian, S., Rohila, J.S., Reese, R.N., Benson, B.V., 2013. Identification, nomenclature, and evolutionary relationships of mitogen-activated protein kinase (MAPK) genes in soybean. Evol. Bioinform. Online 9, 363–386.

Nuruzzaman, M., Sharoni, A.M., Kikuchi, S., 2013. Roles of NAC transcription factors in the regulation of biotic and abiotic stress responses in plants. Front. Microbiol. 4, 248.

Oh, S.J., Song, S.I., Kim, Y.S., Jang, H.J., Kim, S.Y., Kim, M., Kim, Y.K., Nahm, B.H., Kim, J.K., 2005. *Arabidopsis CBF3/DREB1A* and *ABF3* in transgenic rice increased tolerance to abiotic stress without stunting growth. Plant Physiol. 138, 341–351.

Olsson, P.A., Thingstrup, I., Jakobsen, I., Bååth, E., 1999. Estimation of the biomass of arbuscular mycorrhizal fungi in a linseed field. Soil Biol. Biochem. 31, 1879–1887.

Osorio, M.B., Bücker-Neto, L., Castilhos, G., Turchetto-Zolet, A.C., Wiebke-Strohm, B., Bodanese-Zanettini, M.H., Margis-Pinheiro, M., 2012. Identification and *in silico* characterization of soybean trihelix-GT and bHLH transcription factors involved in stress responses. Genet. Mol. Biol. 35, 233–246.

Osundina, M., 1995. Responses of seedlings of *Parkia biglobes* (African locust bean) to drought and inoculation with vesicular-arbuscular mycorrhiza. Niger. J. Bot. 8, 1–10.

Pandey, S.P., Somssich, I.E., 2009. The role of WRKY transcription factors in plant immunity. Plant Physiol. 150, 1648–1655.

Peleg, Z., Blumwald, E., 2011. Hormone balance and abiotic stress tolerance in crop plants. Curr. Opin. Plant Biol. 14, 290–295.

Petersen, M., Brodersen, P., Naested, H., Andreasson, E., Lindhart, U., Johansen, B., Nielsen, H.B., Lacy, M., Austin, M.J., Parker, J.E., 2000. *Arabidopsis* MAP kinase 4 negatively regulates systemic acquired resistance. Cell 103, 1111–1120.

Pinheiro, G.L., Marques, C.S., Costa, M.D.B.L., Reis, P.A.B., Alves, M.S., Carvalho, C.M., Fietto, L.G., Fontes, E.P.B., 2009. Complete inventory of soybean NAC transcription factors: sequence conservation and expression analysis uncover their distinct roles in stress response. Gene 444, 10–23.

Porcel, R., Ruiz-Lozano, J.M., 2004. Arbuscular mycorrhizal influence on leaf water potential, solute accumulation, and oxidative stress in soybean plants subjected to drought stress. J. Exp. Bot. 55, 1743–1750.

Porcel, R., Barea, J.M., Ruiz-Lozano, J.M., 2003. Antioxidant activities in mycorrhizal soybean plants under drought stress and their possible relationship to the process of nodule senescence. New Phytol. 157, 135–143.

Puranik, S., Sahu, P.P., Srivastava, P.S., Prasad, M., 2012. NAC proteins: regulation and role in stress tolerance. Trends Plant Sci. 17, 369–381.

Rao, S.S., El-Habbak, M.H., Havens, W.M., Singh, A., Zheng, D., Vaughn, L., Haudenshield, J.S., Hartman, G.L., Korban, S.S., Ghabrial, S.A., 2014. Overexpression of *GmCaM4* in soybean enhances resistance to pathogens and tolerance to salt stress. Mol. Plant Pathol. 15, 145–160.

Ruiz-Lozano, J.M., Collados, C., Barea, J.M., Azcón, R., 2001. Arbuscular mycorrhizal symbiosis can alleviate drought-induced nodule senescence in soybean plants. New Phytol. 151, 493–502.

Safir, G.R., Boyer, J.S., Gerdemann, J.W., 1971. Mycorrhizal enhancement of water transport in soybean. Science 172, 581–583.

Safir, G.R., Boyer, J.S., Gerdemann, J.W., 1972. Nutrient status and mycorrhizal enhancement of water transport in soybean. Plant Physiol. 49, 700–703.

Sakuma, Y., Liu, Q., Dubouzet, J.G., Abe, H., Shinozaki, K., Yamaguchi-Shinozaki, K., 2002. DNA-binding specificity of the ERF/AP2 domain of *Arabidopsis* DREBs, transcription factors involved in dehydration-and cold-inducible gene expression. Biochem. Biophys. Res. Commun. 290, 998–1009.

Sanders, D., Brownlee, C., Harper, J., 1999. Communicating with calcium. Plant Cell 11, 691–706.

Sandhu, D., Tasma, I.M., Frasch, R., Bhattacharyya, M.K., 2009. Systemic acquired resistance in soybean is regulated by two proteins, orthologous to *Arabidopsis NPR1*. BMC Plant Biol. 9, 105.

Saravanakumar, D., Samiyappan, R., 2007. ACC deaminase from *Pseudomonas fluorescens* mediated saline resistance in groundnut (*Arachis hypogea*) plants. J. Appl. Microbiol. 102, 1283–1292.

Schellenbaum, L., Müller, J., Boller, T., Wiemken, A., Schüepp, H., 1998. Effects of drought on non-mycorrhizal and mycorrhizal maize: changes in the pools of non-structural carbohydrates, in the activities of invertase and trehalase, and in the pools of amino acids and imino acids. New Phytol. 138, 59–66.

Schreiner, R.P., Mihara, K.L., McDaniel, H., Bethlenfalvay, G.J., 1997. Mycorrhizal functioning influence plant and soil functions and interactions. Plant Soil 188, 199–209.

Sharma, M.P., Jaisighani, K., Sharma, S.K., Bhatia, V.S., 2012. Effect of native soybean rhizobia and AM fungi in the improvement of nodulation, growth, soil enzymes and physiological status of soybean under microcosm conditions. Agric. Res. 1 (4), 346–351.

Sharma, R., De Vleesschauwer, D., Sharma, M.K., Ronald, P.C., 2013. Recent advances in dissecting stress-regulatory crosstalk in rice. Mol. Plant 6 (2), 250–260.

Sharoni, A.M., Nuruzzaman, M., Satoh, K., Shimizu, T., Kondoh, H., Sasaya, T., Choi, I.-R., Omura, T., Kikuchi, S., 2011. Gene structures, classification and expression models of the AP2/EREBP transcription factor family in rice. Plant Cell Phys. 52, 344–360.

Singh, R.J., Hymowitz, T., 1999. Soybean genetic resources and crop improvement. Genome 42, 605–616.

Song, Q.X., Liu, Y.F., Hu, X.Y., Zhang, W.K., Ma, B., Chen, S.Y., Zhang, J.S., 2011. Identification of miRNAs and their target genes in developing soybean seeds by deep sequencing. BMC Plant Biol. 11, 5.

Srinivasan, T., Kumar, K.R.R., Meur, G., Kirti, P.B., 2009. Heterologous expression of *Arabidopsis* NPR1 (*AtNPR1*) enhances oxidative stress tolerance in transgenic tobacco plants. Biotechnol. Lett. 31, 1343–1351.

Stahl, P.D., Smith, W.K., 1984. Effects of different geographic isolates of *Glomus* on the water relations of *Agropyron smithii*. Mycologia 76, 261–267.

Su, L.T., Li, J.W., Liu, D.Q., Zhai, Y., Zhang, H.J., Li, X.W., Zhang, Q.L., Wang, Y., Wang, Q.Y., 2014. A novel MYB transcription factor, *GmMYBJ1*, from soybean confers drought and cold tolerance in *Arabidopsis thaliana*. Gene 538, 46–55.

Subramanian, S., Fu, Y., Sunkar, R., Barbazuk, W.B., Zhu, J.K., Yu, O., 2008. Novel and nodulation-regulated microRNAs in soybean roots. BMC Genom. 9, 160.

Tholkappian, P., Prakash, M., Sundaram, M.D., 2001. Effect of AM-fungi on proline, nitrogen and pod number of soybean under moisture stress. Indian J. Plant Physiol. 6, 98–99.

Tran, L.S.P., Mochida, K., 2010. Identification and prediction of abiotic stress responsive transcription factors involved in abiotic stress signaling in soybean. Plant Signal. Behav. 5, 255–257.

Tran, L.S.P., Quach, T.N., Guttikonda, S.K., Aldrich, D.L., Kumar, R., Neelakandan, A., Valliyodan, B., Nguyen, H.T., 2009. Molecular characterization of stress-inducible GmNAC genes in soybean. Mol. Genet. Genom. Med. 281, 647–664.

Tylka, G.L., Hussey, R.S., Roncadori, R.W., 1991. Interactions of vesicular-arbuscular mycorrhizal fungi, phosphorus, and Heterodera glycines on soybean. J. Nematol. 23, 122–133.

US Soybean Export Council (USSEC). 2008. http://www.ussec.org.

Valliyodan, B., Nguyen, H., 2008. Genomics of abiotic stress in soybean. In: Stacey, G. (Ed.), Genetics and Genomics of Soybean SE-18, Plant Genetics and Genomics: Crops and Models. Springer, New York, pp. 343–372.

Vannini, C., Iriti, M., Bracale, M., Locatelli, F., Faoro, F., Croce, P., Pirona, R., Di Maro, A., Coraggio, I., Genga, A., 2006. The ectopic expression of the rice Osmyb4 gene in Arabidopsis increases tolerance to abiotic, environmental and biotic stresses. Physiol. Mol. Plant Pathol. 69, 26–42.

Wang, Y., Li, P., Cao, X., Wang, X., Zhang, A., Li, X., 2009. Identification and expression analysis of miRNAs from nitrogen-fixing soybean nodules. Biochem. Biophys. Res. Commun. 378, 799–803.

Wang, Z., Libault, M., Joshi, T., Valliyodan, B., Nguyen, H.T., Xu, D., Stacey, G., Cheng, J., 2010. SoyDB: a knowledge database of soybean transcription factors. BMC Plant Biol. 10, 14.

Wei, W., Huang, J., Hao, Y.J., Zou, H.F., Wang, H.W., Zhao, J.Y., Liu, X.Y., Zhang, W.K., Ma, B., Zhang, J.S., Chen, S.Y., 2009. Soybean GmPHD-type transcription regulators improve stress tolerance in transgenic Arabidopsis plants. PloS One 4, e7209.

Wrather, J.A., Anderson, T.R., Arsyad, D.M., Gai, J., Ploper, L.D., Porta-Puglia, A., Ram, H.H., Yorinori, J.T., 1997. Soybean disease loss estimates for the top 10 soybean producing countries in 1994. Plant Dis. 81, 107–110.

Wu, Y., Zhang, D., Chu, J.Y., Boyle, P., Wang, Y., Brindle, I.D., De Luca, V., Després, C., 2012. The Arabidopsis NPR1 protein is a receptor for the plant defense hormone salicylic acid. Cell Rep. 1, 639–647.

Xiao, J., Cheng, H., Li, X., Xiao, J., Xu, C., Wang, S., 2013. Rice WRKY13 regulates cross talk between abiotic and biotic stress signaling pathways by selective binding to different cis-elements. Plant Physiol. 163, 1868–1882.

Xie, Z.M., Zou, H.F., Lei, G., Wei, W., Zhou, Q.Y., Niu, C.F., Liao, Y., Tian, A.G., Ma, B., Zhang, W.K., Zhang, J.S., Chen, S.Y., 2009. Soybean Trihelix transcription factors GmGT-2A and GmGT-2B improve plant tolerance to abiotic stresses in transgenic Arabidopsis. PloS One 4, e6898.

Yang, X.B., Lundeen, P., Uphoff, M.D., 1999. Soybean varietal response and yield loss caused by Sclerotinia sclerotiorum. Plant Dis. 83, 456–461.

Yorinori, J.T., Paiva, W.M., Frederick, R.D., Costamilan, L.M., Bertagnolli, P.F., Hartman, G.E., Godoy, C.V., Nunes, Jr., J., 2005. Epidemics of soybean rust (Phakopsora pachyrhizi) in Brazil and Paraguay from 2001 to 2003. Plant Dis. 89, 675–677.

Yuan, Y., Zhong, S., Li, Q., Zhu, Z., Lou, Y., Wang, L., Wang, J., Wang, M.s, Li, Q., Yang, D., 2007. Functional analysis of rice NPR1-like genes reveals that OsNPR1/NH1 is the rice orthologue conferring disease resistance with enhanced herbivore susceptibility. Plant Biotechnol. J. 5, 313–324.

Zhai, Y., Wang, Y., Li, Y., Lei, T., Yan, F., Su, L., Li, X., Zhao, Y., Sun, X., Li, J., Wang, Q., 2013. Isolation and molecular characterization of *GmERF7*, a soybean ethylene-response factor that increases salt stress tolerance in tobacco. Gene 513, 174–183.

Zhang, Y., Wang, L., 2005. The WRKY transcription factor superfamily: its origin in eukaryotes and expansion in plants. BMC Evol. Biol. 5, 1.

Zhang, B., Pan, X., Stellwag, E., 2008a. Identification of soybean microRNAs and their targets. Planta 229, 161–182.

Zhang, G., Chen, M., Chen, X., Xu, Z., Guan, S., Li, L.-C., Li, A., Guo, J., Mao, L., Ma, Y., 2008b. Phylogeny, gene structures, and expression patterns of the ERF gene family in soybean (*Glycine max* L.). J. Exp. Bot. 59, 4095–4107.

Zhang, G., Chen, M., Li, L., Xu, Z., Chen, X., Guo, J., Ma, Y., 2009a. Overexpression of the soybean *GmERF3* gene, an AP2/ERF type transcription factor for increased tolerances to salt, drought, and diseases in transgenic tobacco. J. Exp. Bot. 60, 3781–3796.

Zhang, M., Liu, W., Bi, Y.P., 2009b. Dehydration-responsive element-binding (DREB) transcription factor in plants and its role during abiotic stresses. Yi Chuan 31, 236–244.

Zhang, G., Chen, M., Chen, X., Xu, Z., Li, L., Guo, J., Ma, Y., 2010. Isolation and characterization of a novel EAR-motif-containing gene *GmERF4* from soybean (*Glycine max* L.). Mol. Biol. Rep. 37, 809–818.

Zhou, D.X., 1999. Regulatory mechanism of plant gene transcription by GT-elements and GT-factors. Trends Plant Sci. 4, 210–214.

Zhou, Q.Y., Tian, A.G., Zou, H.F., Xie, Z.M., Lei, G., Huang, J., Wang, C.M., Wang, H.W., Zhang, J.S., Chen, S.Y., 2008. Soybean WRKY-type transcription factor genes, *GmWRKY13*, *GmWRKY21*, and *GmWRKY54*, confer differential tolerance to abiotic stresses in transgenic *Arabidopsis* plants. Plant Biotechnol. J. 6, 486–503.

Zhou, G.A., Chang, R.Z., Qiu, L.J., 2010. Overexpression of soybean ubiquitin-conjugating enzyme gene *GmUBC2* confers enhanced drought and salt tolerance through modulating abiotic stress-responsive gene expression in *Arabidopsis*. Plant Mol. Biol. 72, 357–367.

Zhu, Q., Zhang, J., Gao, X., Tong, J., Xiao, L., Li, W., Zhang, H., 2010. The *Arabidopsis* AP2/ERF transcription factor RAP2.6 participates in ABA, salt and osmotic stress responses. Gene 457, 1–12.

Enhancing soybean response to biotic and abiotic stresses

Mohammad Miransari
Department of Book & Article, AbtinBerkeh Ltd. Company, Isfahan, Iran

Introduction

The production of crop plants is subject to different stresses worldwide, and accordingly different stresses have been evaluated singly or combined (Boyer, 1982; Cushman and Bohnert, 2000). Although plants are subjected to stresses individually, it is also possible that plants are affected by a combination of stresses under field conditions (Figure 3.1). For example, in areas subjected to drought stress, plants may also encounter salinity and heat stress. However, the response of a plant to some of such combinations, for example drought and heat, may be unique and different from the plant response to each stress individually (Craufurd and Peacock, 1993; Heyne and Brunson, 1940; Jiang and Huang, 2001).

Soybean (*Glycine max* (L.) Merr.) plants are important legume crops, used as a major source of food, protein, and oil, worldwide. The production of world soybean in 2007 was 219.8 million metric tons, including USA (70.4), Brazil (61), Argentina (47), and China (14.3), which had the highest rate of production (Soystats, 2008). However, in 2012 soybean production increased to 268 million metric tons, including Brazil (83.5), USA (82.1), Argentina (51.5), China (12.6), and India (11.5), which had the highest rate of soybean yield production (Soystats, 2012).

Soybean is able to develop a symbiotic association with its specific rhizobium, *Bradyrhizobium japonicum*, and acquires most of its essential nitrogen (N) for its growth and yield production thereby. The bacteria can fix atmospheric N and make it available for the use of their host plant. Till now, it has not been possible to enhance the plant and bacteria response in such a way that the plant can acquire its entire essential N by the process of biological N fixation. Hence, when planting soybean, the rate of biological N fixation by bacteria must be determined and, accordingly, the appropriate rate of N chemical fertilization be used so that the plant produces an optimum rate of yield (Miransari, 2011a, b).

The process of biological N fixation is initiated by the production of a signaling molecule, genistein, by host plant roots, which is able to activate N-fixation genes in *Rhizobium*. In response, and as a result of gene activation in *B. japonicum*, some biochemicals, called lipochitooligosaccharides (LCOs) are produced by the bacteria. Such molecules induce morphological and physiological alterations in their host plant roots and eventually the bacteria enter the cells of the root cortex, where they fix atmospheric N for the use of the host plant (Long, 2001).

Under optimal conditions the symbiotic association between soybean and *B. japonicum* results in the fixation of high amounts of N, which is usable by soybean

Abiotic and Biotic Stresses in Soybean Production. http://dx.doi.org/10.1016/B978-0-12-801536-0.00003-7

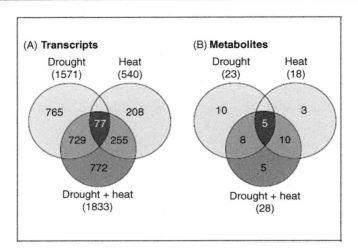

Figure 3.1 The molecular properties of drought and heat stress, singly or combined.
Adapted from Mittler (2006), with permission from Elsevier, license number:
3626500670447.

and can provide most of the plant's essential N. However, under stress activities the host plant and the bacteria are adversely affected, and hence the rate of biological N fixation by bacteria and the growth of soybean decreases. This is because soybean and its symbiotic *B. japonicum* are not tolerant under stress (Wang et al., 2006).

Using biotechnological and molecular methods and techniques, it may be possible to enhance plant and bacteria responses under stress. Accordingly, to develop appropriate methods and techniques, the plant and bacteria must be subjected to stress, and their responses, including the activation of different molecular and signaling pathways, must be determined under field and greenhouse conditions. In the next stage the right gene, which may enhance plant and bacterial response under stress, must be placed so that the plant and/or bacteria can behave appropriately under stress. The use of molecular methods including the proper signaling molecule may also be suitable to improve plant and bacterial response under stress (Fukushima, 1994; Libault et al., 2008; Miransari et al., 2013).

For the development of tolerant strains of bacteria, it is suggested that the bacteria are isolated from stress conditions. This is because such species have adapted to the stress conditions by developing the right mechanisms and signaling pathways. However, this may not be the case for the host plant because the plant itself is sensitive under stress and cannot survive the stress if suitable methods, which make the plant more tolerant under stress are not used (Graham, 1992).

Some of the most important details related to the use of methods and techniques, which may improve the tolerance of soybean and *B. japonicum* under stress, are presented and analyzed. Such details can be used for the production of tolerant soybean and *Rhizobium* species under stress.

Soybean and stress

Soybean is a sensitive plant under stress and its growth and yield decreases under such conditions. The most sensitive stage of biological N fixation is the initial stage of N fixation, including the process of signal communication between soybean and *B. japonicum*. Under stress, production of signaling molecules, specifically genistein, by the roots of host plant reduces and hence the response of the bacterial population in the soil decreases. Different stages of symbiotic association between soybean and *B. japonicum* including the production of signal molecules, attachment of rhizobia, curling of root hairs, production of infection threads, and production of nodules are among the most sensitive stages of nodulation under stress (Zhang and Smith, 1996).

The following stages indicate the development of symbiosis fixation between the host plant, soybean, and the symbiotic bacteria *B. japonicum*.

- Production of the signal molecule, genistein, by the host plant.
- Recognition of the host plant presence by the bacteria and its chemotactic approach toward the roots of the host plant.
- Production of LCO by bacteria resulting in the morphological (bulging and curving) in the root hairs of the host plant.
- Entrance of the bacteria into the cortex cells of the host plant.
- Morphological and physiological alterations in the cortex cells.
- Production of bacteroid, which is able to fix atmospheric N.

Among the most important effects of stress on plant growth is the allocation of carbon to different plant parts. Under stress, a decreased photosynthesis (PN) rate reduces the production and allocation of carbon, and hence crop yield decreases (Berman and Dejong, 1996). Regulated stomatal activities under stress are an important parameter affecting plant growth and yield production. The activities of enzymes, which regulate carbon allocation to different plant parts are also of significance under stress (Alscher and Cumming, 1990; Kluge, 1976).

It has been indicated that under stress the activities of enzymes, such as invertase, which are able to mobilize the sucrose in the vacuole, increase (Roitsch, 1999). The level of PN and carbohydrates are also affected by salinity stress, influencing the types of carbohydrates, which are produced and exported to different plant tissues. Such a response may be used by the plant as a signaling pathway to regulate the activities of sources and sink (Gilbert et al., 1997).

Although under stress plant response including the morphological and physiological alterations is a result of gene expression, more details have yet to be indicated related to the molecular and signaling pathway affecting plant response under stress. Using proteomics may be a suitable method for determining such a pathway. The proteins produced under stress, especially in tolerant plant species, may indicate which molecular and signaling pathways are activated (Yang et al., 2007; Zhen et al., 2007; Zhou et al., 2009a, b).

Chiera et al. (2002) investigated the effects of phosphorus (P) deficiency on the growth of young soybean plants. They evaluated apical meristems and leaf structure in soybean to ascertain the role of cell division and expansion in the regulation of leaf size and number. The seeds were germinated and then were subjected to P deficiency for 32 days. The rates of leaf initiation decreased significantly although the size of the apical dome and the number of cells were similar to those in the control plants. The authors accordingly indicated that under P stress the limitation of cell division numbers restricts soybean growth as regulated by a plant factor in the leaf canopy.

Another important factor affected by stress is crop quality including protein, nonstructural carbohydrate, lipid, minerals, antioxidants, feed value, and the physical properties of food. The protein content of crop products is only 10–30% of the total weight; however, it is an important parameter determining the quality of crop plants including legumes. The protein content of crop plants is affected by the environmental and genetic parameters. Under environmental stresses, and most of the time, the protein concentration of the harvested crop increases, for example, in soybean, wheat, and rice (Feng et al., 2008; Flagella et al., 2010; Mulchi et al., 1988).

Related to the effects of other stresses, the effect of ozone on the quality of crop plants is more complicated and is determined by plant species, ozone concentration, and the effects of other environmental parameters including CO_2. However, the majority of studies have shown that ozone increases the concentration of protein in soybean and wheat. For lipids, the effects of stresses including salinity, drought, and heat have been investigated for oil crop and medicinal plants (Feng and Kobayashi, 2009; Mulchi et al., 1988).

Although several research studies have investigated soil nutrients affecting the nutrient concentration of plant, only a few have evaluated the effects of different stresses on the mineral concentration of plants. However, the results have not been constant. Under stress the antioxidant content of plants including phenolic, ascorbate, and caretonoid increases, affecting the plant response including the uptake of nutrients under stress (Petersen et al., 1998).

Although the effect of stress on the lipid content of harvested crop plants is determined by factors such as plant species, experimental conditions, and type of stress, most research has indicated that under stress the lipid content of harvested crop plants decreases. This was especially the case for drought stress as almost all research has indicated that under such a stress the rate of crop lipid decreases (Feng and Kobayashi, 2009; Mulchi et al., 1988). Starch and sugar are two important nonstructural carbohydrates; in grain yields such as wheat and corn, the stresses of drought, heat, and ultraviolet often decrease the grain starch content. However, for sugar the responses of crop plants under stress were inconsistent (Petersen et al., 1998).

Using genistein and LCO, it is possible to enhance plant and bacteria responses under stress. Enhancing the plant response to stress by activating the related signaling pathway is a strong and promising tool (Cushman and Bohnert, 2000; Kovtun et al., 2000; Miransari et al., 2006). There might be cross talk between different signaling pathways at different levels including mitogen-activated protein kinase (MAPK), plant hormones, calcium signaling pathway, reactive oxygen species (ROS), receptors, and signaling complexes (Bowler and Fluhr, 2000; Casal, 2002; Mittler, 2006; Xiong and Yang, 2003).

Biotic stresses

Under biotic stresses such as pathogenic activities, the plant's systemic resistance is induced and results in the production of biochemicals such as phytoalexins, which are able to reduce the pathogen. Phytoalexins are products with little molecular weight, produced by plants, when subjected to biotic and abiotic stresses; hence they help the plant to control the invading microorganisms. By recognizing the molecular pathways that result in the production of phytoalexins, it is possible to produce plants with a higher tolerance under stress (Conrath et al., 2002; Gill and Tuteja, 2010).

To date, most data that have been gathered on phytoalexins have involved the plant families Fabaceae, Leguminosae, and Solanaceae (Ingham, 1982; Kuc, 1982). However, recent research has investigated plant families of economic significance including Vitaceae and Poaceae (maize and rice) (Schmelz et al., 2014). Phytoalexin production and its pathogen tolerance are regulated by different signaling pathways including plant hormonal signaling, calcium pathway, MAPK, the fungal perception response, and expression of related genes. Hence, details on the production of phytoalexins in plants can be applicable for the production of resistant plants under biotic stresses (Schmelz et al., 2014).

The interactions between a host plant and pathogens are determined by the ability of a pathogen to metabolize the phytoalexins produced by the host plant. Modifying the related genes in the pathogenic fungi, which result in the detoxification of phytoalexins by the fungi, may provide more details related to the interactions between the fungi and the host plant. Examples of phytoalexins in legumes include isoflavans, phaseollin, and glyceollin (Jeandet, 2015).

Plant responses may be similar when the plant is subjected to biotic stresses such as herbivores, pathogens, and abiotic stresses, especially heavy metals. Mithöfer et al. (2004) indicated that under such stresses ROS are activated, adversely affecting cellular structure and resulting in the activation of the related signaling pathway. Although ROS are produced under biotic and abiotic stresses, the related enzymatic and nonenzymatic pathway may be different. Alteration of the cellular structure brought about by the production of ROS determines the plant response under biotic and abiotic stresses (Mithöfer et al., 2004).

Soybean seeds contain a range of antioxidant products including phenolic, tannin, and flavonoid compounds, which are essential for the healthy growth of seed and plant. Malenčić et al. (2007) examined the antioxidant activities of 24 soybean genotypes. The genotypes, which contained the highest level of phenolic, indicated the highest antioxidant activity. However, the genotypes with the lowest level of phenolic compounds indicated the minimum level of scavenging activity. In conclusion, the authors considered that besides protein and oil, phenolic compounds, as antioxidant products, are also essential for seed growth and development.

Zernova et al. (2014) indicated a technique by which they were able to increase the resistance of soybean when it was exposed to the pathogen *Rhizoctonia solani*. They strengthened the root hairs of soybean versus the pathogen by inserting the genes resveratrol-O-methyltransferase (*ROMT*) and resveratrol synthase 3 (*AhRS3*) from peanut. As a result the modified plant indicated few symptoms of necrosis related to the wild types, which indicated 84% necrosis. If the fungal pathogen can resist the toxic effects of phytoalexins produced by the host plant it will be able to invade the plant and hence adversely affect its growth while feeding on it (Jeandet, 2015).

Polyphenols are among the most prevalent biochemicals produced by flowering plants. The antioxidant activities of polyphenols make the host plant resistant to biotic stresses such as pathogens (Ferrazzano et al., 2011). The activities of polyphenols in alleviating plant diseases depends on their antimicrobial activities including: (1) stopping the activity of bacterial replicate enzymes, (2) inactivating the bacterial toxins, and (3) strengthening the plant's cellular structure and antifungal activities (Ferrazzano et al., 2011).

Polyphenols such as catechins can affect different bacterial species including *Pseudomonas aeruginosa, Salmonella choleraesis, Serratia marcescens, Escherichia coli, Bordetella bronchiseptica, Klebsiella pneumoniae, Staphylococcus aureus,* and *Bacillus subtilis* through altering the properties and permeability of cell membranes and the production of hydrogen peroxide (Ferrazzano et al., 2011).

The response of microbes affected by polyphenols includes upregulating the proteins that can strengthen the bacteria under such a stress by adjusting their metabolic activities related to the synthesis of proteins, phospholipids, and carbon products. Polyphenols can also disrupt the process of quorum senescing by bacteria, which results in the regulation of different bacterial activities such as their exponential growth (Dobretsov et al., 2011), and hence can be used as a controlling mechanism for the activities of pathogens.

Abiotic stresses

Salinity

Different research works have been conducted to investigate the effects of salinity on the growth and N-fixation ability of soybean. For example, using greenhouse and field experiments, Miransari and Smith (2007, 2009) evaluated soybean response under salinity. Under field conditions, favorable rates of salinity were established using sodium chloride and soybean was planted. The hypothesis was that the initial stages of the N-fixation process, including the signal exchange between the two symbionts, are the most sensitive stages. Accordingly, bacterial inocula were produced using *B. japonicum* and were treated with different concentrations of the signal molecule, genistein.

At seeding, soybean seeds were inoculated with *B. japonicum* and genistein and planted in the salinized field. At different growth stages, sampling of plants and soil was carried out and different parameters were determined including the leaf properties and nodulation (weight and number). At the final harvest, soybean yield was determined. According to the results genistein was able to significantly alleviate the stress of salinity on soybean growth and yield including the number and weight of nodules, the rate of N fixation, and the amount of yield (Miransari and Smith, 2007, 2014a).

The experiment was also conducted under greenhouse conditions and soybean seeds were inoculated with *B. japonicum*, pretreated with different genistein concentrations at 1 ml per seed. Using Hoagland nutrient solution, which had been treated with the favorable levels of salinity the pots were fertilized and watered. The medium was a mixture of turface and sand, sterilized with heat and pressure. Interestingly, at day 14, following inoculation the pots treated with B. *japonicum* and genistein turned an intense green color indicating that both under nonstressed and stressed conditions the bacteria pretreated with genistein were able to fix N and alleviate the stress of salinity. The sampling of plants was done three times, at 20, 40, and 60 days following the inoculation of soybean seeds with pretreated inocula. Genistein was able to alleviate the stress by increasing the number and the weight of nodules, rate of biological N-fixation, and soybean growth (Miransari and Smith, 2009, 2014a).

The balance between the production of ROS and polyamines under stress is the important factor determining plant response under stress. Polyamines are biochemical products that are accumulated under stress and make the plant regulate the homeostasis of ROS. The active forms of ROS in plants include single oxygen (O_2), superoxide (O_2^-), hydroxyl radical (HO), and hydrogen peroxide (H_2O_2) with the highest stability (Moller et al., 2007; Saha et al., 2015).

A large number of subcellular compartments are able to produce ROS; however, the major ones are chloroplasts, peroxisomes, and mitochondria (Mittler, 2002). The chloroplast is the site that produces the highest level of ROS both under nonstressed and stressed conditions in the thylakoid membrane. The production of ROS increases during salt stress, due to the alteration of membrane fluidity and production of protein complex disrupting electron passage from water to photosystem II (Saha et al., 2015).

Mitochondria are the other important sites of reactive oxygen production. Salt stress adversely affects cellular activity controlled by mitochondria due to the passage of electrons to the molecular oxygen, resulting in the production of O_2^- (Miller et al., 2010). The enzymes, from the other cellular compartments, which may also have a role in the synthesis and production of ROS, are plasma membrane nicotinamide adenine dinucleotide phosphate (NADPH) oxidases, diamine oxidase, oxalate oxidases, cytosolic polyamine oxidase, and cell wall peroxidases (Saha et al., 2015).

Polyamines are also able to enhance plant tolerance under stress by regulating the homeostasis of cellular nutrients. For example, polyamines and H_2O_2 are able to regulate calcium channels under salt stress, which rapidly results in the enhanced concentration of calcium in the plant cells, and hence ROS are produced by the activity of NADPH oxidase in the cellular membrane (Gupta et al., 2013; Mansour, 2013).

Plant hormones such as abscisic acid (ABA), ethylene, jasmonate, and auxin, as well as their signaling, interactions, and cross talk, are important factors affecting the plant response under stress. However, the interactions of plant hormones with the other signaling molecules such as ROS, reactive N species, nitric oxide, cyclic nucleotides, and calcium can also result in the establishment of a complex signaling network affecting the plant response under stress (Hasanuzzaman et al., 2013; Zhang et al., 2006).

For example, it has been indicated that under salt stress the activities of polyamines result in the production of nitric oxide. The molecule acts as a signal and enhances plant resistance under stress by increasing the ratio of potassium to sodium, as a result of stimulating the expression of H^+–ATPase and the antiport of Na^+/H^+ in the tonoplast (Li et al., 2013; Shi et al., 2012; Zagorchev et al., 2013).

Different methods and techniques have been used to alleviate the stress of salinity on the growth and yield of soybean. For example, Yoon et al. (2009) investigated the alleviating effects of methyl jasmonate on the growth of soybean under salinity stress induced by sodium chloride in a hydroponic medium. The stress was established by exposing soybean seedlings to 60 mM sodium chloride 24 h after using methyl jasmonate at 20 and 30 μM. Under salinity stress the following parameters were adversely affected: soybean growth, PN, water loss from stomata, and the production of gibberellins by seedlings. However, under stress the production of ABA and proline increased. Use of methyl jasmonate significantly increased the growth of soybean seedlings by increasing the production of different hormones including ABA, proline, and chlorophyll content, the rate of leaf PN, and stomatal activities.

In another experiment, Lu et al. (2009) evaluated the effects of salinity on different plant growth factors in two soybean genotypes including sensitive and tolerant. In the salt-tolerant genotype the net rate of PN decreased related to the salt-sensitive genotype. The decrease of PN in the salt-tolerant genotype was related to the decreased stomatal activity and CO_2 concentration. The other important factor, which decreased under salinity stress in the salt-sensitive genotype, was the activity of Rubisco. The authors, in conclusion, indicated that the two important parameters affecting salt resistance in the tolerant and salt-sensitive soybean genotypes were PN (resulting from decreased stomatal activity) and Rubisco activity, respectively.

Munns and Tester (2008) reviewed the mechanisms used by plants under salinity stress at the plant, organ, and cellular level. Plant responses under salinity include two different phases including the rapid phase in which the growth of young leaf is disrupted and the longer phase, which results in the more rapidly senescence of old leaf. Three different adaptation mechanisms are used by plants including: (1) the resistance under osmotic stress, (2) exclusion of sodium or chloride, and (3) tissue tolerance under the conditions of sodium and chloride accumulation. The researchers, however, indicated that the details related to plant tolerance under salinity stress need to be further investigated, especially at molecular and physiological levels using functional genomics and molecular genetics.

Similarly, Phang et al. (2008) indicated that salt stress significantly decreases soybean growth, nodulation, and the quality and quantity of seed, and hence soybean yield. However, to decrease the adverse effects of salt stress, soybean has developed certain mechanisms including: (1) response regulation under osmotic stress, (2) maintaining ion homeostasis, (3) establishing the osmotic balance, and (4) other related metabolic and physiological adaptations.

Recognition of genes in crop plants that improve their salinity tolerance is of great significance. Accordingly, Guan et al., (2014) indicated a salt tolerance gene in soybean, *GmSALT3*, which is able to increase soybean tolerance under salinity stress, mainly by decreasing the concentration of sodium in the aerial parts of the plant. The authors speculated that such a gene has a great potential for improving soybean tolerance under salinity stress.

Drought

Interestingly, more than 30% of the world's population resides in places with water deficiency. With respect to elevated rates of CO_2 and climate changes in the future, the stress of drought becomes even more severe and frequent. The tolerance of legumes at the time of grain filling and at early frost under stresses such as salinity, drought, flooding, cold, heat, and heavy metal, can determine tolerance to climate changes in the future (Cutforth et al., 2007; Turner et al., 2001).

Decrease in yield due to drought stress is 40%; with respect to the genotype properties, the amount of water used by soybean is 450–700 mm during the growing season. The most sensitive stage of soybean growth under water stress is flowering and the stages following flowering. Soybean properties including developmental plasticity, leaf area, sensitivity of photoperiod, tolerance under heat stress, adjustment of osmotic potential, root depth and densit y, stomatal activities, and leaf reflectance determine its tolerance under stress (Desclaux et al., 2000).

The morphology and plasticity of soybean roots are among the most important factors affecting soybean resistance under stress. However, with respect to the properties of soybean roots, the plant sometimes faces water stress because the water is available at a greater depth. If the plant genotype grows its roots deeper into the soil, or produces an extensive root network with a high rate of root hairs, it will be able to resist water stress, though research in this respect has been limited (Benjamin and Nielsen, 2006).

The other important parameter affecting plant response under stress is root plasticity. When the plant is subjected to drought stress, more carbon is allocated to the roots and hence the ratio of root to aerial part increases. Root parameters including weight, total length, total number of root hairs, and root volume are significantly correlated with soybean tolerance under stress (Liu et al., 2005).

According to research one of the genotypes of soybean is able to produce an extensive network of root hairs (Carter and Rufty, 1993), which is also able to resist the drought stress by controlling stomatal activities and hence increasing plant water efficiency. Due to its high rate of root surface, the genotype can also fix atmospheric N at a high rate. The roots of this genotype are also able to grow in compacted soil. Although the root properties were positively correlated with the rate of seed proteins, there was no correlation between the root properties and the drought resistance genotype.

With respect to the above mentioned details, it is possible to enhance soybean drought tolerance by modifying its genetic and molecular properties. This can be fulfilled if the related physiological mechanisms and genetic response of soybean and hence the related genes are recognized under drought stress. Such soybean responses must be determined under a wide range of fields using the most responsive genotypes (Figure 3.2).

The process of biological N fixation by soybean and the symbiotic *B. japonicum* is sensitive to the stress of drought. Accordingly, soybean parameters including the production and loss of CO_2, leaf area, and the process of biological N fixation are affected. As a result, plant seed protein and yield decreases (Sinclair and Serraj, 1995). The initial stages of soybean symbiosis with *B. japonicum* are among

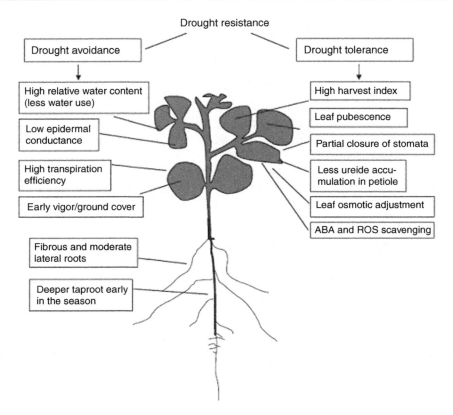

Figure 3.2 Traits of drought tolerance in soybean.
Adapted from Manavalan et al. (2009), with permission from Oxford University Press, license number 3626510189984.

the most sensitive stages of biological N fixation by the two symbionts under stress (Miransari and Smith, 2007, 2009).

Under drought stress the process of N fixation is affected by: (1) decreasing the availability of oxygen, (2) reducing the rate of carbon uptake into the nodules, (3) decreasing the activity of sucrose synthase in nodules, and (4) increasing the amount of ureide and free amino acid. Nodule water potential is also a function of water potential in plant leaf. While the activity of N-fixation enzyme decreased by 70%, the rate of PN reduced by only 5% indicating that the adverse effects of water stress on the process of N fixation are independent of the PN process. Under water stress nodule activity is affected by the increased resistance of oxygen diffusion into the bacteroid (Serraj et al., 2001).

Due to the following, the metabolic activities of bacteroid significantly decreases under drought stress: (1) increased resistance to oxygen diffusion, (2) decreased activities of N-fixation enzyme, (3) accumulation of unusable substrates and oxidized biochemicals, and (4) the activation of antioxidant genes. As a result the adverse effects of oxidative stress in nodules are detectable prior to the reduction in the synthesis of leghemoglobin and sucrose. Water deficit can also result in the accumulation of ureide in plant leaf and hence adversely affect the process of nodulation (Sinclair and Serraj, 1995).

The level of ureide in petioles is a suitable indicator of soybean tolerance under drought stress. It has yet to be indicated if under drought stress the process of biological N fixation is regulated by the whole plant or is done just at a nodule level. However, there is evidence that under drought stress the N content of the aerial part can regulate the process of N fixation by activating the related signals. Some contrasting results indicate that more research work is essential to determine if the process of N fixation under stress is regulated systematically by the plant or not (Serraj et al., 2001).

Water use efficiency, which is the unit of water used for the production of yield, is an important parameter affecting plant growth and yield. High water use efficiency can help the plant survive under drought stress and hence, if genetically improved, can significantly affect plant production under drought stress, for example by modifying the stomatal conductance. The other important parameter influencing plant behavior under stresses such as drought and salinity is osmotic adjustment, indicating the active concentration of solutes in plant cells in response to stress. It has been shown that under water deficit conditions, osmotic adjustment can maintain PN and stomatal conductance, delay leaf and flower senescence, and enhance root growth and water uptake from the soil as the drought stress intensifies. The osmotic potential in soybean ranges from 0.3 MPa to 1.0 MPa, which is less than in other legumes (Turner et al., 2001).

Under water stress, the relative water content of different soybean genotypes, which was significantly correlated ($r = 0.98$) with osmotic adjustment, remained more constant for the genotypes with higher osmotic adjustment. This indicates that such genotypes are able to survive the drought stress more efficiently (James et al., 2008). Accordingly, improving soybean ability for osmotic adjustment may be a suitable method for enhancing soybean tolerance under stress.

Under limited water conditions genetic improvement to enhance plant resistance may be an effective method. However, due to the following reasons such efforts have not been done as fast as expected: (1) selecting the most efficient genotype of soybean tolerant to drought stress is a difficult and time consuming process, which must be done with care and patience, (2) most emphasis has been on the selection of genotypes with a higher tolerance to biotic stresses rather than to abiotic stresses, and (3) drought tolerance is a complicated trait, controlled by different parameters including plant genotype, environment, and the interactions of the two. Selection on the basis of genotype traits can be a favorable method to improve soybean response significantly under stress (Serraj and Sinclair, 1998; Sinclair, 2011; Vadez and Sinclair, 2001).

Plant functional genomics is among the fastest growing techniques used for the investigation of genes and their behavior and hence improving plant tolerance under stress. Developing the total sequence of soybean (975 Mb) (http://www.phytozome. net/soybean) is among the most important progresses, which have been made so far. Accordingly, the functions of genes can be improved under different conditions including stress.

Because the genome sequence of soybean has been determined it is possible to find the functional activity of each gene. Accordingly, Le et al. (2012) investigated the functional response of different soybean genes under drought stress. Of 46,093 soybean genes, 41,059 genes have a known function. They found that during the stages of V6 and R2 under drought stress in plant leaf, 1458, 1818, 1582, and 1688 genes were up- and downregulated, respectively. Most upregulated genes resulted in the activation of regular proteins such as transcription factors and kinases or functional proteins including late embryogenesis abundant proteins.

Gene expression also indicated that during stages V6 and R2 under drought stress, the activity of hormone related genes was affected. Downregulation of PN genes, which can decrease the growth rate of plant under drought stress, indicated that the plant response can be a survival mechanism under drought stress. According to this research, excellent drought-responsive genes have been recognized, which can be used for further analysis and development of drought tolerant soybean genotypes.

Gil-Quintana et al. (2013) conducted an interesting experiment to examine the effects of drought stress on N fixation and nodule metabolism in soybean. The activity of N-fixing enzyme decreased under drought stress relative to the control treatment. The amount of amino acid and ureide increased in soybean root and the related components. The amount of ureide increased because its degradation decreased and not because of its increased biosynthesis. The use of proteomic analyses indicated that under stress the activities of nodules which are affected the most are the metabolism of plant carbon, metabolism of amino acids, synthesis of protein, and cell growth. Accordingly, the authors stated that their results support the hypothesis that the process of biological N fixation is locally regulated in soybean nodules and the process is not adversely affected by ureide.

In relation to proteomic analyses, investigating the metabolic activity of plant cells can reveal the functions of such cells under different conditions. It has been known for a long time that under stress osmolytes including ABA, sugars, proline, and glycine betaine are accumulated in plant. If the development of tolerant soybean genotypes is desirable, the related physiological and biochemical responses of soybean as well as the functional genes under stress must be understood in great detail (Gil-Quintana et al., 2013).

A large number of cultivated and wild type soybean genotypes are available in germplasm collected worldwide; however, only a few germplasms have been analyzed for their tolerance under stress. There has not been much effort placed on the investigation of soybean physiological properties affecting plant tolerance under stress including osmotic adjustment, plant cell properties, depression of canopy heat, and production of antioxidants (Qi et al., 2010; Serraj, 2003).

With respect to the importance of soybean in the world economy, recognition of desirable traits and genotypes which may enhance soybean tolerance under stress is of great significance. In relation to the soybean genome sequence, combining a suitable trait/gene related to drought stress has become increasingly feasible. If genomics, proteomics, and biology are combined, finding new genes, pathways, and physiological responses that can significantly allow the control of the soybean response under stress will be come more likely (Deshmukh et al., 2014).

To acquire favorable results related to the precise soybean response under drought stress, the conditions of the drought experiments must match the field conditions. The length of the experiment, the severity of drought stress, the way the experiment is conducted, and the parameters which are determined must closely follow to the actual field conditions. The environment of the experiments must be manageable in different places and for years. Accordingly, the knowledge of breeding, physiology, and molecular biology must be combined so that the soybean response under stress can be evaluated precisely and the development of more tolerant plant species becomes more feasible (Cominelli et al., 2013).

Lei et al. (2006) investigated the effects of drought and rewetting on the physio-ecological properties of soybean in China. They found that the activity of stomata and hence water loss from soybean is more sensitive to drought stress than the process of PN. However, rewetting the soil resulted in a higher water potential in soybean under drought stress.

Benjamin and Nielsen (2006) investigated the effects of drought stress on the root distribution of soybean. Soybean plants were subjected to different water treatments including natural precipitation and irrigation. Root distribution was determined at 0.23-m soil intervals to a depth of 1.12 m directly underneath the plant at two different stages including late bloom and midpod fill growth. The relative root distribution was not affected by water deficit, and 97% of total soybean roots were found at a 0.23-m depth at both sampling times and under both treatments. Such results indicate that soybean may not be a suitable legume for planting under drought stress unless its morphological and physiological properties are adjusted.

Silvente et al. (2012) examined the effects of water stress on the physiological response of two soybean genotypes including sensitive and the tolerant ones. Plant growth, the rate of biological N fixation, the dynamics of proline and ureide, and metabolic changes in both genotypes under water stress were evaluated. The results indicated that there were significant differences between the two genotypes with respect to their drought tolerance. The authors were also able to elucidate the signaling pathway activated under stress to regulate plant response by affecting plant metabolic activities and physiology.

Plants utilize different mechanisms to alleviate the adverse effects of stress. For example, under drought stress, proline is produced in plants to adjust water behavior and efficiency and hence make the plant absorb water. Soybean nodules are able to produce proline under drought stress, and as a result a high rate of NADP/NADPH is maintained in nodules. The authors also suggested that proline shifts the redox potential from plant cytoplasm to the bacteroid (Kohl et al., 1990). Soybean production in South Africa is subjected to drought stress. However, using the *P5CR* gene in soybean resulted in improved tolerance of soybean under drought and heat stress related to wild genotypes. This indicates the possibility of commercializing drought-tolerant soybean able to produce proline (Verbruggen and Hermans, 2008).

Heavy metals

Chen et al. (2003) determined the effects of cadmium stress on the nodulation, biological N fixation, soybean growth, and distribution of cadmium in soybean plants in contaminated soils. Soybean nodulation was adversely and significantly affected by cadmium stress, especially at the level of 10–20 mg kg^{-1} soil. Root growth also decreased with increasing levels of cadmium and the symptoms of stress appeared in the roots. Cadmium stress also reduced the weight ratio of roots compared to the aerial parts, which can also be the reason for the decreased weight of root nodules.

Although low concentrations of cadmium moderately stimulated the process of N fixation by soybean plants, at the higher levels of cadmium the rate of N-fixation significantly decreased. At high cadmium concentrations, there were also some significant changes in the ultrastructure of root nodules and the effective area of N fixation as the number of N fixing cells decreased. The distribution of cadmium in different parts of the plant was as follows: root > aerial part > seed, which is also the reason for the adverse effects on plant roots (Chen et al., 2003).

Pawlowska and Charvat (2004) investigated the effects of heavy metal stress on the activities of different mycorrhizal species. Different growth stages of mycorrhizal fungal ecotypes were evaluated under the stress of heavy metals including the germination of spores, presymbiotic hyphal growth, spore production, and mycelium growth. Although the two species of mycorrhizal fungi including *Glomus intraradices* and *G. etunicatum* were grown for a long time under a low level of heavy metal stress, their sensitivity to the stress was different as *G. etunicatum* was more sensitive.

The hyphal growth of *G. etunicatum* increased under heavy metal stress and the fungi were able to produce spores under stress. The production of spores was more sensitive to the stress of heavy meals than the hyphal growth. Evaluating the different growth stages of the fungus *G. etunicatum* indicated that the fungi are able to survive the stress of heavy metals by using a metal avoidance strategy (Pawlowska and Charvat, 2004). Such results can be useful for planting soybean under heavy metal stress by selecting the correct and most fungal-resistant species.

Soybean seeds were planted in a nutrient solution treated with cadmium and nickel. Plant growth and seed production were significantly decreased by the metals and the highest concentrations of heavy metals were found in the roots. Because of higher mobility, nickel was found with a higher concentration in the higher parts of the plant including the seeds. In the seeds the higher content of nickel was found in the axis and testa (Malan and Farrant, 1998).

The highest cadmium concentrations were found in the cotyledon and testa and the lowest was related to the axis. Metal content of seeds increased with plant age. Nickel content of seeds was higher than in the pods, although for cadmium the difference was not significant. Due to the decreased levels of lipid, protein, and carbohydrate, cadmium decreased seed yield. Although nickel reduced the number of seeds, the seed yield was not affected (Malan and Farrant, 1998).

Although the BURP plant proteins are important in growth, development, and plant response under stress, most of their functions in plants have not yet been indicated. The properties of these proteins indicate that these proteins are able to interact with heavy metals. Tang et al. (2014) found that the soybean BURP protein, SALI3-2, is able to bind soft heavy metals such as cadmium, zinc, cobalt, nickel, and copper but not hard metals such as calcium and magnesium. Such proteins are localized in vacuole. The expression of *sali3-2* in *Arabidopsis thaliana* enhanced its tolerance under the copper and cadmium stress by increasing their concentration in the roots rather than the aerial parts of the plant.

According to research, heavy metals may delay the process of N fixation in soybean. For example, the presence of arsenic in the nutrient solution delayed the time essential for the nodulation of soybean roots and decreased the number of nodules per plant. The stress also reduced the number of root hairs and decreased the growth of roots and the aerial parts of soybean (Ahmad et al., 2012).

The unfavorable effects of aluminum on plant growth and yield production adversely affects root growth in aluminum-sensitive plants. The effects, which cause toxicity, include: (1) decreased uptake of water and nutrients and their subsequent allocation to different plant parts, (2) disruption of calcium homeostasis and cellular structure, (3) adverse effects on cellular behavior including division and growth, (4) callose deposition in apoplast adversely affecting the cellular crossing of substances, and (5) peroxidation of lipid, which results in the production of ROS. Such effects impinge on the morphological properties of plant roots (Duressa et al., 2011; Horst, 1995; Yamamoto et al., 2002; Zhou et al., 2009a, b).

Plant cellular function is also disrupted by aluminum through its binding to phosphate and carbonyl components in the apoplast and symplast. Plants produce organic products, containing phosphates, phenolics, and polypeptides, which bind aluminum and ameliorate its unfavorable effects on plant growth. In a wide range of plant species, the production of citrate, oxalate, and malate is among the most important mechanisms used by plants to mitigate the unfavorable effects of aluminum. In soybean, detoxification of aluminum is a trait and as a result plant roots produce citrate and phenolic compounds (Pellet et al., 1996; Yang et al., 2007; Zhou et al., 2009a, b).

Duressa et al. (2011) investigated the effects of aluminum toxicity on the related molecular mechanism for the survival of different soybean genotypes under unfavorable pH conditions. They analyzed the root hairs of aluminum tolerant and sensitive genotypes using proteomic. In the tolerant genotype, aluminum resulted in the induction of aluminum tolerance related enzymes and proteins; however, in the sensitive genotype, only the production of general stress proteins was induced.

Aluminum upregulated the production of different antioxidation and detoxification enzymes such as enolase, malate, malate oxidoreductase, and others in the tolerant genotype and not in the sensitive genotype (Begum et al., 2009). These activity of the enzymes resulted in the production of citrate, which is able to alleviate the unfavorable effects of aluminum on the growth of soybean by the process of detoxification. The authors accordingly suggested that if such enzymes are overexpressed simultaneously in soybean, it is possible to produce soybean tolerant genotypes under aluminum toxicity.

Suboptimal root heat

Suboptimal root heat is a stress that can adversely affect plant growth and yield production. Under cold-included chilling (<20°C) and freezing (<0°C), different metabolic and genetic functions of plants are negatively affected resulting in a cold-induced osmotic potential, decreased water and nutrient uptake, oxidative stress, etc. However, during the process of cold acclimation plants may acquire resistance before being subjected to cold stress. Such ability can be obtained by most temperate plants. Crop plants that are planted during the winter must be vernalized helping the plant to overcome winter stress as seedlings. However, after the vernalization period and before the reproductive period the acclimation capability of plants decreases (Mittler, 2006; Mahajan and Tuteja, 2005).

Seed germination is a complex process that is affected by different environmental factors, especially heat and water. This is also the case for germinating soybean seeds and, if they are subjected to environmental stresses, the production of seedlings and plant growth and yield production may be adversely affected (Miransari and Smith, 2014b). Using proteomic techniques Swigonska and Weidner (2013) investigated the long-term effects of cold and osmotic stress on the roots of germinating soybean. Seeds were subjected to the cold (10°C, H_2O) and osmotic (25°C/ −0.2 MPa) stresses and a combination of the two (10°C and −0.2 MPa).

Proteomic analyses showed the role of different proteins under long time stress in metabolic pathway, synthesis of protein, seed germination and growth, production of signals, and cellular behavior. The authors accordingly indicated that the differences in the expression patterns of proteins under stress can be used to illustrate soybean response to long-term stress and the similarities and the differences, which may be present in soybean response under osmotic and cold stresses (Swigonska and Weidner, 2013).

MYB-type transcription factors have MYB–DNA binding for 50 amino acids, which are able to regulate different plant activities including growth, development, and stress response. Liao et al. (2008) found 156 *GmMYB* genes for soybean and indicated the same, when treated with stresses such as drought, salinity, and cold. The authors selected three *GmMYB* genes including *GmMYB76, GmMYB92*, and *GmMYB177*. The first two indicated transcriptional activity. It was also shown that the three genes are able to bind the *cis* elements TAT AAC GGT TTT TT and CCG GAA AAA AGG AT, however, with different affinities.

The overexpression of *GmMYB76* or *GmMYB177* by transgenic *Arabidopsis* indicated better tolerance under salt and cold stress, related to *GmMYB9*. But the sensitivity of seeds treated with ABA decreased in such plants, relative to the wild ones. Besides their regular activities, the genes were also able to regulate the activities of a subset of responsive genes. In conclusion, the authors indicated that the genes are able to differentially affect the soybean response under stress by regulating the activities of stress-responsive genes.

Miransari and Smith (2008) investigated the effects of suboptimal root heat on the growth of soybean by stimulating the conditions of field in the greenhouse. Accordingly, intact soil pots were collected from the field with different soil textures, using aluminum cylinders, which were then placed in the greenhouse. Using a compressor and thermostat different soil heats were applied to the planted seeds, which had been inoculated with *B. japonicum*. The bacterial inocula had been pretreated with different genistein concentrations. The results indicated that genistein was able to alleviate the stress of suboptimal heat and increase the growth of soybean plants and the effects of soil textures were also different on growth of soybean and activity of genistein. The effects of suboptimal heat on the growth and yield of soybean and its alleviation by using the signal molecule genistein have been investigated both under field and greenhouse conditions (Bai et al., 2003; Lynch and Smith, 1993; Pan et al., 1998; Zhang and Smith, 1995, 1996).

B. japonicum and stress

B. japonicum is not tolerant under stress and its population, growth, and activity decrease under stress. The bacterium may use different mechanisms to tolerate stress. For example, under iron (Fe) deficiency, it is able to produce the specific ligands of Fe such as citrate as a siderophore, and hence increase Fe uptake. The *hemA* gene is induced by the availability of Fe (Page et al., 1994). If used in combination with plant growth promoting rhizobacteria (PGPR), such a process may be useful for providing a plant with its essential Fe (Miransari, 2014).

Fujihara and Yoneyama (1993) investigated the effects of pH and salinity on the growth and activity (including cellular polyamine) of *B. japonicum* A1017 and *Rhizobium fredii* P220. At unfavorable pH, the growth of *B. japonicum* was adversely affected, although the polyamine content of bacterial cells remained unaffected. However, *R. fredii* showed more tolerance under a wide range of pH from 4.0 to 9.5; and with decreasing the medium pH, the content of cellular polyamine and magnesium increased.

While *R. fredii* was able to grow in a medium with a salinity concentration of up to 0.4 M, *B. japonicum* was not able to grow even in a medium with a salinity of 0.14 M. The glutamine and potassium contents of *R. fredii* and the salt concentrations were positively correlated; however, the polyamine and magnesium content of *R. fredii* were negatively related. Such results indicate that the polyamine content of *R. fredii* and the stresses of unfavorable pH and salinity are correlated (Fujihara and Yoneyama, 1993).

Narberhaus et al. (1997) indicated three genes in *B. japonicum* including $rpoH_1$, $rpoH_2$, and $rpoH_3$, which have, for the first time, been recognized in an organism as a family with a *rpoH* multi gene. Each gene is regulated by a different mechanism. The activity rates of these genes in affecting the growth and activity of *B. japonicum* are different. With respect to their results, the authors indicated that *rpoH2* is essential for cellular protein synthesis under different physiological conditions; however, *rpoH1* and probably *rpoH3* are just essential for the growth and activities of bacteria under stress.

It has been indicated that manganese is essential for the nonstressed growth of *B. japonicum*; however, it is also able to induce responses in *B. japonicum* under stress. The mutant of *B. japonicum* without the *mntH* gene (as a transporter of Mn) indicated some severe symptoms of stressed growth under Mn deficiency indicating that Mn must be available for the suitable growth of *B. japonicum* under nonstressed conditions. Hohle and O'Brian (2012) found that the activities of superoxide dismutase and glycolytic pyruvate kinase were not present in a *mntH* mutant of *B. japonicum* under Mn deficient conditions. They accordingly indicated that Mn is essential for the oxidative metabolism of *B. japonicum* under nonstressed conditions.

The bacterial membrane controls the uptake of elements such as heavy metals. The membrane is permeable to most small solutes because of the presence of membrane pores. Hohle et al. (2011) revealed the related genes in the membrane as *mnoP* and *mntH*, regulating the uptake of Mn by *B. japonicum*. *mnoP* is especially expressed under the limitation of Mn and its cellular uptake is adjusted by bacteria. An *mnoP* mutant is deficient in the uptake of Mn and hence the symptoms of deficiency appear. The authors indicated that the bacterial membrane regulates the nutritional requirement of *B. japonicum* by the presence of the related membrane proteins.

Under different conditions such as stress and anaerobic conditions, ROS are produced. Different bacteria including *B. japonicum* are able to produce enzymes which detoxify such unfavorable stress products, and their production is regulated by the activity of redox sensing transcription factors. The other method used by bacteria to mitigate the unfavorable effects of stress products such as ROS is metal homeostasis. The presence of ROS results in the alteration of metal distribution in the cytosol. The concentration of, for example, Fe is controlled by different microbial metabolic activities present in proteins and metalloenzymes, which is also the case for Mn and Zn. Among the most important mechanisms used by bacteria to alleviate the adverse effects of ROS under high metal cytosolic concentration is a high ratio of Mn:Fe (Faulkner and Helmann, 2011).

If soybean plants are to be planted under Fe deficient conditions not previously used, the establishment of symbiosis between *B. japonicum* and soybean plants must be ensured with respect to N fertilization for the use of the host plant. However, N fertilization increases the pH of soil, enhancing Fe deficiency. Accordingly, Wiersma (2010) conducted a three-year experiment to determine the response of different soybean genotypes with different Fe deficiency tolerance to N fertilization. In conclusion, and in line with the results, the author indicated that N fertilization can result in Fe deficiency, especially in genotypes that are inefficient in relation to Fe uptake and utilization.

The response of soybean plants to the use of LCO is similar to the response to pathogens. Accordingly, to test the response of soybean plants to the use of LCO molecules under control and stress conditions, soybean plants grown under control conditions were subjected to the stress of suboptimal root zone temperature, and after a 5-day acclimation period, the trifoliate leaves were treated with 10^{-7} M LCO (NodBj-V [C18:1, Me-Fuc]) isolated from a *B. japonicum* culture, treated with genistein (Wang et al., 2012).

Plant leaves were harvested at 0 and 48 h following the LCO treatment and the expression of different genes was evaluated. A total of 147 genes including some stress responsive genes were activated on treatment with LCO. The authors suggested that soybean is a responsive plant with regard to the use of LCO and hence such a response can be used for treating soybean plants under stress with LCO (Wang et al., 2012).

In a similar experiment Miransari et al. (2006) tested the effects of LCO on the root hair properties of soybean seeds subjected to the stress of unfavorable pH. LCO was able to alleviate the stress of pH resulting in the bulging and curling of soybean root hairs related to the control treatment. Such results can be of practical use under stress conditions to enhance the process of biological N fixation by *B. japonicum* and soybean.

Sánchez-Pardo and Zornoza, (2014) investigated the effects of *B. japonicum* on the growth of soybean and white lupin under contamination with copper. *B. japonicum* resulted in a higher growth of root and aerial part in both soybean and white lupin, with a higher concentration of Cu in the aerial part of plants related to the control treatments. *B. japonicum* enhanced plant tolerance in both plants as they were able to absorb higher rate of Cu, while their growth was less affected, related to the control treatments.

A higher rate of Cu was indicated in the presence or absence of *B. japonicum*. White lupin resulted in the induction of a greater malondialdehyde rate and total thiols, when contaminated with copper in the presence or absence of *B. japonicum*. However, in soybean the increase in the concentration of malondialdehyde and total thiols was a little higher in the presence of *B. japonicum* in comparison with the absence of *B. japonicum*. The aerial parts of the plant were the most sensitive ones in both plants in the presence and absence of *B. japonicum*. In conclusion, the authors indicated that *B. japonicum* enhances tolerance of both plants to the stress of copper contamination, with a higher effect on white lupin, and the N fixation of *B. japonicum* can effectively increase the Cu tolerance of both plants.

Conclusion and future perspectives

Soybean and its symbiotic *Rhizobium*, *B. japonicum*, are not tolerant under stress and hence soybean growth and yield as well as the process of N biological fixation decrease under such conditions. Different parts of the world are subjected to different types of stresses such as salinity, unfavorable pH, drought, suboptimal root heat, and heavy metals. Under stress, the growth and activities of both soybean and *B. japonicum* are adversely affected. Hence, for the more efficient planting of soybean under stress suitable methods and techniques must be used to alleviate the unfavorable effects of stress on the growth and activity of soybean and *B. japonicum*. Under stress, the most sensitive stages of the process of biological N fixation are the initial ones including the process of signaling exchange between the two symbionts. Accordingly, among the most successful techniques that have been tested and proved to be effective is the use of signal molecule genistein with the bacterial inoculums of *B. japonicum*. Pretreatment of *B. japonicum* with genistein has been shown to significantly alleviate the effects of different stresses including salinity unfavorable pH, and suboptimal root heat on the growth, biological N fixation and the yield of soybean. Using different molecular and biotechnological techniques the response of soybean to different stresses has been illustrated. Accordingly, the related genes and proteins which may result in such responses have been indicated and they also have been used for the enhanced tolerance of soybean under different types of stresses. However, more details have yet to be indicated related to the response of soybean and *B. japonicum* under stress and how such responses may be improved. Future research may focus more on the responses of soybean and bacteria under stress by indicating the related molecular and signaling pathways, and by illustrating how such responses may be improved under stress.

References

Ahmad, E., Zaidi, A., Khan, M., Oves, M., 2012. Heavy metal toxicity to symbiotic nitrogen-fixing microorganism and host legumes. In: Zaidi, A., Wani, P.A., Khan, M.S. (Eds.), Toxicity of Heavy Metals to Legumes and Bioremediation. Springer, Vienna, pp. 29–44.

Alscher, R., Cumming, J., 1990. Stress Responses in Plants: Adaptation and Acclimation Mechanisms. Wiley-Liss, Inc., USA.

Bai, Y., Zhou, X., Smith, D.L., 2003. Enhanced soybean plant growth resulting from coinoculation of strains with *Bradyrhizobium japonicum*. Crop Sci. 43, 1774–1781.

Begum, H.H., Osaki, M., Watanabe, T., Shinano, T., 2009. Mechanisms of aluminum tolerance in phosphoenolpyruvate carboxylase transgenic rice. J. Plant Nutr. 32, 84–96.

Benjamin, J.G., Nielsen, D.C., 2006. Water deficit effects on root distribution of soybean, field pea and chickpea. Field Crops Res. 97, 248–253.

Berman, M.E., Dejong, T.M., 1996. Water stress and crop load effects on fruit fresh and dry weights in peach (*Prunus persica*). Tree Physiol. 16, 859–864.

Bowler, C., Fluhr, R., 2000. The role of calcium and activated oxygens as signals for controlling cross-tolerance. Trends Plant Sci. 5, 241–246.

Boyer, J.S., 1982. Plant productivity and environment. Science 218, 443–448.

Carter, T.E., Rufty, T.W., 1993. A soybean plant introduction exhibiting drought and aluminum tolerance. In: Kuo, G. (Ed.), Adaptation of Vegetables and other Food Crops to Temperature and Water Stress. Asian Vegetable Research and Development Center, Shanhua, Taiwan, pp. 335–346.

Casal, J.J., 2002. Environmental cues affecting development. Curr. Opin. Plant Biol. 5, 37–42.

Chen, Y.X., He, Y.F., Yang, Y., Yu, Y.L., Zheng, S.J., Tian, G.M., Luo, Y.M., Wong, M.H., 2003. Effect of cadmium on nodulation and N_2-fixation of soybean in contaminated soils. Chemosphere 50, 781–787.

Chiera, J., Thomas, J., Rufty, T., 2002. Leaf initiation and development in soybean under phosphorus stress. J. Exp. Bot. 53, 473–481.

Cominelli, E., Conti, L., Tonelli, C., Galbiati, M., 2013. Challenges and perspectives to improve crop drought and salinity tolerance. New Biotechnol. 30, 355–361.

Conrath, U., Pieterse, C., Mauch-Mani, B., 2002. Priming in plant–pathogen interactions. Trends Plant Sci. 7, 210–216.

Craufurd, P.Q., Peacock, J.M., 1993. Effect of heat and drought stress on sorghum. Exp. Agric. 29, 77–86.

Cushman, J.C., Bohnert, H.J., 2000. Genomic approaches to plant stress tolerance. Curr. Opin. Plant Biol. 3, 117–124.

Cutforth, H.W., McGinn, S.M., McPhee, K.E., Miller, P.R., 2007. Adaptation of pulse crops to the changing climate of the northern great plains. Agron. J. 99, 1684–1699.

Desclaux, D., Huynh, T.T., Roumet, P., 2000. Identification of soybean plant characteristics that indicate the timing of drought stress. Crop Sci. 40, 716–722.

Deshmukh, R., Sonah, H., Patil, G., Chen, W., Prince, S., Mutava, R., Vuong, T., Valliyodan, B., Nguyen, H., 2014. Integrating omic approaches for abiotic stress tolerance in soybean. Front. Plant Sci. 5, 244.

Dobretsov, S., Teplitski, M., Bayer, M., Gunasekera, S., Proksch, P., Paul, V.J., 2011. Inhibition of marine biofouling by bacterial quorum sensing inhibitors. Biofouling 27, 893–905.

Duressa, D., Soliman, K., Taylor, R., Senwo, Z., 2011. Proteomic analysis of soybean roots under aluminum stress. Int. J. Plant Genomics, 2011:282531.

Faulkner, M.J., Helmann, J.D., 2011. Peroxide stress elicits adaptive changes in bacterial metal ion homeostasis. Antioxid. Redox Signal. 15, 175–189.

Feng, Z.Z., Kobayashi, K., 2009. Assessing the impacts of current and future concentrations of surface ozone on crop yield with meta-analysis. Atmos. Environ. 43, 1510–1519.

Feng, Z.Z., Kobayashi, K., Ainsworth, E.A., 2008. Impact of elevated ozone concentration on growth, physiology, and yield of wheat (Triticum aestivum L.): a meta-analysis. Glob. Chang. Biol. 14, 2696–2708.

Ferrazzano, G.F., Amato, I., Ingenito, A., Zarrelli, A., Pinto, G., Pollio, A., 2011. Plant polyphenols and their anti-cariogenic properties: a review. Molecules 16, 1486–1507.

Flagella, Z., Giuliani, M.M., Giuzio, L., Volpi, C., Masci, S., 2010. Influence of water deficit on durum wheat storage protein composition and technological quality. Eur. J. Agron. 33, 197–207.

Fujihara, S., Yoneyama, T., 1993. Effects of pH and osmotic stress on cellular polyamine contents in the soybean rhizobia Rhizobium fredii P220 and Bradyrhizobium japonicum A1017. Appl. Environ. Microbiol. 59, 1104–1109.

Fukushima, D., 1994. Recent progress on biotechnology of soybean proteins and soybean protein food products. Food Biotechnol. 8, 83–135.

Gilbert, G.A., Wilson, C., Madore, M.A., 1997. Root-zone salinity alters raffinose oligosaccharide metabolism and transport in coleus. Plant Physiol. 115, 1267–1276.

Gill, S.S., Tuteja, N., 2010. Reactive oxygen species and antioxidant machinery in abiotic stress tolerance in crop plants. Plant Physiol. Biochem. 48, 909–930.

Gil-Quintana, E., Larrainzar, E., Seminario, A., Díaz-Leal, J., Alamillo, J., Pineda, M., Arrese-Igor, C., Wienkoop, S., González, E., 2013. Local inhibition of nitrogen fixation and nodule metabolism in drought-stressed soybean. J. Exp. Bot. 64, 2171–2182.

Graham, P., 1992. Stress tolerance in Rhizobium and Bradyrhizobium, and nodulation under adverse soil conditions. Can. J. Microbiol. 38, 475–484.

Guan, R., Qu, Y., Guo, Y., Yu, L., Liu, Y., Jiang, J., Chen, J., et al., 2014. Salinity tolerance in soybean is modulated by natural variation in GmSALT3. Plant J. 80, 937–950.

Gupta, K., Dey, A., Gupta, B., 2013. Plant polyamines in abiotic stress responses. Acta Physiol. Plant. 35, 2015–2036.

Hasanuzzaman, M., Gill, S., Fujita, M., 2013. Physiological role of nitric oxide in plants grown under adverse environmental conditions. In: Tuteja, A., Singh Gill, S. (Eds.), Plant Acclimation to Environmental Stress. Springer, New York, pp. 269–322.

Heyne, E.G., Brunson, A.M., 1940. Genetic studies of heat and drought tolerance in maize. J. Am. Soc. Agron. 32, 803–814.

Hohle, T.H., O'Brian, M.R., 2012. Manganese is required for oxidative metabolism in unstressed Bradyrhizobium japonicum cells. Mol. Microbiol. 84, 766–777.

Hohle, T.H., Franck, W.L., Stacey, G., O'Brian, M.R., 2011. Bacterial outer membrane channel for divalent metal ion acquisition. Proc. Natl. Acad. Sci. USA 108, 15390–15395.

Horst, W.J., 1995. The role of the apoplast in aluminum toxicity and resistance of higher plants. Z. Pflanzeneráh. Bodenk. 158, 419–428.

Ingham, J.L., 1982. Phytoalexins from the Leguminosae. In: Bailey, J.A., Mansfield, J.W. (Eds.), Phytoalexins. Blackie, Glasgow/London, UK, pp. 21–80.

James, A.T., Lawn, R.J., Cooper, M., 2008. Genotypic variation for drought stress response traits in soybean. I. Variation in soybean and wild Glycine spp. for epidermal conductance, osmotic potential, and relative water content. Aust. J. Agric. Res. 59, 656–669.

Jeandet, P., 2015. Phytoalexins: current progress and future prospects. Molecules 20, 2770–2774.

Jiang, Y., Huang, B., 2001. Drought and heat stress injury to two cool season turfgrasses in relation to antioxidant metabolism and lipid peroxidation. Crop Sci. 41, 436–442.

Kluge, M., 1976. Carbon and nitrogen metabolism under water stress. Water and Plant Life. Springer, Berlin–Heidelberg, pp. 243–252.

Kohl, D.H., Lin, J.J., Shearer, G., Schubert, K.R., 1990. Activities of the pentose phosphate pathway and enzymes of proline metabolism in legume root nodules. Plant Physiol. 94, 1258–1264.

Kovtun, Y., Chiu, W.L., Tena, G., Sheen, J., 2000. Functional analysis of oxidative stress-activated mitogen-activated protein kinase cascade in plants. Proc. Natl. Acad. Sci. USA 97, 2940–2945.

Kuc, J., 1982. Phytoalexins from the Solanaceae. In: Bailey, J.A., Mansfield, J.W. (Eds.), Phytoalexins. Blackie, Glasgow/London, UK, pp. 81–105.

Le, D.T., Nishiyama, R., Watanabe, Y., Tanaka, M., Seki, M., Ham, L., Yamaguchi-Shinozaki, K., Shinozaki, K., Lam, T., 2012. Differential gene expression in soybean leaf tissues at late developmental stages under drought stress revealed by genome-wide transcriptome analysis. PLoS One 7, e49522.

Lei, W., Tong, Z., Shengyan, D., 2006. Effect of drought and rewatering on photosynthetic physioecological characteristics of soybean. Acta Ecol. Sinica 26, 2073–2078.

Li, X., Jiang, H., Liu, F., Cai, J., Dai, T., Cao, W., Jiang, D., 2013. Induction of chilling tolerance in wheat during germination by pre-soaking seed with nitric oxide and gibberellin. J. Plant Growth Regul. 71, 31–40.

Liao, Y., Zou, H., Wang, H., Zhang, W., Ma, B., Zhang, J., Chen, S., 2008. Soybean *GmMYB76*, *GmMYB92*, and *GmMYB177* genes confer stress tolerance in transgenic *Arabidopsis* plants. Cell Res. 18, 1047–1060.

Libault, M., Thibivilliers, S., Bilgin, D.D., Radwan, O., Benitez, M., Clough, S.J., Stacey, G., 2008. Identification of four soybean reference genes for gene expression normalization. Plant Genome 1, 44–54.

Liu, F., Anderson, M.N., Jacobson, S.E., Jensen, C.R., 2005. Stomatal control and water use efficiency of soybean (*Glycine max* L. Merr.) during progressive soil drying. Environ. Exp. Bot. 54, 33–40.

Long, S.R., 2001. Genes and signals in the *Rhizobium*-legume symbiosis. Plant Physiol. 125, 69–72.

Lu, K.X., Cao, B.H., Feng, X.P., He, Y., Jiang, D.A., 2009. Photosynthetic response of salt-tolerant and sensitive soybean varieties. Photosynthetica 47, 381–387.

Lynch, D.H., Smith, D.L., 1993. Soybean (*Glycine max*) modulation and N_2-fixation as affected by exposure to a low root-zone temperature. Physiol. Plant. 88, 212–220.

Mahajan, S., Tuteja, N., 2005. Cold, salinity and drought stresses: an overview. Arch. Biochem. Biophys. 444, 139–158.

Malan, H.L., Farrant, J.M., 1998. Effects of the metal pollutants cadmium and nickel on soybean seed development. Seed Sci. Res. 8, 445–453.

Malenčić, D., Popović, M., Miladinović, J., 2007. Phenolic content and antioxidant properties of soybean (*Glycine max* (L.) Merr.) seeds. Molecules 12, 576–581.

Manavalan, L., Guttikonda, K., Tran, L., Nguyen, H., 2009. Physiological and molecular approaches to improve drought resistance in soybean. Plant Cell Physiol. 50, 1260–1276.

Mansour, M., 2013. Plasma membrane permeability as an indicator of salt tolerance in plants. Biol. Plant. 57, 1–10.

Miller, G., Suzuki, N., Ciftci-Yilmaz, S., Mittler, R., 2010. Reactive oxygen species homeostasis and signaling during drought and salinity stresses. Plant Cell Environ. 33, 453–467.

Miransari, M., 2011a. Soil microbes and plant fertilization. Review article. Appl. Microbiol. Biotechnol. 92, 875–885.

Miransari, M., 2011b. Interactions between arbuscular mycorrhizal fungi and soil bacteria. Appl. Microbiol. Biotechnol. 89, 917–930.

Miransari, M., 2014. Plant growth promoting rhizobacteria. J. Plant Nutr. 37, 2227–2235.

Miransari, M., Smith, D.L., 2007. Overcoming the stressful effects of salinity and acidity on soybean [*Glycine max* (L.) Merr.] nodulation and yields using signal molecule genistein under field conditions. J. Plant Nutr. 30, 1967–1992.

Miransari, M., Smith, D.L., 2008. Using signal molecule genistein to alleviate the stress of suboptimal root zone temperature on soybean-*Bradyrhizobium* symbiosis under different soil textures. J. Plant Interact. 3, 287–295.

Miransari, M., Smith, D.L., 2009. Alleviating salt stress on soybean (*Glycine max* (L.) Merr.)-*Bradyrhizobium japonicum* symbiosis, using signal molecule genistein. Eur. J. Soil Biol. 45, 146–152.

Miransari, M., Smith, D.L., 2014a. Improving Soybean (*Glycine max* L.) N_2 Fixation under Salinity Stress. LAP LAMBERT. Academic Publishing, Germany, ISBN: 978-3-659-53717-2.

Miransari, M., Smith, D.L., 2014b. Plant hormones and seed germination. Environ. Exp. Bot. 99, 110–121.

Miransari, M., Balakrishnan, P., Smith, D.L., Mackenzie, A.F., Bahrami, H.A., Malakouti, M.J., Rejali, F., 2006. Overcoming the stressful effect of low pH on soybean root hair curling using lipochitooligosaccharides. Commun. Soil Sci. Plant Anal. 37, 1103–1110.

Miransari, M., Riahi, H., Eftekhar, F., Minaie, A., Smith, D.L., 2013. Improving soybean (*Glycine max* L.) N_2 fixation under stress. J. Plant Growth Regul. 32, 909–921.

Mithöfer, A., Schulze, B., Boland, W., 2004. Biotic and heavy metal stress response in plants: evidence for common signals. FEBS Lett. 566, 1–5.

Mittler, R., 2002. Oxidative stress, antioxidants and stress tolerance. Trends Plant Sci. 7, 405–410.

Mittler, R., 2006. Abiotic stress, the field environment and stress combination. Trends Plant Sci. 11, 15–19.

Moller, I.M., Jensen, P.E., Hansson, A., 2007. Oxidative modifications to cellular components in plants. Annu. Rev. Plant Biol. 58, 459–481.

Mulchi, C.L., Lee, E., Tuthill, K., Olinick, E.V., 1988. Influence of ozone stress on growth processes, yields and grain quality characteristics among soybean cultivars. Environ. Pollut. 53, 151–169.

Munns, R., Tester, M., 2008. Mechanisms of salinity tolerance. Annu. Rev. Plant Biol. 59, 651–681.

Narberhaus, F., Krummenacher, P., Fischer, H.M., Hennecke, H., 1997. Three disparately regulated genes for σ^{32}-like transcription factors in *Bradyrhizobium japonicum*. Mol. Microbiol. 24, 93–104.

Page, K., Connolly, E., Guerinot, M., 1994. Effect of iron availability on expression of the *Bradyrhizobium japonicum* hemA gene. J. Bacteriol. 176, 1535–1538.

Pan, B., Zhang, F., Smith, D.L., 1998. Genistein addition to the rooting medium of soybean at the onset of nitrogen fixation increases nodulation. J. Plant Nutr. 21, 1631–1639.

Pawlowska, T.E., Charvat, I., 2004. Heavy-metal stress and developmental patterns of arbuscular mycorrhizal fungi. Appl. Environ. Microbiol. 70, 6643–6649.

Pellet, D.M., Papernik, L.A., Kochian, L.V., 1996. Multiple aluminum-resistance mechanisms in wheat roles of root apical phosphate and malate exudation. Plant Physiol. 112, 591–597.

Petersen, K.K., Willumsen, J., Kaack, K., 1998. Composition and taste of tomatoes as affected by increased salinity and different salinity sources. J. Hortic. Sci. Biotechnol. 73, 205–215.

Phang, T.H., Shao, G., Lam, H.M., 2008. Salt tolerance in soybean. J. Integr. Plant Biol. 50, 1196–1212.

Qi, X., Wang, X., Xu, J., Zhang, J., Mi, J., 2010. Drought-resistance evaluation of Flax Germplasm at adult plant stage. Scientia Agric. Sinica 43, 3076–3087.

Roitsch, T., 1999. Source-sink regulation by sugar and stress. Curr. Opin. Plant Biol. 2, 198–206.

Saha, J., Brauer, E.K., Sengupta, A., Popescu, S.C., Gupta, K., Gupta, B., 2015. Polyamines as redox homeostasis regulators during salt stress in plants. Front. Environ. Sci. 3, 21.

Sánchez-Pardo, B., Zornoza, P., 2014. Mitigation of Cu stress by legume–*Rhizobium* symbiosis in white lupin and soybean plants. Ecotoxicol. Environ. Saf. 102, 1–5.

Schmelz, E.A., Huffaker, A., Sims, J.W., Christensen, S.A., Lu, X., Okada, K., Peters, R.J., 2014. Biosynthesis, elicitation and roles of monocot terpenoid phytoalexins. Plant J. 79, 659–678.

Serraj, R., 2003. Effects of drought stress on legume symbiotic nitrogen fixation: physiological mechanisms. Indian J. Exp. Biol. 41, 1136–1141.

Serraj, R., Sinclair, T.R., 1998. Soybean cultivar variability for nodule formation and growth under drought. Plant Soil 202, 159–166.

Serraj, R., Vadez, V., Sinclair, T.R., 2001. Feedback regulation of symbiotic N_2 fixation under drought stress. Agronomie 21, 621–626.

Shi, H., Li, R., Cai, W., Liu, W., Fu, Z.W., Lu, Y., 2012. *In vivo* role of nitric oxide in plant response to abiotic and biotic stress. Plant Signal. Behav. 7, 437–439.

Silvente, S., Sobolev, A.P., Lara, M., 2012. Metabolite adjustments in drought tolerant and sensitive soybean genotypes in response to water stress. PLoS One 7, e38554.

Sinclair, T.R., 2011. Challenges in breeding for yield increase for drought. Trends Plant Sci. 16, 289–293.

Sinclair, T.R., Serraj, R., 1995. Legume nitrogen fixation and drought. Nature 378, 344.

Soystats, 2008. http://www.soystats.com/.

Soystats, 2012. http://www.soystats.com/.

Swigonska, S., Weidner, S., 2013. Proteomic analysis of response to long-term continuous stress in roots of germinating soybean seeds. J. Plant Physiol. 170, 470–479.

Tang, Y., Cao, Y., Qiu, J., Gao, Z., Ou, Z., Wang, Y., Zheng, Y., 2014. Expression of a vacuole-localized BURP-domain protein from soybean (SALI3-2) enhances tolerance to cadmium and copper stresses. PLoS One 9, e98830.

Turner, N.C., Wright, G.C., Siddique, K.H.M., 2001. Adaptation of grain legumes (pulses) to water limited environments. Adv. Agron. 71, 123–193.

Vadez, V., Sinclair, T.R., 2001. Leaf ureide degradation and N_2 fixation tolerance to water deficit in soybean. J. Exp. Bot. 52, 153–159.

Verbruggen, N., Hermans, C., 2008. Proline accumulation in plants: a review. Amino Acids 35, 753–759.

Wang, L., Zhang, T., Ding, S., 2006. Effect of drought and rewatering on photosynthetic physioecological characteristics of soybean. Acta Ecol. Sinica 26, 2073–2078.

Wang, N., Khan, W., Smith, D.L., 2012. Changes in soybean global gene expression after application of lipo-chitooligosaccharide from *Bradyrhizobium japonicum* under sub-optimal temperature. PLoS One 7, e31571.

Wiersma, J.V., 2010. Nitrate-induced iron deficiency in soybean varieties with varying iron-stress responses. Agron. J. 102, 1738–1744.

Xiong, L., Yang, Y., 2003. Disease resistance and abiotic stress tolerance in rice are inversely modulated by an abscisic acid-inducible mitogen-activated protein kinase. Plant Cell 15, 745–759.

Yamamoto, Y., Kobayashi, Y., Devi, S., Rikiishi, S., Matsumoto, H., 2002. Aluminum toxicity is associated with mitochondrial dysfunction and the production of reactive oxygen species in plant cells. Plant Physiol. 128, 63–72.

Yang, Q., Wang, Y., Zhang, J., Shi, W., Qian, C., Peng, X., 2007. Identification of aluminum-responsive proteins in rice roots by a proteomic approach: cysteine synthase as a key player in Al response. Proteomics 7, 737–749.

Yoon, J.Y., Hamayun, M., Lee, S., Lee, I., 2009. Methyl jasmonate alleviated salinity stress in soybean. J. Crop Sci. Biotechnol. 12, 63–68.

Zagorchev, L., Seal, C., Kranner, I., Odjakova, M., 2013. A central role for thiols in plant tolerance to abiotic stress. Int. J. Mol. Sci. 14, 7405–7432.

Zernova, O.V., Lygin, A.V., Pawlowski, M.L., Hill, C.B., Hartman, G.L., Widholm, J.M., Lozovaya, V.V., 2014. Regulation of plant immunity through modulation of phytoalexin synthesis. Molecules 19, 7480–7496.

Zhang, F., Smith, D.L., 1995. Preincubation of *Bradyrhizobium japonicum* with genistein accelerates nodule development of soybean at suboptimal root zone temperatures. Plant Physiol. 108, 961–968.

Zhang, F., Smith, D.L., 1996. Genistein accumulation in soybean (*Glycine max* (L.) Merr.) root systems under suboptimal root zone temperatures. J. Exp. Bot. 47, 785–792.

Zhang, Y.Y., Wang, L.L., Liu, Y.L., Zhang, Q., Wei, Q.P., Zhang, W.H., 2006. Nitric oxide enhances salt tolerance in maize seedlings through increasing activities of proton-pump and Na$^+$/H$^+$ antiport in the tonoplast. Planta 224, 545–555.

Zhen, Y., Qi, J.L., Wang, S.S., et al., 2007. Comparative proteome analysis of differentially expressed proteins induced by Al toxicity in soybean. Physiol. Plant. 131, 542–554.

Zhou, S., Sauve, R., Thannhauser, T.W., 2009a. Proteome changes induced by aluminum stress in tomato roots. J. Exp. Bot. 57, 4201–4213.

Zhou, S., Sauve, R., Thannhauser, T.W., 2009b. Aluminum induced proteome changes in tomato cotyledons. Plant Signal. Behav. 4, 769–772.

Use of proteomics to evaluate soybean response under abiotic stresses

Nacira Muñoz*,**,†, Man-Wah Li*, Sai-Ming Ngai*, Hon-Ming Lam*
*Centre for Soybean Research of the Partner State Key Laboratory of Agrobiotechnology and School of Life Sciences, The Chinese University of Hong Kong, Shatin, Hong Kong; **Instituto de Fisiología y Recursos Genéticos Vegetales, Centro de Investigaciones Agropecuarias – INTA, Córdoba, Argentina; †Cátedra de Fisiología Vegetal, FCEFyN – UNC, Córdoba, Argentina

Introduction

Soybean is one of the most important crops in the world. This unique high-protein oilseed contributes to diverse demands of human foods and animal feeds. Abiotic stresses are a primary cause of crop loss worldwide (Vinocur and Altman, 2005; Wang et al., 2003). Likewise, soybean production is vulnerable to the global environmental challenges. For instance, there are extended cultivated areas suffering from salinity, drought, and flooding problems, which are aggravated by the processes of desertification and climate change (Ahuja et al., 2010; Ciais et al., 2005).

The mechanisms of tolerance to abiotic stresses have attracted much attention in the scientific community (Mittler, 2006; Rai and Takabe, 2006). These tolerance mechanisms are often complex and dependent on the expression of a cascade of genes involved in specific signaling and regulatory pathways, as well as other cellular mechanisms. The advent of next generation sequencing has expedited the exploration of genomes by reducing the cost and increasing the throughput of the genome sequencing data production.

High-throughput sequencing technologies are important tools to discover genomic loci that control responses of stress tolerance. Strategies such as restriction site DNA–associated sequencing or genotyping-by-sequencing are useful for the discovery of genetic determinants of characters that are controlled by a few genes. However, in some cases, multiplicity of characters was involved. Furthermore, the complexity of epistatic interactions also makes it difficult to formulate a strategy for improvement against stresses (Cabello et al., 2014; Deshmukh et al., 2014).

The fascinating complexity and diversity of stress responses suggest that an integration of "omics" information is essential to achieve a comprehensive understanding of stress responses. In this chapter, we provide an integrated update on the soybean proteomic studies related to abiotic stresses. Here, we summarize the frequently used technologies and experimental systems, including information on the types and forms of stress conditions applied. We place special emphasis on the concepts of stress and tolerance in relationship to the experimental approaches and result interpretation. We also

Abiotic and Biotic Stresses in Soybean Production. http://dx.doi.org/10.1016/B978-0-12-801536-0.00004-9

summarize the major findings of soybean proteomic studies in conditions of flooding, salinity, drought, osmotic stress, cold, ultraviolet (UV)-B, and metals. Finally, we give an overall perspective on integrating the proteomic approach with other "omics" studies.

Stress in plants: general concepts

Stress could be defined as any unfavorable condition or substance that negatively affects a plant's metabolism, growth, or development. A sequence or phases of responses could be induced in plants after stress exposure (Lichtenthaler, 1996). In the initial alarm phase, stress is perceived as a shock to nonacclimated plants and the tolerance level is low, leading to destabilization of functions and excess in turnover of metabolites over biosynthesis. In the second phase that could last for several days, a new homeostasis will be established via initiating the repair processes and implementing the adaptation mechanisms, leading to an increased stress tolerance level. The final exhaustion phase will come with a reduced stress tolerance level, when the plant is not able to maintain metabolic homeostasis, as a result of high-stress intensity and the weakened adaptation capacity.

Tolerance can be defined as those mechanisms allowing a plant to sustain a physiological state under suboptimal conditions. For example, acute damage may occur rapidly in sensitive plants with a low level of tolerance or restricted tolerance mechanisms. During the alarm phase, these plants will activate their stress responding mechanisms by the rapid acclimation of metabolic fluxes as well as the repair processes and long-term metabolic and morphological changes (Lichtenthaler, 1996).

Acclimation can be defined as a set of reversible physiological changes occurring in the individual plant under stress (Baxter et al., 2014). The ability of a plant to acclimate is also often referred as phenotypic plasticity (Shao et al., 2007). The acclimation is distinguished from classical definition of adaptation because the latter refers to the genetically determined level of tolerance that has been acquired by selection after several generations (Bohnert and Sheveleva, 1998; Lichtenthaler, 1996, 1998). Both acclimation and adaptation to stressful environments are complex processes involving different levels of control.

The concept of phases of stress is useful to differentiate between short-term and long-term stress effects, as well as between events of low stress that can be partially alleviated by acclimation, adaptation, repair mechanisms, and strong stress or chronic stress events that cause significant damage, and eventually plant death.

Proteomics: definition and importance

Proteomics is the study of protein profiles. It allows us to identify and categorize proteins according to their putative functions. It has become an important research tool for "omics" study in crop plants. An additional advantage of proteomics over other large-scale studies is its ability to reveal posttranslational modifications. These are keys in the determination of functional impacts of proteins in a biological system (Komatsu et al., 2013b).

Proteomics has evolved from cataloging proteins to understanding how proteins are altered in abundance, location, activity, etc. These changes have been subjected to the development of new technologies and have redefined the concept of comparative quantitative studies in understanding the dynamics of the proteome (Thelen and Peck, 2007). The total proteome within a given cell is incredibly variable and responds to multiple factors such as metabolic, physiological, nutritional, and environmental ones. In other words, the proteome within a cell is highly dynamic and varies according to the stages of development and the stress conditions.

Dynamic changes in the abundance and types of proteins are part of the adaptive responses in plants (Kosová et al., 2011). How much we could achieve using this tool depends on factors such as the throughput of the available technology, capacity of data processing, and the experimental design. Important advances in proteomic technologies have given researchers new analytical capacities to investigate questions related to stress tolerance in crop plants.

The first challenge in proteomic approaches is sample extraction. Detailed methodologies for total soybean protein extraction from different plant tissues under different stress conditions have been developed (Hossain and Komatsu, 2014). Nevertheless, it is necessary to separate individual proteins in the crude extract. The most popular separation approaches are based on electrophoresis: conventional two-dimensional (2D) gel electrophoresis and 2D difference gel electrophoresis (DIGE). Alternatively, the total peptides from a sample in solution (gel-free) or in gel can also be separated and characterized by liquid chromatography–mass spectrometry/mass spectrometry (LC–MS/MS). Separation allows the identification of a higher number of peptides in complex samples. However, such analysis demands both high-performance chromatography and high-end mass spectrometry. The sample quality could also affect the downstream data characterization.

In general, the 2D approach allows the separation of complex protein extracts, that is, in the first dimension according to the protein's isoelectric point and, in the second dimension, according to the protein's molecular size. The presence of proteins in a gel can hence be revealed by different methods and each spot reflects the abundance of the corresponding separated protein. Nonetheless, it is often the case that a single protein spot contains more than one protein species.

The 2D DIGE technique is an alternative to the conventional 2D method, in which up to three batches of total protein samples can be individually prelabeled with fluorescent dyes (fluorophores) before separation by 2D electrophoresis. The 2D separated proteins from the labeled protein sample could be visualized as spectrally distinguishable fluorescent spots. Therefore, both protein identification and quantification can be compared within the labeled protein sample. Since the experimental and control protein samples are loaded onto the same 2D gel, errors resulting from variation between gels can be minimized. Thus, it improves the experimental reproducibility and allows quantitative comparison.

After 2D gel separation, in-gel digestion of the protein spot using a site-specific protease, such as trypsin, can generate a subset of peptides for analysis. This step could be operated automatically or manually. Masses of the digested peptides can be identified by either matrix-assisted laser desorption ionization (MALDI) or use of the electrospray ionization (ESI) mass spectrometer.

In the MALDI mode, peptide samples are mixed with organic acid which is known as matrix. The three most commonly used organic acids are 3,5-dimethoxy-4-hydroxycinnamic acid (sinapinic acid), α-cyano-4-hydroxycinnamic acid (also known as CHCA, α-cyano, or α-matrix), and 2,5-dihydroxybenzoic acid. It is necessary to introduce "charges" into the peptide sample so it can fly along the flight tube of the MS machine upon laser excitation.

Alternatively, charging of the peptide sample could be achieved using ESI mode, in which peptide ions are transferred from solution into the gaseous phase via the electrical energy before being subjected to mass spectrometric analysis. Ionic species in solution can be analyzed by ESI–MS with increased sensitivity (no requirement of matrix which, by itself, could contribute noise to the mass spectral measurement).

Either the MALDI or ESI mode could be coupled to a time-of-flight (TOF) mass spectrometer and allow the generation of the peptide mass finger print (PMF) profile of the corresponding protein. The target protein can then be identified through comparison of the mass list from the PMF profile with public databases, subject to the availability of existing information. If the submitted mass list cannot be identified with available databases, a *de novo* sequencing approach (tandem MS) would be the alternative method. This requires high-fidelity configuration of the mass spectrometer and expertise in protein/peptide sample handling. A general flow chart for proteomic analysis by MS (adapted from Van Wijk, 2001) is shown in Figure 4.1.

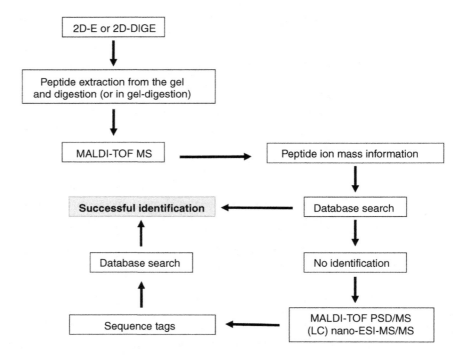

Figure 4.1 Flow chart for proteomic analysis by MS. 2D-E, bidimensional gel electrophoresis; 2D-DIGE, difference gel electrophoresis; LC, liquid chromatography; MALDI, matrix-assisted laser desorption ionization; PSD, post-source decay; TOF, time of flight; ESI, electrospray ionization. Adapted from Van Wijk (2001).

The mass spectrometry-based quantification methods could be divided into two groups such as the following:

1. Label-based protein quantification: The ^{15}N metabolic labeling *in vivo* labeling method can give good results in quantitative comparative proteomics of plants, although this method requires long labeling exposure periods. The isobaric tags for relative and absolute quantitation (iTRAQ) labeling methods labels samples with iTRAQ reagents during sample preparation and can be used to analyze eight samples simultaneously. The labeled peptides are separated by LC coupled to ESI and subsequently analyzed by mass spectrometry. Comparative analysis using this technique requires an analysis system that is able to examine the low-molecular-weight range. Other labeling methods are also available, such as isotope-coded affinity tag, ^{18}O labeling, and stable isotope labeling by amino acids in cell culture.
2. Label-free protein quantification: Direct quantification and comparison of the spectral information generated by independent LC–MS/MS acquisitions were involved.

The advantages and disadvantages of applying these strategies to study stress responses in crops have been previously discussed (Nanjo et al., 2011a). Table 4.1 summarizes the most commonly used methods in soybean stress proteomic studies and Figure 4.1 provides a general flow chart for proteomic analysis by MS.

In the subsequent section, we will summarize the major contributions of soybean proteomic studies to the understanding of stress responses.

Responses of soybean proteome under different stresses

Cell division, expansion, and growth

The growth and survival of plants is affected by stress conditions. Stress can affect the growth of plants by causing differences in the capacity of water acquisition and transport, or drastic changes in the metabolism (Chaves et al., 2002).

Important types of stress conditions such as drought, salinity, high temperature, or cold induce a reduction in water potential and affect the plant water status. The plant hormone abscisic acid (ABA) plays a major role in plant responses to stress. It is produced after dehydration is perceived by plant cells, resulting in the activation of the stress signal transduction pathway among other plant responses to stresses (Zhang et al., 2006).

Under dehydration conditions, rapid ABA signaling will induce a partial closure of stomata in order to avoid water loss. This response accompanies the reduction in photosynthesis given by the reduction of CO_2 diffusion in the leaf and carbon assimilation. A more complex scenario arises because ABA itself is a regulator in soybean, regulating shoot growth and development under water stress. Variation in endogenous ABA levels and differing sensitivity to ABA in roots and hypocotyls can result in the modulation of root/shoot growth ratios under low water potentials in soybean (Creelman et al., 1990).

Under different stress conditions, proteomic approaches in soybean characterized proteins that are linked with ABA signaling pathway. During the stage of soybean seed germination under salinity, two cultivars, Lee68 (salt-tolerant) and N2899 (salt-sensitive), have shown an increase in ABA contents. Moreover, the ABA level in Lee68 was much higher than that in N2899. Some proteins characterized in this work

Table 4.1 **Details of soybean proteomic studies under different stress conditions**

Stress conditions	Cultivars of *Glycine max*	Tissues	Stress applications	Methods	Proteins characterized	References
Aluminum	Baxi10 (aluminum tolerant line)	Root	50 µM AlCl$_3$, applied to 1 week seedlings for 24, 48, and 72 h	2D-E MALDI–TOF	39 proteins	Zhen et al. (2007)
Cadmium	Harosoy, Fuku-yutaka and their RIL CDH-80	Root and unifoliate leaves	10-day-old seedlings treated with 100 µM CdCl$_2$ for 3 days	2D-E LC-MS/MS	12 proteins	Hossain et al. (2012)
UV-B	Two isolines of Clark	First trifoliate leaf	12 days seedlings under field conditions with and without natural levels of UV-B	2D MALDI–TOF LC-MS/MS	67 proteins	Xu et al. (2008)
Salinity	Enrei	Hypocotyl and root	100 mM NaCl applied from germination, 3 days seedlings	2D ESI–Q/TOF–MS/MS	7 proteins	Aghaei et al. (2009)
Salinity	Enrei	Leaves, hypocotyl, and root	40 mM NaCl applied from germination for 7 days	2DE MALDI–TOF	19 proteins leaves, 22 hypocotyl, and 14 root	Sobhanian et al. (2010)
Salinity	Lee68 tolerant N2899 sensitive	Germinated seeds	100 mM NaCl from germination. Samples harvested when the radicle protruded by 2–3 mm from the seed coat	2D MALDI–TOF-MS	18 proteins	Xu et al. (2011)
SA, JA, ACC, H$_2$O$_2$, SNP vitamin B1, and *Phytophthora sojae* zoosperm	Xinyixiaohei-dou	Leaves	Seedlings treated for 24 h	2D MALDI–TOF/TOF	21 proteins	Zhao et al. (2013)

Ozone	Pioneer 93B15	Leaves and root	85 and 40 ppb 9 h per day from seedlings to reproductive stage	2D-E LC-MS/MS	277 Proteins	Galant et al. (2012)
Waterlogging	Asogari	Root	3 and 7 days after treatment water saturated substrate	2D-E MALDI-TOF ESI-MS/MS	24 proteins	Alam et al. (2010)
Flooding	Enrei	Cotyledon and root	2-day-old seedlings treated for 2 days	2D-E LC-MS/MS	73 proteins root and 28 cotyledon	Komatsu et al. (2013d)
Flooding	Enrei	Cell wall of root and hypocotyl	2-day-old germinated seeds subjected to flooding for 2 days	2D-E MALDI TOF/MS	16 proteins	Komatsu et al. (2010a)
Flooding	Enrei	Leaves hypocotyl root	2-day-old germinated seeds subjected to flooding for 5 days	2D-E LC-MS/MS	51 root 66 hypocotyl 51 leaves	Khatoon et al. (2012)
Flooding	Enrei	Mitochondrial fractions from root and hypocotyl	4-day-old seedlings subjected to flooding for 2 days	2D-E LC-MS/MS	52 proteins	Komatsu et al. (2011b)
Flooding	Enrei	Hypocotyl and root	2-day-old germinated seeds subjected to flooding for 2 days	2D-E MALDI-TOF MS	28 proteins	Komatsu et al. (2010b)
Flooding	Enrei	Root and hypocotyl	2-day-old seedlings were flooded for 12 h	2D-E MALDI-TOF MS	34 proteins	Komatsu et al. (2009b)
Flooding	Enrei	Plasma membrane proteins from root and hypocotyl	2-day-old seedlings were flooded for 24 h	2D-E MALDI-TOF MS	14 proteins	Komatsu et al. (2009a)
Water stress	Magellan	Root (four different regions from the root tip)	15 mm root length transfer to -1.6 MPa for 48 h (control -0.1 MPa)	2D-E MALDI-TOF MS/MS	35 proteins	Yamaguchi et al. (2010)

(Continued)

Table 4.1 Details of soybean proteomic studies under different stress conditions (*cont.*)

Stress conditions	Cultivars of *Glycine max*	Tissues	Stress applications	Methods	Proteins characterized	References
Osmotic and comparisons with salinity, water stress, and cold	Enrei	Root	2-day-old seedlings treated with PEG 6000 (−0.01, −0.3, or −0.49 MPa) from 1 to 4 days. 100 Mm NaCl, 5°C and drought	2D-E MALDI-TOF	37 proteins	Toorchi et al. (2009)
Drought	Enrei	Leaf, hypocotyl, and root	3-day-old seedlings subjected to drought-stress by water restriction. Osmotic control using 10% PEG 6000	2D-E LC-MS/MS	49 root, 37 hypocotyl, 37 leaves	Mohammadi et al. (2012)
Drought	Taegwang	Root	2-week-old plants treated with drought for 5 days (gradual water shortage, 79, 62, 46, 29, and 23% of field capacity)	2D-E MALDI-TOF MS	28 proteins	Alam et al. (2010)
Osmotic stress	Enrei	Plama membrane proteins from root and hypocotyl	2-day-old germinated seeds subjected to 10% PEG for 2 days	2D-E LC-MS/MS	86 proteins	Nouri and Komatsu (2010)
Cold and osmotic stress	Aldana	Root	10 °C/−0.2 MPa applied from germination for 72 h	2D-E ESI-LTQ-FTICR	59 proteins	Swigonska and Weidner (2013)
Cold stress	Chilling-resistant soybean genotype Z22	Embryonic axes	Seeds exposed to chilling temperature 4 °C for 24 h	2D-E MALDI-TOF	40 proteins	Cheng et al. (2010)

2D-E, bidimensional gel electrophoresis; 2D DIGE, difference gel electrophoresis; LC, liquid chromatography; MALDI, matrix-assisted laser desorption ionization; TOF, time of flight; ESI, electrospray ionization; SA, salicylic acid, JA, methyl jasmonate; ACC, 1-amino cyclopropane-1-carboxylic acid; SNP, sodium nitroprusside.

are reported to be involved in or regulated by the ABA signaling pathway. Ferritin is an iron storing protein and appears to be a key protein in the soybean response to salt stress conditions (Xu et al., 2011). Induction of ferritin by ABA has been documented at both transcript and protein levels (Lobreaux et al., 1993). In the roots of water stressed soybean, the increased abundance of ferritin proteins was shown to effectively sequester more iron, and thereby prevent excess free iron throughout the root elongation zone. A protective role against Fenton reactive oxygen species (ROS) production has been proposed (Yamaguchi et al., 2010).

Glutathione S-transferases (GSTs) characterized by proteomic approaches have been reported in soybean roots under aluminum toxicity using an aluminum-tolerant cultivar (Zhen et al., 2007) and associated with salt responses in soybean (Chan and Lam, 2014; Xu et al., 2011). GSTs play important roles in oxidative stress control (Edwards et al., 2000). GST expression is induced by phytohormones including ABA. Interestingly, a new function was cited for a specific GST in *Arabidopsis* with a negative role in drought and salt stress tolerance through a combined effect of glutathione (GSH) and ABA (Chen et al., 2012). This highlights the importance of GSTs in response to stress beyond their role of oxidative protection and detoxification.

Analysis of label-free quantitative proteomic indicated the effect of ABA in early soybean stage under flooding. This study characterized a number of protein involved and concluded that ABA might enhance the flooding tolerance of soybean through the control of energy conservation via a glycolytic system and the regulation of zinc finger proteins, cell division cycle 5 protein, and transducin (Komatsu et al., 2013a).

Cell division and cell expansion are the two main processes that sustain organ growth. In higher plants, both processes take place in growth zones that can be subdivided into specialized regions, particularly extensively characterized in root tips. From the tip, a plant root can be divided into three major zones (Beemster et al., 2003): (1) apical meristem – found in two root areas, closest to the tip where cells divide and expand at approximately equal rates, and next to it, where the cells maintain the division rate but the expansion rate rapidly increases; (2) elongation zone – there is no cell division, but rapid cell expansion; (3) mature part of the root – largest part of the root system, where cells no longer divide or expand. Researchers have generated a protein map of apex and differentiated root of soybean and describe a classification of proteins involved in this growth zone (Mathesius et al., 2011).

The root system demonstrated plastic responses to some abiotic stresses. Kinematic characterization of the spatial patterns of cell expansion within the zone of root elongation in soybean and other plants showed that at low water potentials, elongation rates are preferentially maintained towards the root tip, but are progressively inhibited at more basal locations, resulting in a shortened growth zone (Yamaguchi and Sharp, 2010).

The same study also analyzed the spatial profile of soluble proteins in two regions of the soybean root (Yamaguchi and Sharp, 2010). Within the first 4 mm from the root tip, the root elongation remains similar between plants with or without a water supply. On the other hand, in the region 4–8 mm from the tip root, the elongation is progressively inhibited under drought conditions. Proteomic analysis showed a possible differential and specific spatial regulation in the phenylpropanoids pathway, which could explain the observed difference in the two regions.

It was also found that several enzymes related to the biosynthesis of isoflavonoids were increased in the region closest to the root tip, correlated with the increased levels of flavonoids. Meanwhile, two important enzymes involved in lignin biosynthesis were upregulated in the more distal region from the root tip under water stress, correlated with an increment in the lignin content that could be a strategy to prevent loss of water in roots. Such change in cell wall composition will strengthen the tissue and lead to a reduction of cell wall flexibility that is necessary to keep the process of cell expansion during plant growth. Proteins that were common to both regions are those involved in the protection against oxidative damage and could also be involved in the control of soybean cell wall extensibility. It has also been found in other plants (Bustos et al., 2008; Müller et al., 2009; Yamaguchi and Sharp, 2010) and reported in a comparative proteome of soybean cell wall under flooding stress (Komatsu et al., 2010a).

The root growth is also affected in flooding. The proteome of different regions of the root revealed responses, such as induction of cell death in the root tip, which may negatively affect the plant's overall growth and development (Nanjo et al., 2011b). Furthermore, there seems to be a reduction of cell wall lignification in soybean roots and hypocotyl subjected to flooding stress (Komatsu et al., 2010a).

It is known that the physical properties of the cell wall could change at low water potentials (Nonami and Boyer, 1990). Lignins are crucial for structural integrity of the cell wall and stiffness and strength of the stem. They are complex racemic aromatic heteropolymers derived mainly from three monomers of hydroxycinnamyl alcohol, differing in their degree of methoxylation. The amount and composition of lignins is a function of taxa, cell types, and individual cell wall layers. They are also influenced by developmental and environmental cues (Boerjan et al., 2003).

O-methyltransferases are key enzymes in the biosynthesis of lignin because they control the production of differentially methylated monolignols, as the precursors of lignins. Caffeoyl-CoA-3-O-methyltransferase (CCoAOMT) converts caffeoyl-CoA into feruloyl-CoA (Boudet, 2000). The expression of CCoAOMT is coincident with lignin deposition and has been involved in the response to salinity and water deficit stresses in *Arabidopsis* and tobacco (Senthil-Kumar et al., 2010). Interestingly, the CCoAOMT gene promoter is stress responsive (Chen et al., 2000).

Simulations of osmotic stress using polyethylene glycol (PEG) led to a decrease in CCoAOMT and S-adenosyl methionine synthetase, suggesting that the reduction of CCoAOMT could confer a reduction of lignification in roots under osmotic stress (Toorchi et al., 2009). Downregulation of caffeoyl-CoA-O-methyltransferase in soybean roots from an organ specific proteomic study under saline conditions was also reported (Sobhanian et al., 2010), suggesting that lignification is related to salt response.

Methionine synthase (MS) was reduced in leaf, hypocotyl, and root under drought conditions. This enzyme was expressed in tissues undergoing lignification in non-stressful environments because lignin monomers should be methylated before polymerization. MS is also involved in a number of important functions in plants such as the synthesis of ethylene and polyamines (Ingram and Bartels, 1996). The decrease of MS in soybean is specific to drought and associated with the process of cell wall lignification (Mohammadi et al., 2012).

The cytoskeleton is involved in a variety of cellular functions such as cell division, morphogenesis, and signal transduction (Wang et al., 2011). The importance of the cytoskeleton in plants under biotic and abiotic stresses becomes an important topic for plant science (Lin et al., 2014). Proteomic approaches have revealed the role of cytoskeleton-associated proteins in soybean, such as tubulin, responding under stress and acting during flooding stress indicated by an analysis of plasma membrane proteome (Komatsu et al., 2009a).

The other important cytoskeleton-associated proteins in soybean, such as actin depolymerizing factor, tubulin, and profiling, have been characterized and shown to be downregulated under osmotic stress (Nouri and Komatsu, 2010). Actins and kinesins are suggested to be related with growth suppression in soybean root under drought stress (Alam et al., 2010). However, it was also reported that actin isoform B is upregulated in the leaf, hypocotyl, and root of soybean under drought, and in leaf and hypocotyl of soybean under osmotic stress (Mohammadi et al., 2012). How the cytoskeleton dynamics are regulated under stress in soybean is an important yet unexplored field. Their roles in acclimation, adaptation, and tolerance are essential to the understanding of soybean responses to stress.

Dehydration and osmotic adjustment

At some stage in the life cycle, most plants encounter transient decreases in relative water content, and many also produce highly desiccation-tolerant structures such as seeds, spores, or pollen. Indeed, physiological drought also occurs during cold and salt stresses, when the main damage caused to the living cell can be related to water deficit (Ingram and Bartels, 1996).

Plant water potential can be maintained by osmotic adjustment, brought about by the presence of sugars or other compatible solutes. Osmotic adjustment is the plant's response to environmental changes related to any stress that could be perceived and results in the induction of dehydration by, for example, drought, salinity, or cold. The mechanism of osmotic adjustment under drought and salinity could help plants avoid ion toxicity and maintain water uptake by accumulating large quantities of osmolytes. Such products could be differentiated into two types: organic solutes including amino acids, sugars, or low molecular weight metabolites, and inorganic ions including mainly Na^+, K^+, Ca^{2+}, and Cl^- (Chen and Jiang, 2010).

In organic solute type-osmolytes, proline, an amino acid, seems to play a key role during osmotic adjustment. Remarkable accumulation of proline due to enhanced synthesis and decreased degradation under a range of stress conditions such as salt, drought, and metal exposure has been documented in many plants (Kishor et al., 2005). The accumulation of proline normally occurs in the cytosol and contributes substantially to the cytoplasmic osmotic adjustment in response to drought or salinity stress (Ashraf and Foolad, 2007).

In soybean roots, the accumulation of proline has been reported (Alam et al., 2010; Delauney and Verma, 1993). A proteomic approach in soybean root subjected to short-term drought stress showed that glutamine synthetase (GS) is upregulated (Alam et al., 2010) and in plants this enzyme has been implicated in the regulation of levels of proline derived from glutamate (Chen and Jiang, 2010; Kishor et al., 2005). So far, it is unknown whether the decreased degradation of proline in soybean is a response that promotes its accumulation under dehydration, although the proteomic evidence supports the increment of its synthesis. Whatever the cause of proline accumulation, this response provides an adaptive advantage to soybean under low water potentials.

Betaines, another organic solute type-osmolyte, are quaternary ammonium compounds. Glycine betaine is synthesized by several plant families in response to saline or drought stress (Chen and Jiang, 2010; Munns, 2002). The main function is to balance the osmotic potential of intracellular and extracellular compartments together with important roles as a compatible solute to stabilize proteins, and to protect membrane structures and the photosynthetic apparatus (Chen and Jiang, 2010).

Glycine betaine is synthesized in chloroplasts and several enzymes play important roles in the pathway of glycine betaine synthesis such as betaine aldehyde dehydrogenase (BADH). Proteomic analysis in soybean under long-term stress revealed that BADH is upregulated under cold and osmotic stress as well as a combination of both stresses in roots of germinating soybean seeds (Swigonska and Weidner, 2013), highlighting the potential importance of this pathway in soybean osmoregulation. Interestingly, it has been suggested that the introduction of this enzyme in plants may enhance salt and drought tolerances (Zhang et al., 2008).

Glycerol is an important osmolyte in plants, which is synthesized from glucose and has been linked with salinity stress (Chen et al., 2009). It has been proposed that glycerol may be an effective osmotic element at high salinity for many reasons, for instance, high solubility, nontoxicity, and low energy cost of synthesis (Chen and Jiang, 2010). Nevertheless, reports on a possible function of glycerol as an osmolyte in soybean are yet to be found; even proteomic analysis has yet to discover any important role of enzymes related to its synthesis. Further study on this alternative possibility of osmoregulation in soybean is needed.

Proteomic studies using different tissues also identified late embryogenesis abundant (LEA) proteins, which are often induced under stress conditions in soybean, such as cold (Cheng et al., 2010), salt (Aghaei et al., 2009), and drought (Alam et al., 2010). These proteins are highly hydrophilic and may act as osmotic protectors. A specific group of LEA II proteins, known as dehydrins, have been repeatedly reported in drought-stressed plants (Bray, 1997; Goyal et al., 2005; Hong-Bo et al., 2005). Under dehydrating conditions, dehydrins may confer stability to the cytoplasm (Alam et al., 2010).

Sodium ion, K^+, and Ca^{2+} are the main inorganic ions under osmotic and saline stresses. These ions could prevent plants from harm caused by drought by absorbing water into cells or as a component of general responses to dehydration like stomata opening. In this section, we refer to the role of inorganic ions specifically in the context of saline stress. Plant growth responses to salinity include two phases: a rapid osmotic phase and a slower ionic phase.

Sodium ions are particularly toxic to leaves. Therefore, most Na^+ that is delivered to the shoot remains in the shoot since the recirculation of Na^+ can occur on a small scale. The key processes controlling the net delivery of Na^+ into the root xylem are regulated by four components (Munns and Tester, 2008). Briefly, Na⁺ enters roots passively via nonselective cation channels. Once Na⁺ is inside the outer part of the root, it can be pumped back out to the soil via plasma membrane Na⁺/H⁺ antiporters (SOS1 in *Arabidopsis*).

Na⁺ remaining in the root can be sequestered in vacuoles via tonoplast Na⁺/H⁺ antiporters (NHX for Na⁺/H⁺ exchanger) or moved to the xylem through the symplast across the endodermis and stellar apoplast. Once in the xylem, it moves with the transpiration stream to the shoot. The plasma membrane SOS1 type transporter in stellar cells could be involved in the efflux of Na⁺ from these cells into the xylem. In order to avoid sodium ion delivery to the leaves, use of the process called retrieval of Na⁺ from the xylem has great importance. A high-affinity K⁺ transporter seems to be also involved.

In the context of soybean physiology, Cl⁻ homeostasis has long been implicated as a major mechanism of soybean tolerance to NaCl. However, new evidence supports that both Cl⁻ and Na⁺ homeostasis are important for soybean salt tolerance (El-Samad and Shaddad, 1997). Another study characterized two putative soybean transporters to be involved in soybean NaCl tolerance: a chloride channel gene (*GmCLC1*) and a putative Na⁺/H⁺ antiporter gene (*GmNHX1*) (Li et al., 2006).

Both transporters are localized on tonoplasts and evidence suggests they may help to sequester ions from the cytoplasm into the vacuole to reduce their toxic effects and enhance tolerance. GmCAX1 is a cation/proton antiporter, which is also characterized in soybean. A study of the plasma membrane localization suggests that it could be involved in ion homeostasis under salinity (Luo et al., 2005).

GmCHX1, a novel cation/proton exchange transporter, which is highly tolerant to salt stress, has been identified in soybean using an interesting approach involving use of a wild soybean genetic background. GmCHX1 was discovered when researching a major quantitative trait locus for salt tolerance using a unique RI population, genotyped by sequencing and generated from wild and cultivated soybeans (Qi et al., 2014). It has been suggested that GmCHX1 could be involved in Na⁺ transportation. However, the role of these transporters in salt tolerance remains unknown.

Some characterized functions in *Arabidopsis*, including K⁺ transport across the plasma membrane and endomembrane system, have been correlated with K⁺ and pH homeostasis (Chanroj et al., 2011; Zhao et al., 2008). Plant adaptations to salinity through the above-mentioned mechanisms are essential; however, the limited information available on the soybean proteome under saline stress and the absence of ion transporters in soybean proteomic profiles underlines the importance of strengthening the proteomic approaches, especially focusing on specific characterization of membrane proteins.

It is known that the active transport of Na⁺ across membranes involves H⁺–ATPase activity. In soybean, H⁺–ATPase is upregulated during osmotic stress using a plasma membrane proteomic approach, highlighting the importance of this strategy in soybean proteomics under salinity and others types of stress (Nouri and Komatsu, 2010).

Protein synthesis and turnover

Protein synthesis from amino acids involves a number of concatenated processes, highly regulated and essential for life. The relative abundance of different proteins involved in protein biosynthesis often appears in proteomic approaches. For example, ribosomal proteins are directly related to translation process and key proteins are involved in associated processes, such as recruitment of specific mRNA.

Different ribosomal proteins are differentially expressed in proteomic studies of soybean under different stress conditions (60S, 50S, ribosomal protein S10, 40S ribosomal S3 protein, 40S ribosomal S4 protein, 30S ribosomal protein S5, ribosomal protein L30, 50S ribosomal protein L9). Reports on different soybean tissues during flooding (Khatoon et al., 2012), chilling (Cheng et al., 2010), and osmotic stress induced by PEG (Mohammadi et al., 2012; Nouri and Komatsu, 2010) and salt (Sobhanian et al., 2010) showed downregulation of ribosomal proteins related with protein synthesis, suggesting that this process is affected by stresses.

Translation initiation factor (eIF5A) is a protein that can recruit specific mRNA for translation and has been reported to be involved in a set of plant responses to abiotic stress (Ren et al., 2013; Wang et al., 2012). eIF5A is downregulated in soybean roots under drought stress, suggesting an early level of negative stress effect on protein synthesis (Alam et al., 2010).

As well as protein synthesis, proteolysis is essential in plant physiology and development. Proteolysis controls the degradation of abnormal or misfolded proteins, contributing to remobilization of amino acids and controlling the last step of posttranscriptional events. Not only does protein degradation control protein recycling, but it is a key step that regulates metabolism and development to reduce the abundance of important proteins such as receptors, second messengers, protein-controlling cell cycle, and transcription factors (Vierstra, 1996).

In plants, protein degradation is a complex process that involves multiple proteolytic pathways in different subcellular compartments. The ubiquitin-dependent proteolytic pathway is one of the most important pathways in cytoplasm and nucleus, where proteins are "labeled" with ubiquitin and subsequently degraded by a multisubunit protein complex named 26S proteasome. The labeling step is complex and involves a number of ubiquitin-dependent enzymes (E1, E2, E3) that will recognize the target protein and prepare the ubiquitin to conjugate for proteasome degradation. The 26S proteasome is composed of two major subcomplexes (20S and 19S). It is interesting that the 26S proteasome can also degrade proteins in a ubiquitin independent pathway.

Vacuoles contain proteases that are responsible for degradation of storage protein (cathepsin class of cystein proteases). Vacuole proteases also seem to be important in remobilization processes during senescence, starvation, or stress conditions (Müntz, 2007). Others subcellular compartments also contain specific proteases. Their regulation and functions are fundamental for the control of imported (cleavage of transit peptide) or misfolded proteins (Schaller, 2004).

Different proteins in the ubiquitin proteasome pathway have been characterized in soybean by proteomic approaches. Protein components of ubiquitin/proteasome-mediated proteolysis are involved in response to flooding (Khatoon et al., 2012; Yanagawa and Komatsu, 2012), drought (Mohammadi et al., 2012), salinity (Xu et al., 2011), chilling (Cheng et al., 2010), and osmotic stress (Toorchi et al., 2009). The generation and characterization of flooding-tolerant soybean mutants has been indicated (Komatsu et al., 2013c). In these mutants, ubiquitin/proteasome-mediated cell death is absent and a possible contribution of phytohormones, especially ABA, has been suggested. The 26S proteasome base protein is involved in ABA responses and has a specific function in ABA signaling (Smalle et al., 2003).

The information in soybean suggests that 26S proteasome is involved in the maintenance and control of protein under different stress conditions. More in-depth studies are needed to unravel the details of these degradation pathways in order to understand their impact and importance beyond a general response. Other proteins involved in degradation or its control have been characterized in soybean under different stresses, for instance, cystein proteinase precursors, serine carboxypeptidases, and proteinase inhibitors (Cheng et al., 2010; Komatsu et al., 2009a; Mohammadi et al., 2012; Nouri and Komatsu, 2010; Toorchi et al., 2009; Yamaguchi et al., 2010).

Typically characterized proteins from proteomic approaches are chaperones. Molecular chaperones are a diverse group of proteins involved in the folding, assembly and transport of proteins. They are ubiquitous in the cell. Their expression and activity increase under stress and specific chaperones can perform specific tasks (Feder and Hofmann, 1999; Gupta and Tuteja, 2011). The Hsp70 molecular chaperone is often upregulated under stress conditions in soybean proteomic approaches (Alam et al., 2010; Hossain et al., 2012; Komatsu et al., 2009a, b, 2010; Mohammadi et al., 2012; Nouri and Komatsu, 2010; Toorchi et al., 2009; Xu et al., 2008).

This chaperone corresponds to a large family of proteins, widely distributed in different organisms. It may work alone or together with other cochaperones facilitating the folding, transport, or labeling of proteins. It was reported that Hsp70 induced by stress participates in the aggregation of stress-denatured proteins so helping to restore protein function through refolding (Sung et al., 2001).

Another important chaperone in soybean is Hsp90. Hsp90 is a key conserved chaperone involved in the folding, maturation, stabilization, and activation of important proteins such as kinases, transcription factors, and receptors. The regulation function of key proteins involved in metabolism, defense responses, and development highlights the importance of this chaperone protein in plants (Kadota and Shirasu, 2012).

A comparative analysis of soybean plasma membrane proteins under osmotic stress shows that Hsp90 is differentially expressed with PEG treatment (Nouri and Komatsu, 2010). A dependence of Hsp90 protein function under osmotic stress was also reported (Swigonska and Weidner, 2013). A recent report identifies 12 soybean Hsp90 proteins to be responsive to abiotic stresses and play a protective role in Arabidopsis, suggesting it may have an important role in soybean stress responses (Xu et al., 2013).

Calreticulin was found to be downregulated in soybean leaves under salt treatment (Sobhanian et al., 2010) and under osmotic stress (Nouri and Komatsu, 2010). Calreticulin is an important chaperone with calcium binding activity involved in the protein folding and homeostasis at the endoplasmic reticulum under stress (Gupta and Tuteja, 2011). This endoplasmic reticulum chaperone plays an important role in response to abiotic stress (including binding protein (BiP), immunophilins, calnexin, etc.) Soybean BiP isoform B and a calnexin homolog have been reported under osmotic stress (Nouri and Komatsu, 2010). Further study of these proteins in soybean is important.

Antioxidant system

A common response to oxidative stress conditions is characterized by an exacerbated generation of ROS. Certain levels of ROS can induce damage to macromolecules such as proteins, lipids, and nucleic acids, affecting their functions and possibly leading to the death of cells, tissues, or even the organism itself (Foyer and Noctor, 2005; Trippi et al., 1989).

In order to control ROS levels, living organisms have developed a complex enzymatic and nonenzymatic antioxidant system. Key enzymes that remove ROS are superoxide dismutase (SOD), ascorbate peroxidase (APX), and catalase (CAT) (Scandalios et al., 1980, 1984). The activities of these three enzymes are crucial in maintaining the levels of superoxide radical (O_2^-) and hydrogen peroxide (H_2O_2). Nonenzymatic antioxidants such as ascorbic acid (ASC) and GSH are also crucial on the control of ROS. A set of important enzymes are involved in a cycle that maintains the reduction pool of these compounds: glutathione reductase (GR), glutathione peroxidase (GPX), monodehydroascorbate reductase (MDAR), and dehydroascorbate reductase (DAR) (Asada, 1999).

There is a positive correlation between antioxidant enzyme activity and tolerance to different stress conditions (Gill and Tuteja, 2010; Lascano, 2003; Mittler, 2002). Oxidative homeostasis highly depends on the type and intensity of the stress. The proteomic approaches, which showed responses related to up- and downregulations of antioxidant enzymes, should only be considered for the specific stress, tissue, and ontogenetic stage analyzed.

Soybean SOD and CAT in leaf (Khatoon et al., 2012) and APX in root and hypocotyl (Komatsu et al., 2010) have been reported to be downregulated during flooding. However, specific plasma membrane proteome analysis showed that other isoforms of SOD in soybean are upregulated under flooding, suggesting a putative specific role of this enzyme at the cell wall or plasma membrane (Komatsu et al., 2009a). Similarly, SOD was upregulated in soybean root under osmotic stress, but specific plasma membrane proteomes showed that three major antioxidant enzymes, APX, MDAR, and CuZnSOD, are downregulated.

This comparative approach and the differentiated subcellular compartments of the proteome demonstrate the importance of specific protein analysis. In a complete organ-specific proteomic analysis, it was reported that APX and MDAR are upregulated in leaves and hypocotyl under drought stress (Mohammadi et al., 2012). MDAR was also reported in soybean root under drought stress (Yamaguchi et al., 2010). The level of antioxidant enzymes such as SOD, APX, and CAT also increased in soybean under heavy metal stress (Hossain et al., 2012). These findings suggest antioxidant enzymes may play a putative role in soybean's response to stresses.

Energy production

Metabolism optimization in terms of energy production is fundamental to plants under stress. Sustaining growth and inducing defense systems are two basic responses related to the acclimation and adaptation, leading to the establishment of new homeostatic states. These states may depend on the type and intensity of the stress conditions as well as the developmental stage of the plant.

The decline in oxygen levels induced by flooding results in a metabolic shift from aerobic to anaerobic metabolism (Komatsu et al., 2009b). Most proteomic studies were conducted on roots because it is the organ directly subjected to this stress. The metabolism of soybean roots under flooding is redirected from oxidative phosphory-lation to the induction of glycolytic and fermentation pathways, which contribute to the regeneration of NADH through fermentative pathways. Enzymes such as UDP-glucose pyrophosphorylase and fructose biphosphate aldolase are usually affected (Hashiguchi et al., 2009; Komatsu et al., 2009b).

The increase in other key enzymes such as alcohol dehydrogenase (ADH) revealed an activation of alcohol fermentative pathways (Komatsu et al., 2011a). These changes in the proteome are accompanied by glucose degradation and sucrose accumulation in roots (Nanjo et al., 2010). In the mitochondrial proteome there is a severe effect of flood-ing stress on mitochondrial electron transport pathways and the pathways for NADH pro-duction are activated (Komatsu et al., 2011b). Interestingly, an organ-specific proteomic study showed that while this metabolism-related protein is increased in flooded soybean root, decreases were recorded in hypocotyl and leaf, showing clearly different metabolic responses in the tissues analyzed (Khatoon et al., 2012).

Glyceraldehyde 3-phosphate dehydrogenase (GAPDH) is a key enzyme in the glycolytic pathway, and participates in the degradation of glucose to obtain energy and carbon molecules. Soybean proteomic studies under salt stress showed that this enzyme is commonly downregulated under stress in leaf, hypocotyl (Sobhanian et al., 2010), and seeds (Xu et al., 2011), suggesting that it could be a target enzyme regulated under salt. Similar results were observed in soybean hypocotyl and root under osmotic or drought stress (Mohammadi et al., 2012) and in leaf under UV-B radiation (Xu et al., 2008).

Interestingly, GAPDH has also been observed to be clearly upregulated in soybean leaves under cadmium stress and involved in energy production, which is necessary to meet the high-energy demand to sustain some of the costly metabolic pathways involved in heavy-metal tolerance, such as the synthesis of phytochelatins (PCs) (Hossain et al., 2012). Under cadmium stress, GS, a key player in nitrogen assimila-tion, accumulated in all cultivars but maximum expression was observed in the high-accumulating cultivar (Hossain et al., 2012).

The enzyme glutathione synthetase is involved in the synthesis of GSH through the glutamate biosynthesis pathway. PCs are oligomers of GSH that act as chelators and are important for heavy-metal detoxification. PCs are synthesized in the cytosol where they bind to metal and form complexes (Cobbett and Goldsbrough, 2002; Grill et al., 1985). It has been proposed that the transport and accumulation of PCs and metal inside the vacu-ole contributes to sequestering of heavy metals. This detoxification pathway is common to cultivars, although they may differ in the ability to accumulate cadmium.

Another important target protein under stress in soybean characterized by pro-teomic approaches is RuBisCo activase. RuBisCo activase is a chaperone that pro-motes and maintains the catalytic activity of RuBisCo (Portis, 2003). Downregulation of RuBisCo activase under different stress conditions in soybean seems to be critical for maintenance of photosynthesis and may be involved in the inhibitory effect of stress on soybean photosynthesis (Mohammadi et al., 2012; Sobhanian et al., 2010).

However, specific stress on leaves such as UV-B radiation was reported to upregulate this enzyme, suggesting a stress-specific response (Xu et al., 2008).

The upregulation of ADH in soybean under flooding seems to be an important way to obtain energy (Komatsu et al., 2010). ADH has been shown to be upregulated under salt stress in the hypocotyl (Sobhanian et al., 2010) and in chilling during germination (Cheng et al., 2010). However, long-term experiments using stressed soybean plants showed that ADH is downregulated under cold and/or osmotic stress. These findings suggest that the response of this enzyme could depend on stress duration (Swigonska and Weidner, 2013).

Other important proteins associated with energy production have also been reported. During germination, when embryos were subjected to low temperature, it was found that some responsive proteins (ATP synthase α-subunit, malate dehydrogenase, phosphoenolpyruvate (PEP) carboxylase, and triose phosphate isomerase) are associated with energy generation (Cheng et al., 2010).

General conclusions and perspectives

Proteomic studies have contributed greatly to the understanding of soybean stress responses, characterizing a large number of proteins involved in specific metabolic pathways and relevant under different stress conditions. Soybean as an experimental system faces a large number of limitations but taking advantage of the special complexity of proteomic studies can result in their complementing each other in advancing our understanding, as well as the larger general scientific question of how plants respond to stress.

Our understanding of tolerance mechanisms in plants has increased exponentially in the last 20 years. In general, we may say that tolerance mechanisms are diverse but have interesting common pathways (Bohnert et al., 2006; Fujita et al., 2006; Knight and Knight, 2001; Mittler, 2002; Shinozaki et al., 2003). This observation applies to soybean proteomic studies.

Figure 4.2 shows some important metabolic pathways with proteins that have been reported up- or downregulated in soybean proteomic studies under cold, drought, flooding, osmotic, salt, UV-B, and cadmium stresses. The diagram is a simplified representation and cannot cover all aspects due to the complexity of proteomic approaches and lack of available information. For example, although it is not a common response, a single protein in a particular metabolic pathway could be up- or downregulated depending on the tissue in which it is analyzed. Therefore, in this chapter, we have selected and discussed only proteins that are commonly found in the protein profile of soybean.

Soybean stress responses such as upregulation of chaperones of different types (Hsps, LEA proteins, etc.), osmolyte accumulation, antioxidant enzymes, or key metabolic shift (like activation of glycolysis-related enzymes) are often seen in the protein profile under different stress conditions. Beyond the range of experimental systems tested (Table 4.1), common responses between types of stresses and tissues could be particularly important because proteomic approaches involve putative posttranslational targets.

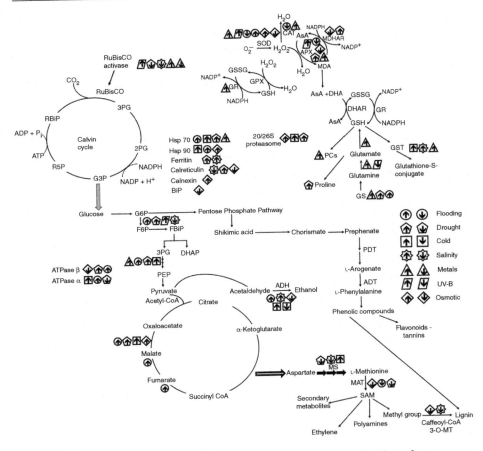

Figure 4.2 General important metabolic pathways with proteins that have been reported to be upregulated (↑) or downregulated (↓) in soybean proteomic studies under cold, drought, flooding, osmotic, salt, UV-B, and metal stresses.
RuBisCo, ribulose-1,5-bisphosphate carboxylase; RBiP, ribulose-1,5-bisphosphate; 3PG, 3-phosphoglyceric acid; 2PG, 1,3-bisphosphoglyceric acid; G3P, glyceraldehyde 3-phosphate; R5P, ribose 5-phosphate; GSSG, oxidized glutathione; GSH, reduced glutathione; SOD, superoxide dismutase; CAT, catalase; APX, ascorbate peroxidase; AsA, ascorbic acid; MDHAR, monodehydroascorbate reductase; MDA, monodehydroascorbate; GR, glutathione reductase; DHAR, dehydroascorbate reductase; PCs, phytochelatins; GST, glutathione S-transferases; G6P, glucose 6-phosphate; F6P, fructose 6-phosphate; FbiP, fructose-1,6-diphosphate; DHAP, dihydroxyacetone phosphate; PEP, phosphoenolpyruvate; MS, methionine synthase; SAM, S-adenosylmethionine; MAT, methionine adenosyltransferase; ADH, alcohol dehydrogenase.

In addition to common responses, specific changes have been also characterized such as synthesis of PCs under cadmium stress (Hossain et al., 2012), induction of fermentative pathways on root under flooding (Nanjo et al., 2011b), or proline accumulation under drought (Alam et al., 2010). The complexity of specific responses increases when different tissues, stresses, stress intensity, etc. are involved and could influence the interpretation of results as a "general" soybean stress response.

Likewise, in the context of proteomic approaches, the majority of studies focus on general soluble protein profile characterization. Those involving specific subcellular proteomic experiments have explained a variety of important stress responses. For example, in the mitochondrial proteome during flooding, it has been characterized that electron transport chains are affected but the reduction power could be maintained through the tricarboxylic acid cycle (Komatsu et al., 2011b). Specific protein profile analysis from cell walls also showed that the lignification process could be affected in flooded soybean plants (Komatsu et al., 2010). More in-depth subcellular proteomic analysis certainly will contribute to the understanding of soybean stress responses.

We also wanted to include some general stress concepts like acclimation, adaptation, or tolerance in this chapter because such concepts are related to basic and general plant responses that can enrich the discussions in others fields of plant science.

Stress acclimation is often linked with proteomic approaches (Hossain et al., 2013). The concepts of the common tolerance response pathways and cross-tolerance to various types of stress are fundamental in this context. Acclimation phenomena could be understood as the ability of the plant to adapt to environmental changes (Umezawa et al., 2000). In this regard, not all physiological changes that we can observe will necessarily lead to an adaptive stage. Concepts such as phase of stress are useful when interpreting the results. The stress response is sufficiently different at the stress shock phase (were the plant is clearly not acclimated) and the exhaustion phase, even when both have an important putative impact on the overall adaptive responses. Dissecting these phases and examining the relevance of the players in each of them help to form novel acclimation concept on plant. For example, the role of ROS has been defined as systemic acquired acclimation (SAA) in *Arabidopsis*. SAA activates defense or acclimation mechanisms in systemic or nonchallenged tissues (Baxter et al., 2014). As a component of stress tolerance responses, acclimation in this sense is an important unexplored field in soybean and a new possible challenge to proteomic approaches.

Likewise, the complexity of the stress tolerance mechanism increases when we understand that tolerance and the acclimation response to abiotic stresses not only depend on the plant genetic background, but also on the type of stress, the dose, stage of development when stressed, and even the combination of stresses that may occur under natural conditions (Knight and Knight, 2001; Lichtenthaler, 1996, 1998). Given this complexity, the discussion in terms of how much we can understand about stress responses should be considered in a continuous and dynamic context. The concept of phases of stress in the context of proteomic approaches has been discussed by Kosová et al. (2011).

Useful nondestructive tools have been developed in order to follow plant stress responses in a time line (Lichtenthaler and Miehe, 1997) and their use has been widely successful (Ehlert and Hincha, 2008; Flexas et al., 2002; Kao et al., 2003; Maxwell and Johnson, 2000; Musser et al., 1984; O'Neill et al., 2006; Ohashi et al., 2006). The usefulness of these tools could be powerful together with the analysis of protein profiles in leaves, leading to a more complete space–time understanding of the stress responses in soybean.

"Omics" is a powerful tool used to characterize the stress responsive pathways and proteomics is especially important in its ability to reveal posttranslational modifications. Every "omics" faces its own challenges, which are sometimes limited by the availability of technologies and capacity of data processing. Beyond technical limitations, the biggest challenge that we face is more conjunctural to the understanding of stress responses: to bring together efforts using these tools with the goal of comprehensively unraveling complex biological processes. Data integration has been pointed out as one increasing demand from the life research community (Gomez-Cabrero et al., 2014). Cooperative efforts in soybean via multidisciplinary and international consortia will certainly contribute to carry out coordinated projects and to generate homogenous data to improve our knowledge on soybean biology.

Acknowledgments

Nacira Muñoz's fellowship at the Chinese University of Hong Kong was funded by Instituto Nacional de TecnologiaAgropecuaria, INTA, Argentina.

This work was financially supported by the Hong Kong RGC Collaborative Research Fund (CUHK3/CRF/11G), the Lo Kwee-Seong Biomedical Research Fund, and Lee Hysan Foundation.

The authors also thank Ms Lydia K. Siu for copy editing.

References

Aghaei, K., Ehsanpour, A., Shah, A., Komatsu, S., 2009. Proteome analysis of soybean hypocotyl and root under salt stress. Amino Acids 36, 91–98.

Ahuja, I., de Vos, R.C., Bones, A.M., Hall, R.D., 2010. Plant molecular stress responses face climate change. Trends Plant Sci. 15, 664–674.

Alam, I., Sharmin, S.A., Kim, K.-H., Yang, J.K., Choi, M.S., Lee, B.-H., 2010. Proteome analysis of soybean roots subjected to short-term drought stress. Plant Soil 333, 491–505.

Asada, K., 1999. The water-water cycle in chloroplast: scavenging of active oxygens and dissipation of excess photons. Annu. Rev. Plant Physiol. Plant Mol. Biol. 50, 601–639.

Ashraf, M., Foolad, M., 2007. Roles of glycine betaine and proline in improving plant abiotic stress resistance. Environ. Exp. Bot. 59, 206–216.

Baxter, A., Mittler, R., Suzuki, N., 2014. ROS as key players in plant stress signalling. J. Exp. Bot. 65, 1229–1240.

Beemster, G.T., Fiorani, F., Inzé, D., 2003. Cell cycle: the key to plant growth control? Trends Plant Sci. 8, 154–158.

Boerjan, W., Ralph, J., Baucher, M., 2003. Lignin biosynthesis. Annu. Rev. Plant Biol. 54, 519–546.

Bohnert, H.J., Sheveleva, E., 1998. Plant stress adaptations – making metabolism move. Curr. Opin. Plant Biol. 1, 267–274.

Bohnert, H.J., Gong, Q.Q., Li, P.H., Ma, S.S., 2006. Unraveling abiotic stress tolerance mechanisms – getting genomics going. Curr. Opin. Plant Biol. 9, 180–188.

Boudet, A.-M., 2000. Lignins and lignification: selected issues. Plant Physiol. Biochem. 38, 81–96.

Bray, E.A., 1997. Plant responses to water deficit. Trends Plant Sci. 2, 48–54.

Bustos, D., Lascano, R., Villasuso, A.L., Machado, E., Senn, M.E., Cordoba, A., Taleisnik, E., 2008. Reductions in maize root-tip elongation by salt and osmotic stress do not correlate with apoplastic O_2^*-levels. Ann. Bot. 102, 551–559.

Cabello, J.V., Lodeyro, A.F., Zurbriggen, M.D., 2014. Novel perspectives for the engineering of abiotic stress tolerance in plants. Curr. Opin. Biotechnol. 26, 62–70.

Chan, C., Lam, H.-M., 2014. A putative lambda class glutathione s-transferase enhances plant survival under salinity stress. Plant Cell Physiol. 55, 570–579.

Chanroj, S., Lu, Y., Padmanaban, S., Nanatani, K., Uozumi, N., Rao, R., Sze, H., 2011. Plant-specific cation/H^+ exchanger 17 and its homologs are endomembrane K^+ transporters with roles in protein sorting. J. Biol. Chem. 286, 33931–33941.

Chaves, M.M., Pereira, J.S., Maroco, J., Rodrigues, M.L., Ricardo, C.P.P., Osório, M.L., Carvalho, I., Faria, T., Pinheiro, C., 2002. How plants cope with water stress in the field? Photosynthesis and growth. Ann. Bot. 89, 907–916.

Chen, H., Jiang, J.-G., 2010. Osmotic adjustment and plant adaptation to environmental changes related to drought and salinity. Environ. Rev. 18, 309–319.

Chen, C., Meyermans, H., Burggraeve, B., De Rycke, R.M., Inoue, K., De Vleesschauwer, V., Steenackers, M., Van Montagu, M.C., Engler, G.J., Boerjan, W.A., 2000. Cell-specific and conditional expression of caffeoyl-coenzyme A-3-O-methyltransferase in poplar. Plant Physiol. 123, 853–868.

Chen, H., Jiang, J.-G., Wu, G.-H., 2009. Effects of salinity changes on the growth of *Dunaliella salina* and its isozyme activities of glycerol-3-phosphate dehydrogenase. J. Agric. Food Chem. 57, 6178–6182.

Chen, J.-H., Jiang, H.-W., Hsieh, E.-J., Chen, H.-Y., Chien, C.-T., Hsieh, H.-L., Lin, T.-P., 2012. Drought and salt stress tolerance of an *Arabidopsis* glutathione S-transferase U17 knockout mutant are attributed to the combined effect of glutathione and abscisic acid. Plant Physiol. 158, 340–351.

Cheng, L., Gao, X., Li, S., Shi, M., Javeed, H., Jing, X., Yang, G., He, G., 2010. Proteomic analysis of soybean [*Glycine max* (L.) Merr.] seeds during imbibition at chilling temperature. Mol. Breed. 26, 1–17.

Ciais, P., Reichstein, M., Viovy, N., Granier, A., Ogée, J., Allard, V., Aubinet, M., Buchmann, N., Bernhofer, C., Carrara, A., 2005. Europe-wide reduction in primary productivity caused by the heat and drought in 2003. Nature 437, 529–533.

Cobbett, C., Goldsbrough, P., 2002. Phytochelatins and metallothioneins: roles in heavy metal detoxification and homeostasis. Annu. Rev. Plant Biol. 53, 159–182.

Creelman, R.A., Mason, H.S., Bensen, R.J., Boyer, J.S., Mullet, J.E., 1990. Water deficit and abscisic acid cause differential inhibition of shoot versus root growth in soybean seedlings analysis of growth, sugar accumulation, and gene expression. Plant Physiol. 92, 205–214.

Delauney, A.J., Verma, D.P.S., 1993. Proline biosynthesis and osmoregulation in plants. Plant J. 4, 215–223.

Deshmukh, R.K., Sonah, H., Patil, G., Chen, W., Prince, S., Mutava, R., Vuong, T., Valliyodan, B., Nguyen, H.T., 2014. Integrating omic approaches for abiotic stress tolerance in soybean. Plant Genet. Genom. 5, 244.

Edwards, R., Dixon, D.P., Walbot, V., 2000. Plant glutathione S-transferases: enzymes with multiple functions in sickness and in health. Trends Plant Sci. 5, 193–198.

Ehlert, B., Hincha, D.K., 2008. Chlorophyll fluorescence imaging accurately quantifies freezing damage and cold acclimation responses in *Arabidopsis* leaves. Plant Method. 4, 1–7.

El-Samad, H.A., Shaddad, M., 1997. Salt tolerance of soybean cultivars. Biol. Plant. 39, 263–269.

Feder, M.E., Hofmann, G.E., 1999. Heat-shock proteins, molecular chaperones, and the stress response: evolutionary and ecological physiology. Annu. Rev. Physiol. 61, 243–282.

Flexas, J., Escalona, J.M., Evain, S., Gulías, J., Moya, I., Osmond, C.B., Medrano, H., 2002. Steady-state chlorophyll fluorescence (Fs) measurements as a tool to follow variations of net CO_2 assimilation and stomatal conductance during water-stress in C_3 plants. Physiol. Plant. 114, 231–240.

Foyer, C.H., Noctor, G., 2005. Oxidant and antioxidant signalling in plants: a re-evaluation of the concept of oxidative stress in a physiological context. Plant Cell Environ. 28, 1056–1071.

Fujita, M., Fujita, Y., Noutoshi, Y., Takahashi, F., Narusaka, Y., Yamaguchi-Shinozaki, K., Shinozaki, K., 2006. Crosstalk between abiotic and biotic stress responses: a current view from the points of convergence in the stress signaling networks. Curr. Opin. Plant Biol. 9, 436–442.

Galant, A., Koester, R.P., Ainsworth, E.A., Hicks, L.M., Jez, J.M., 2012. From climate change to molecular response: redox proteomics of ozone-induced responses in soybean. New Phytol. 194, 220–229.

Gill, S.S., Tuteja, N., 2010. Reactive oxygen species and antioxidant machinery in abiotic stress tolerance in crop plants. Plant Physiol. Biochem. 48, 909–930.

Gomez-Cabrero, D., Abugessaisa, I., Maier, D., Teschendorff, A., Merkenschlager, M., Gisel, A., Ballestar, E., Bongcam-Rudloff, E., Conesa, A., Tegnér, J., 2014. Data integration in the era of omics: current and future challenges. BMC Syst. Biol. 8, I1.

Goyal, K., Walton, L., Tunnacliffe, A., 2005. LEA proteins prevent protein aggregation due to water stress. Biochem. J. 388, 151–157.

Grill, E., Winnacker, E.-L., Zenk, M.H., 1985. Phytochelatins: the principal heavy-metal complexing peptides of higher plants. Science 230, 674–676.

Gupta, D., Tuteja, N., 2011. Chaperones and foldases in endoplasmic reticulum stress signaling in plants. Plant Signal. Behav. 6, 232–236.

Hashiguchi, A., Sakata, K., Komatsu, S., 2009. Proteome analysis of early-stage soybean seedlings under flooding stress. J. Proteome Res. 8, 2058–2069.

Hong-Bo, S., Zong-Suo, L., Ming-An, S., 2005. LEA proteins in higher plants: structure, function, gene expression and regulation. Colloids Surf. B 45, 131–135.

Hossain, Z., Komatsu, S., 2014. Potentiality of soybean proteomics in untying the mechanism of flood and drought stress tolerance. Proteomes 2, 107–127.

Hossain, Z., Hajika, M., Komatsu, S., 2012. Comparative proteome analysis of high and low cadmium accumulating soybeans under cadmium stress. Amino Acids 43, 2393–2416.

Hossain, Z., Khatoon, A., Komatsu, S., 2013. Soybean proteomics for unraveling abiotic stress response mechanism. J. Proteome Res. 12, 4670–4684.

Ingram, J., Bartels, D., 1996. The molecular basis of dehydration tolerance in plants. Annu. Rev. Plant Biol. 47, 377–403.

Kadota, Y., Shirasu, K., 2012. The HSP90 complex of plants. Biochim. Biophys. Acta 1823, 689–697.

Kao, W.-Y., Tsai, T.-T., Shih, C.-N., 2003. Photosynthetic gas exchange and chlorophyll *a* fluorescence of three wild soybean species in response to NaCl treatments. Photosynthetica 41, 415–419.

Khatoon, A., Rehman, S., Hiraga, S., Makino, T., Komatsu, S., 2012. Organ-specific proteomics analysis for identification of response mechanism in soybean seedlings under flooding stress. J. Proteomics 75, 5706–5723.

Kishor, P.K., Sangam, S., Amrutha, R., Laxmi, P.S., Naidu, K., Rao, K., Rao, S., Reddy, K., Theriappan, P., Sreenivasulu, N., 2005. Regulation of proline biosynthesis, degradation, uptake and transport in higher plants: its implications in plant growth and abiotic stress tolerance. Curr. Sci. 88, 424–438.

Knight, H., Knight, M.R., 2001. Abiotic stress signalling pathways: specificity and cross-talk. Trends Plant Sci. 6, 262–267.

Komatsu, S., Wada, T., Abaléa, Y., Nouri, M.Z., Nanjo, Y., Nakayama, N., Shimamura, S., Yamamoto, R., Nakamura, T., Furukawa, K., 2009a. Analysis of plasma membrane proteome in soybean and application to flooding stress response. J. Proteome Res. 8, 4487–4499.

Komatsu, S., Yamamoto, R., Nanjo, Y., Mikami, Y., Yunokawa, H., Sakata, K., 2009b. A comprehensive analysis of the soybean genes and proteins expressed under flooding stress using transcriptome and proteome techniques. J. Proteome Res. 8, 4766–4778.

Komatsu, S., Kobayashi, Y., Nishizawa, K., Nanjo, Y., Furukawa, K., 2010a. Comparative proteomics analysis of differentially expressed proteins in soybean cell wall during flooding stress. Amino Acid. 39, 1435–1449.

Komatsu, S., Sugimoto, T., Hoshino, T., Nanjo, Y., Furukawa, K., 2010b. Identification of flooding stress responsible cascades in root and hypocotyl of soybean using proteome analysis. Amino Acids 38, 729–738.

Komatsu, S., Thibaut, D., Hiraga, S., Kato, M., Chiba, M., Hashiguchi, A., Tougou, M., Shimamura, S., Yasue, H., 2011a. Characterization of a novel flooding stress-responsive alcohol dehydrogenase expressed in soybean roots. Plant Mol. Biol. 77, 309–322.

Komatsu, S., Yamamoto, A., Nakamura, T., Nouri, M.-Z., Nanjo, Y., Nishizawa, K., Furukawa, K., 2011b. Comprehensive analysis of mitochondria in roots and hypocotyls of soybean under flooding stress using proteomics and metabolomics techniques. J. Proteome Res. 10, 3993–4004.

Komatsu, S., Han, C., Nanjo, Y., Altaf-Un-Nahar, M., Wang, K., He, D., Yang, P., 2013a. Label-free quantitative proteomic analysis of abscisic acid effect in early-stage soybean under flooding. J. Proteome Res. 12, 4769–4784.

Komatsu, S., Mock, H.-P., Yang, P., Svensson, B., 2013b. Application of proteomics for improving crop protection/artificial regulation. Front. Plant Sci. 4, 522.

Komatsu, S., Nanjo, Y., Nishimura, M., 2013c. Proteomic analysis of the flooding tolerance mechanism in mutant soybean. J. Proteomics 79, 231–250.

Komatsu, S., Makino, T., Yasue, H., 2013d. Proteomic and biochemical analyses of the cotyledon and root of flooding-stressed soybean plants. PLoS One 8, e65301.

Kosová, K., Vítámvás, P., Prášil, I.T., Renaut, J., 2011. Plant proteome changes under abiotic stress–contribution of proteomics studies to understanding plant stress response. J. Proteomics 74, 1301–1322.

Lascano, H., 2003. Effect of photooxidative stress induced by paraquat in two wheat cultivars with differential tolerance to water stress. Plant Sci. 164, 841–848.

Li, W.Y.F., Wong, F.-L., Tsai, S.N., Phang, T.H., Shao, G., Lam, H.M., 2006. Tonoplast-located GmCLC1 and GmNHX1 from soybean enhance NaCl tolerance in transgenic bright yellow (BY)-2 cells. Plant Cell Environ. 29, 1122–1137.

Lichtenthaler, H.K., 1996. Vegetation stress: an introduction to the stress concept in plants. J. Plant Physiol. 148, 4–14.

Lichtenthaler, H.K., 1998. The stress concept in plants: an introduction. Ann. NY Acad. Sci. 851, 187–198.

Lichtenthaler, H.K., Miehe, J.A., 1997. Fluorescence imaging as a diagnostic tool for plant stress. Trends Plant Sci. 2, 316–320.

Lin, F., Qu, Y., Zhang, Q., 2014. Phospholipids: molecules regulating cytoskeletal organization in plant abiotic stress tolerance. Plant Signal. Behav. 9 (3), e28337.

Lobreaux, S., Hardy, T., Briat, J., 1993. Abscisic acid is involved in the iron-induced synthesis of maize ferritin. EMBO J. 12, 651–657.

Luo, G.-Z., Wang, H.-W., Huang, J., Tian, A.-G., Wang, Y.-J., Zhang, J.-S., Chen, S.-Y., 2005. A putative plasma membrane cation/proton antiporter from soybean confers salt tolerance in *Arabidopsis*. Plant Mol. Biol. 59, 809–820.

Mathesius, U., Djordjevic, M.A., Oakes, M., Goffard, N., Haerizadeh, F., Weiller, G.F., Singh, M.B., Bhalla, P.L., 2011. Comparative proteomic profiles of the soybean (*Glycine max*) root apex and differentiated root zone. Proteomics 11, 1707–1719.

Maxwell, K., Johnson, G.N., 2000. Chlorophyll fluorescence – a practical guide. J. Exp. Bot. 51, 659–668.

Mittler, R., 2002. Oxidative stress, antioxidants and stress tolerance. Trends Plant Sci. 7, 405–410.

Mittler, R., 2006. Abiotic stress, the field environment and stress combination. Trends Plant Sci. 11, 15–19.

Mohammadi, P.P., Moieni, A., Hiraga, S., Komatsu, S., 2012. Organ-specific proteomic analysis of drought-stressed soybean seedlings. J. Proteomics 75, 1906–1923.

Müller, K., Linkies, A., Vreeburg, R.A., Fry, S.C., Krieger-Liszkay, A., Leubner-Metzger, G., 2009. *In vivo* cell wall loosening by hydroxyl radicals during cress seed germination and elongation growth. Plant Physiol. 150, 1855–1865.

Munns, R., 2002. Comparative physiology of salt and water stress. Plant Cell Environ. 25, 239–250.

Munns, R., Tester, M., 2008. Mechanisms of salinity tolerance. Annu. Rev. Plant Biol. 59, 651–681.

Müntz, K., 2007. Protein dynamics and proteolysis in plant vacuoles. J. Exp. Bot. 58, 2391–2407.

Musser, R.L., Thomas, S.A., Wise, R.R., Peeler, T.C., Naylor, A.W., 1984. Chloroplast ultra-structure, chlorophyll fluorescence, and pigment composition in chilling-stressed soy-beans. Plant Physiol. 74, 749–754.

Nanjo, Y., Skultety, L., Ashraf, Y., Komatsu, S., 2010. Comparative proteomic analysis of early-stage soybean seedlings responses to flooding by using gel and gel-free techniques. J. Proteome Res. 9, 3989–4002.

Nanjo, Y., Nouri, M.-Z., Komatsu, S., 2011a. Quantitative proteomic analyses of crop seedlings subjected to stress conditions; a commentary. Phytochemistry 72, 1263–1272.

Nanjo, Y., Skultety, L., Uváčková, L.u., Klubicová, K.n., Hajduch, M., Komatsu, S., 2011b. Mass spectrometry-based analysis of proteomic changes in the root tips of flooded soybean seedlings. J. Proteome Res. 11, 372–385.

Nonami, H., Boyer, J.S., 1990. Wall extensibility and cell hydraulic conductivity decrease in enlarging stem tissues at low water potentials. Plant Physiol. 93, 1610–1619.

Nouri, M.Z., Komatsu, S., 2010. Comparative analysis of soybean plasma membrane proteins under osmotic stress using gel-based and LC MS/MS-based proteomics approaches. Proteomics 10, 1930–1945.

Ohashi, Y., Nakayama, N., Saneoka, H., Fujita, K., 2006. Effects of drought stress on photosynthetic gas exchange, chlorophyll fluorescence and stem diameter of soybean plants. Biol. Plant. 50, 138–141.

O'Neill, P.M., Shanahan, J.F., Schepers, J.S., 2006. Use of chlorophyll fluorescence assessments to differentiate corn hybrid response to variable water conditions. Crop Sci. 46, 681–687.

Portis, Jr, A.R., 2003. Rubisco activase – Rubisco's catalytic chaperone. Photosyn. Res. 75, 11–27.

Qi, X., Li, M.-W., Xie, M., Liu, X., Ni, M., Shao, G., Song, C., Yim, A.K.-Y., Tao, Y., Wong, F.-L., 2014. Identification of a novel salt tolerance gene in wild soybean by whole-genome sequencing. Nature Commun. 5, 4340.

Rai, A.K., Takabe, T. (Eds.), 2006. Abiotic Stress Tolerance in Plants. Springer, The Netherlands.

Ren, B., Chen, Q., Hong, S., Zhao, W., Feng, J., Feng, H., Zuo, J., 2013. The *Arabidopsis* eukaryotic translation initiation factor eIF5A-2 regulates root protoxylem development by modulating cytokinin signaling. Plant Cell 25, 3841–3857.

Scandalios, G., Tong, F., Roupakias, G., 1980. Cat3, a third gene locus coding for a tissue-specific catalase in maize: genetics, intracellular location, and some biochemical properties. Mol. Gen. Genet. 179, 3341.

Scandalios, G., Tsaftaris, S., Chandlee, M., Skadsen, W., 1984. Expression of the developmentally regulated catalase (cat) genes in maize. Dev. Genet. 4, 281–293.

Schaller, A., 2004. A cut above the rest: the regulatory function of plant proteases. Planta 220, 183–197.

Senthil-Kumar, M., Hema, R., Suryachandra, T.R., Ramegowda, H., Gopalakrishna, R., Rama, N., Udayakumar, M., Mysore, K.S., 2010. Functional characterization of three water deficit stress-induced genes in tobacco and *Arabidopsis*: an approach based on gene down regulation. Plant Physiol. Biochem. 48, 35–44.

Shao, H.-B., Guo, Q.-J., Chu, L.-Y., Zhao, X.-N., Su, Z.-L., Hu, Y.-C., Cheng, J.-F., 2007. Understanding molecular mechanism of higher plant plasticity under abiotic stress. Colloids Surf. B 54, 37–45.

Shinozaki, K., Yamaguchi-Shinozaki, K., Seki, M., 2003. Regulatory network of gene expression in the drought and cold stress responses. Curr. Opin. Plant Biol. 6, 410–417.

Smalle, J., Kurepa, J., Yang, P., Emborg, T.J., Babiychuk, E., Kushnir, S., Vierstra, R.D., 2003. The pleiotropic role of the 26S proteasome subunit RPN10 in *Arabidopsis* growth and development supports a substrate-specific function in abscisic acid signaling. Plant Cell 15, 965–980.

Sobhanian, H., Razavizadeh, R., Nanjo, Y., Ehsanpour, A.A., Jazii, F.R., Motamed, N., Komatsu, S., 2010. Proteome analysis of soybean leaves, hypocotyls and roots under salt stress. Proteome Sci. 8, 19.

Sung, D.Y., Kaplan, F., Guy, C.L., 2001. Plant Hsp70 molecular chaperones: protein structure, gene family, expression and function. Physiologia Plant. 113, 443–451.

Swigonska, S., Weidner, S., 2013. Proteomic analysis of response to long-term continuous stress in roots of germinating soybean seeds. J. Plant Physiol. 170, 470–479.

Thelen, J.J., Peck, S.C., 2007. Quantitative proteomics in plants: choices in abundance. Plant Cell 19, 3339–3346.

Toorchi, M., Yukawa, K., Nouri, M.-Z., Komatsu, S., 2009. Proteomics approach for identifying osmotic-stress-related proteins in soybean roots. Peptides 30, 2108–2117.

Trippi, V., Gidrol, J., Pradet, A., 1989. Effect of oxidative stress caused by oxygen and hydrogen peroxide on energy metabolism and senescence in oat leaves. Plant Cell Physiol. 30, 157–163.

Umezawa, T., Shimizu, K., Kato, M., Ueda, T., 2000. Enhancement of salt tolerance in soybean with NaCl pretreatment. Physiol. Plant. 110, 59–63.

Van Wijk, K.J., 2001. Challenges and prospects of plant proteomics. Plant Physiol. 126, 501–508.

Vierstra, R.D., 1996. Proteolysis in plants: mechanisms and functions. Plant Mol. Biol. 32, 275–302.

Vinocur, B., Altman, A., 2005. Recent advances in engineering plant tolerance to abiotic stress: achievements and limitations. Curr. Opin. Biotechnol. 16, 123–132.

Wang, W., Vinocur, B., Altman, A., 2003. Plant responses to drought, salinity and extreme temperatures: towards genetic engineering for stress tolerance. Planta 218, 1–14.

Wang, C., Zhang, L.-J., Huang, R.-D., 2011. Cytoskeleton and plant salt stress tolerance. Plant Signal. Behav. 6, 29–31.

Wang, L., Xu, C., Wang, C., Wang, Y., 2012. Characterization of a eukaryotic translation initiation factor 5A homolog from *Tamarix androssowii* involved in plant abiotic stress tolerance. BMC Plant Biol. 12, 118.

Xu, C., Sullivan, J.H., Garrett, W.M., Caperna, T.J., Natarajan, S., 2008. Impact of solar ultraviolet-B on the proteome in soybean lines differing in flavonoid contents. Phytochemistry 69, 38–48.

Xu, X.-Y., Fan, R., Zheng, R., Li, C.-M., Yu, D.-Y., 2011. Proteomic analysis of seed germination under salt stress in soybeans. J. Zhejiang Univ. Sci. B 12, 507–517.

Xu, J., Xue, C., Xue, D., Zhao, J., Gai, J., Guo, N., Xing, H., 2013. Overexpression of GmHsp90s, a heat shock protein 90 (Hsp90) gene family cloning from soybean, decrease damage of abiotic stresses in *Arabidopsis thaliana*. PloS One 8, e69810.

Yamaguchi, M., Sharp, R.E., 2010. Complexity and coordination of root growth at low water potentials: recent advances from transcriptomic and proteomic analyses. Plant Cell Environ. 33, 590–603.

Yamaguchi, M., Valliyodan, B., Zhang, J., Lenoble, M.E., Yu, O., Rogers, E.E., Nguyen, H.T., Sharp, R.E., 2010. Regulation of growth response to water stress in the soybean primary root. I. Proteomic analysis reveals region-specific regulation of phenylpropanoid metabolism and control of free iron in the elongation zone. Plant Cell Environ. 33, 223–243.

Yanagawa, Y., Komatsu, S., 2012. Ubiquitin/proteasome-mediated proteolysis is involved in the response to flooding stress in soybean roots, independent of oxygen limitation. Plant Sci. 185, 250–258.

Zhang, J., Jia, W., Yang, J., Ismail, A.M., 2006. Role of ABA in integrating plant responses to drought and salt stresses. Field Crops Res. 97, 111–119.

Zhang, J., Tan, W., Yang, X.-H., Zhang, H.-X., 2008. Plastid-expressed choline monooxygenase gene improves salt and drought tolerance through accumulation of glycine betaine in tobacco. Plant Cell Rep. 27, 1113–1124.

Zhao, J., Cheng, N.-H., Motes, C.M., Blancaflor, E.B., Moore, M., Gonzales, N., Padmanaban, S., Sze, H., Ward, J.M., Hirschi, K.D., 2008. AtCHX13 is a plasma membrane K^+ transporter. Plant Physiol. 148, 796–807.

Zhao, J., Zhang, Y., Bian, X., Lei, J., Sun, J., Guo, N., Gai, J., Xing, H., 2013. A comparative proteomics analysis of soybean leaves under biotic and abiotic treatments. Mol. Biol. Rep. 40, 1553–1562.

Zhen, Y., Qi, J.L., Wang, S.S., Su, J., Xu, G.H., Zhang, M.S., Miao, L., Peng, X.X., Tian, D., Yang, Y.H., 2007. Comparative proteome analysis of differentially expressed proteins induced by Al toxicity in soybean. Physiol. Plant. 131, 542–554.

Soybean N fixation and production of soybean inocula

Mohammad Miransari
Department of Book & Article, AbtinBerkeh Ltd. Company, Isfahan, Iran

Introduction

With respect to the nutritional and economical importance of soybean (*Glycine max* (L.) Merr.), worldwide, its production has increased dramatically in different parts of the world including the USA (Ruiz Diaz et al., 2009), Brazil, Argentina, Canada, England, France, Japan, China, India, and many other countries. For example, the production of soybean as the second most important crop in the USA was 10 million tons per year (Mt year^{-1}) in the early 1950s, while it is now 70–75 Mt year^{-1}, accounting for half of global production and 15% of arable crop fields. The growth rate of soybean production has been even faster in Brazil (with a high rate of export) at a production rate of over 60 Mt year^{-1} in 2007 (a 60-fold time growth relative to the early 1950s) followed by Argentina, China, India, and Paraguay at 47, 14.3, 9.3, and 7 Mt year^{-1}, respectively. In Europe, the production rate is only 1 Mt year^{-1}, mostly by Italy, France, and Romania (Rodriguez-Navarro et al., 2011).

Accordingly, it is important to use modern techniques that can increase the amount of crop yield in plants such as soybean as there is a high demand for soybean production and utilization, worldwide (Ruiz Diaz et al., 2009). The rate of soybean yield has steadily increased since the mid-1980s as a result of using new techniques and strategies. For example, the production increase in the world (Wilcox, 2004) and in the USA (Specht et al., 1999) has been equal to 28 and 31 kg ha^{-1}, respectively.

An important point about soybean production is its increased efficiency in hectare, because field development is not much likely. Such a goal may be attained by using suitable strategies related to the production of soybean under different conditions inducing stress. This can be achieved by providing conditions conducive to soybean growth. For example, the rate of soybean production in the USA Corn Belt is estimated to be between 6 t ha^{-1} and 8 t ha^{-1} (Salvagiotti et al., 2008; Specht et al., 1999), which is an acceptable rate. As a result of high photosynthesis rate and high N accumulation in soybean seeds, the rate of soybean-yield production has increased per hectare (Sinclair, 2004).

Some of the most important strategies that can be used for increased production of soybean yield per hectare are: (1) using efficient varieties with a high nodulating capacity, (2) using commercially produced inoculants, (3) combined use of beneficial soil microbes including rhizobia, the other strains of plant growth promoting rhizobacteria (PGPR) and arbuscular mycorrhizal (AM) fungi, and (4) using appropriate amounts of chemical N fertilization to be used with the inoculants especially at

Abiotic and Biotic Stresses in Soybean Production. http://dx.doi.org/10.1016/B978-0-12-801536-0.00005-0

seeding (Afzal et al., 2010; Chebotar et al., 2001; Miransari, 2011a, b; Miransari and Mackenzie, 2011, 2014; Miransari and Smith, 2007, 2008, 2009).

The use of highly efficient varieties is a suitable method to increase soybean yield per hectare. These varieties must be able to produce high yield and develop a strong symbiotic association with efficient strains of *Bradyrhizobium japonicum*. As a result, the most important part of the host plant N is supplied by the bacteria and the remaining part is supplied from the soil or by N chemical fertilization (Hungria et al., 2006; Njira et al., 2013).

Due to the benefits of using commercially produced inocula, their production and use has become widespread. The inocula must be producible on a large scale and be able to compete with the soil native bacteria, and hence effectively inoculate their host plant roots. If the inocula are used under stress they will be especially effective if they are isolated from the prevailing stress conditions. Use of a suitable carrier is also important for the production and use of inocula (Chebotar et al., 2001; Delves et al., 1986).

Sometimes, instead of using just one strain of bacteria such as rhizobia, the combined use of soil microbes may be more applicable. However, they must be compatible and useable in combination. Is such a case, besides an increased rate of plant N availability by the process of symbiotic N fixation, other advantages are also resulted such as higher uptake of other nutrients by the host plant, production of plant growth promoting substances, production of plant hormones, and alleviation of soil stresses (Rebah et al., 2007; Njira et al., 2013).

Although a significant part of soybean N is supplied by the process of N fixation, the remaining part must be presented by the soil organic matter or chemical N fertilization. It is estimated that about 50–60% of the soybean N requirement occurs via the process of N fixation. In particular, mineral N must be available for soybean at seeding. Strains of *B. japonicum* need to develop a symbiotic association with their host plant and hence fix atmospheric N. In the subsequent sections some of the most important and recent details related to the use and importance of inocula for the production of soybean yield are presented and analyzed.

The process of soybean N fixation

Among the most important processes that provide required N for soybean growth and yield production, is biological N fixation. This is the process of symbiotic N fixation between soybean and its specific *Rhizobium, B. japonicum* (Miransari and Smith, 2007, 2008, 2009; Njira et al., 2013; Saito et al., 2014). It has been found that legumes are able to develop a symbiotic association with the bacteria from the Rhizobiaceae family (α-probacteria) including the genera *Bradyrhizobium, Rhizobium, Sinorhizobium, Mesorhizobium*, and *Azorhizobium*.

In the symbiotic association between the host plant and microbes, signals that determine the specificity of symbiosis are exchanged between these two symbionts. Different types of symbiosis have been specified including their specific and nonspecific associations. In a specific symbiotic association, the host plant is only able to undergo symbiosis with specific microbial species (Oldroyd and Downie, 2008).

For example, in the symbiotic association between soybean and *B. japonicum* the related signal molecule, genistein, which also determines the specificity of symbiosis, is produced by the host plant resulting in the activation of bacterial *nod* genes. However, in a nonspecific symbiotic association, the host plant is able to establish symbiosis with a wide range of microbes including bacteria and fungi (Martínez-Romero, 2009; Wang et al., 2012).

After the recognition of the host plant by the rhizobia as the result of signal perception, the bacteria produce lipochitooligosaccharide molecules, which are able to induce morphological and physiological alterations in the root hair of the host plant. Root hair bulging and curling are the plant's response to such molecules, which is followed by the production of infection thread. Accordingly, the bacteria are able to enter the host plant roots and, by altering the morphology and physiology of root cortex cells, this results in the production of root nodules, wherein rhizobial settlement and the process of N fixation by bacteria take place. The ammonia produced by the rhizobia can be used by the host plant as a suitable source of N and is incorporated into the structure of proteins. The legume host plants provide the bacteria with its essential carbon (Long, 2001).

AM fungi are among the most widespread fungal species in the soil that are able to make nonspecific symbiotic associations with their host plant. Although there might be some kind of specificity, mycorrhizal fungi are able to make symbiotic associations with most terrestrial plants. The fungi favor their host plant by enhancing the uptake of water and nutrients in the exchange of carbon. The fungal spore is able to produce an extensive hyphal network, which can significantly increase the host plant's ability to absorb water and nutrients (Miransari, 2010).

Fungal hyphae are able to grow in the soil micro- and macropores, where even the finest root hairs are not able to grow, and absorb water and nutrients. As a result, mycorrhizal plants can grow more efficiently under different conditions including stress, compared to nonmycorrhizal plants. It has been found that the fungi are able to alleviate the adverse effects of stress on the growth of the host plant by the following methods: (1) enhanced uptake of water and nutrients, (2) interaction with other microbes, (3) production of different enzymes, (4) production of plant hormones, (5) controlling pathogens, etc. (Bothe, 2012; Daei et al., 2009; Miransari, 2010; Miransari et al., 2008).

The fungi are able to develop a tripartite symbiosis with their legume host plant in the presence of rhizobia. Such a symbiosis is beneficial to the host plant, the fungi, and the rhizobia. The uptake of the two important nutrients nitrogen and phosphorus is enhanced by the host plant and as a result plant growth and yield increase. The two microbes, AM fungi and rhizobia, can also have positive interactions on their activities, especially in the rhizosphere (Miransari 2011a, b).

Soybean N fixation and N fertilization

Both the processes of biological N fixation and using chemical N fertilizer provide soybean with essential N. However, adverse interactions between these processes may decrease their efficiency, especially if N chemical fertilizer is overused

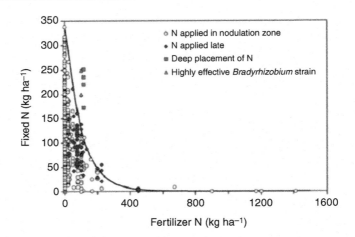

Figure 5.1 The rate of biological N fixation as affected by chemical N fertilizer. The related equation ($y = 337\ e^{-0.0098\times}$) indicates how the maximum rate of N fixation was affected by chemical N fertilization in such experiments. The chemical N fertilizer was used preplant, in the nodulation zone (top 20 cm of soil) during the early growth period, or on the soil surface. The N values are the amounts of N in the total plant biomass of above ground. Adopted from Salvagiotti et al. (2008). With kind permission from Elsevier, license number: 3577411072815.

(Figure 5.1). The maximum rate of N fixation is related to the growth stages R3 and R5, and hence accordingly the soybean N budget must be adjusted so that the plant does not face N deficiency. After the R5 growth stage, which is the time of N pick-demand for soybean, the N-fixation ability of soybean begins to decline and hence addition of N fertilization at this stage may effectively increase soybean growth and yield (Seneviratne et al., 2000; Schmitt et al., 2001).

The other suitable time for N fertilization is at seeding or during the early vegetative stage. Hence, in addition to the process of biological N fixation as the most important process for supplying the N required by soybean, chemical N fertilization and the available soil N can also provide some part of N required for plant growth (Miransari and Mackenzie, 2011; Zhang and Smith, 1995). Soybean inoculation may be more effective if the soil pH is not above 6, in sandy soil, or if there is flooding in the field (Ruiz Diaz et al., 2009); however, further investigations are required in this respect.

Under suitable growing conditions the nodule nitrogenase is able to adjust nodule activities so that the highest rate of N fixation provides the important part of N required for soybean plant growth and yield production. Important factors that determine soybean response to N fertilization include: the amount of yield, the environment (under stress, the rate of yield decreases), and plant growth stage. Accordingly, the role of rhizobia under such conditions becomes clear, determining soybean response to N fertilization, and hence the rate of soybean yields. If soybean plants are well nodulated, with a yield higher than 4.5 t ha^{-1}, a positive response to N fertilization is likely (Mpepereki et al., 2000; Salvagiotti et al., 2008; Zhou et al., 2006).

As previously mentioned, although chemical N fertilization is a way of providing soybean with some of its required N, due to its antagonism with biological N fixation, the process of symbiotic N fixation can be adversely affected. N deficiency during the processes of R5 and R7 may decrease soybean N seed under suitable growing conditions (Liu et al., 2004; Wang et al., 2011). The effects of N fertilization on the growth and yield of soybean become greater when the process of biological N fixation does not supply adequate amounts of N for the use of the host plant. If the soybean plants have been inoculated for several years they may not be responsive to inocula because the bacterial population in the soil might prove adequate to colonize the plants (Barker and Sawyer, 2005; Schmitt et al., 2001).

In this experiment, Saito et al. (2014) investigated the effects of 5 mM nitrate on the root and nodule growth of soybean plants. Addition of nitrate suppressed the growth and hence activity of nodules 7 h later. Although it is speculated that there might be similar mechanisms which control the adverse effects of nitrate on the process of nodulation and the process of autoregulation of nodulation (Delves et al., 1986; Gremaud and Harper, 1989; Ohyama et al., 1993; Reid et al., 2011), certain points have been proposed as the main mechanisms, which are responsible for adverse effects of nitrate upon the process of nodulation: (1) deficiency of carbohydrates in the nodules (Fujikake et al., 2003; Streeter, 1988; Thorpe et al., 1998; Vessey et al., 1988), (2) the adverse effects of nitrate assimilation products including asparagine (Bacanamwo and Harper, 1997) and glutamate (Neo and Layzell, 1997), and (3) decreased diffusion rate of oxygen into the nodules, adversely affecting the activities of bacteroids (Gordon et al., 2002; Schuller et al., 1988; Vessey et al., 1988).

Inoculum microbes and their interactions

B. japonicum is the most important microbe used for the production of soybean inocula. However, other PGPR such as *Pseudomonas* spp. and *Azospirillum* spp. with the ability to fix atmospheric N nonsymbiotically and produce plant growth substances such as plant hormones can also be useful for the production of inocula (Table 5.1). Mycorrhizal fungi can also be used as an effective inoculum for soybean production (Miransari, 2011a; Rodriguez-Navarro et al., 2011). The rhizobial inoculants were first marketed in the USA; however, now the inoculants are being used in most parts of the world with a net production rate of 2000 t year^{-1} (Rebah et al., 2007).

The world of microbes is interesting and fascinating. There is a wide range of soil microbes affecting plant growth and crop production. Plants and microbes are continuously interacting. Plants have significant effects on the rhizosphere by the growth of their roots and production of different compounds including organic products and ions. As a result the growth and activities of the microbes in the rhizosphere are also affected. Depending on the plant and microbial species and strains, interactions between the plant and microbes are determined including symbiotic and nonsymbiotic associations. The most intimate and beneficial interaction is the symbiotic association between host plants and the related microbes (Chisholm et al., 2006; Miransari 2011a, b; Pozo et al., 2005).

Table 5.1 Different strains of soil bacteria and the related mechanisms affecting soybean growth and grain yield

Microbes	Mechanisms	References
B. japonicum	Biological N fixation	Jordan (1982)
Azospirillum	Production of plant hormones such as auxin	Chebotar et al. (2001)
Sinorhizobium fredii	Biological N fixation	Chen et al. (1988)
Bacillus sp.	Production of antibiotics	Pankhurst (1977)
Pseudomonas sp.	Production of siderophore	Fuhrmann and Wollum (1989)
Pseudomonas, Enterobacter, Ralstonia, Pantoea, Acinetobacter	Production of plant hormones such as auxin, P solubilization	Kuklinsky-Sobral et al. (2004)
Serratia sp.	Production of plant growth-promoting substances	Fuhrmann and Wollum (1989)
Streptomyces sp.	Production of antibiotics	Hu et al. (2009)

The interactions between soil microbes are of great importance in the production of inocula. This is because sometimes more than one species of microbe may be used for the production of inocula. Such kinds of interactions are regulated by different signal molecules and can control the following biological processes: (1) the cycling of nutrients and organic matter, (2) establishment of symbiotic association with the host plant, and (3) maintaining the health of plants and the soil. It has been indicated that plant growth promoting rhizobacteria are able to improve the N-fixing ability of efficient rhizobia by increasing the occupancy of rhizobia in the nodules and enhancing the rate of root nodulation (Araujo et al., 2012; Kuiper et al., 2004; Young and Crawford, 2004; Zhang et al., 1996).

Wang et al. (2011) investigated the effects of coinoculation with rhizobia and mycorrhizal fungi on the growth of two different soybean genotypes differing in root architecture under field conditions. They found that there were synergetic effects between the two microbes significantly affecting the growth of soybean, especially the growth of deep root genotype compared with the shallow root genotype at low levels of N and P. Such findings can be of theoretical and practical use for the production and use of rhizobia and mycorrhizal fungi inocula under field conditions.

Rhizobia and PGPR are able to affect the efficiency and survival of each other. Afzal et al. (2010) tested the interactions between *B. japonicum* and *Pseudomonas* in the presence and absence of P_2O_5 fertilizer, affecting soybean N fixation and yield, under field and greenhouse conditions. Coinoculation, in the presence of the P fertilization, resulted in higher soybean yield as well as the survival of both species of bacteria under field and greenhouse conditions.

Production of plant hormones is one of the beneficial effects of PGPR on rhizobial growth and activities affecting nodulation and hence N fixation. Some *Pseudomonas* strains increased the number of nodules and decreased the rate of ethylene production in nodulated soybean plants with *B. japonicum* (Chebotar et al., 2001). Under stress, some PGPR are able to alleviate the adverse effects of stress on the growth of the host plant by producing 1-aminocyclopropane-1-carboxylate (ACC) deaminase. This enzyme can catabolize ethylene (stress hormone) precursor to α-ketobutyrate and ammonia and hence alleviate the stress (Glick, 2005; Jalili et al., 2009).

Azospirillum brasilense and *Azospirillum lipoferum* as α-proteobacteria are also among the PGPR, which are able to increase plant growth by fixing atmospheric N nonsymbiotically and by producing plant growth substances such as plant hormones (auxins). Their alleviating effects on the growth of plants under stress have also been proved (Arzanesh et al., 2011). As a result some morphological and physiological alterations in plant roots make the plant absorb more water and nutrients. An important point about using PGPR is that their positive effects on the uptake of nutrients by the host plant decrease the rate of chemical fertilization, and hence they are chemically and environmentally recommendable (Dobbelaere et al., 2003; Zabihi et al., 2010).

The bacterial endophytes are also able to produce the enzyme ACC deaminase, which does not have any function in the bacteria; however, as previously mentioned, they can catabolize the precursor for ethylene production and hence decrease the production of the stress hormone, ethylene, in the plant and alleviate the stress. In addition, such bacteria can also produce plant hormones and hence affect plant growth, the most important of which is auxin indole-3-acetic acid, produced by *Herbaspirillum*, *Azospirillum*, *Pseudomonas*, *Pantoea*, *Methylobacterium*, *Gluconoacetobacter*, and *Erwinia* (Kuklinsky-Sobral et al., 2004; Sun et al., 2009).

AM fungi are able to interact with different soil microbes including rhizobia. It has been indicated that AM fungi are also present in the soybean nodules significantly affecting the process of nodulation and the activities of rhizobia (Barea et al., 2005). The mechanisms by which the fungi are able to interact with the soil microbes are: (1) the production of different metabolites such as plant hormones, (2) enhancing the solubility of different nutrients including phosphorus, (3) increasing the activities of soil bacteria such as rhizobia, (4) controlling pathogens, (5) affecting the rhizosphere and hence plant root activities and the associative microbes, (6) alleviating soil stresses, (7) improving the structure of soil, (8) the bacteria can reside on the spores of fungi and may enter into the spores, (9) the bacteria may affect the fungi activities by producing some metabolites, and (10) affecting the of fungi cellular wall (De Boer et al., 2005; Miransari, 2011a, b; Miransari et al., 2013).

The other important point related to the use of inocula is their interactions with the soil native bacteria. Microbial inocula may significantly affect the phylogeny and taxonomy of the major soil microbes. Use of inocula is a desirable method of enhancing N fixation and hence plant growth and grain yield. However, if such a procedure has adverse effects on the main communities of soil microbes which are of significance for the ecosystem, it must be reconsidered. In some situations the functions of soil microbes may be replaced by those of other soil microbes (Trabelsi and Mhamdi, 2013).

Production of soybean inocula

The technological production of inocula for soybean inoculation under different conditions has resulted in a considerable increase in the growth and yield of soybean. The successful inoculation of soybean is related to the selection of suitable strain with a good carrier, tested, controlled, formulated, and produced in adequate amounts (Hungria et al., 2006; Khavazi et al. 2007).

Ruiz Diaz et al. (2009) found that if soybean is inoculated with *Bradyrhizobium* spp., the yield increase per hectare is equal to 130 kg. There were also positive effects of rhizobium inoculation on N uptake, accumulation of grain N, and soybean dry matter, while the quality of seed was not affected. Soybean grain yields were not responsive to N fertilization with inoculation or non-inoculation, although plant dry matter increased. Accordingly, the authors suggested that if soybean is planted after use of pasture, the plant response to inoculation with bacteria is high and N fertilization would not affect plant yield.

With respect to the economical and environmental advantages of microbial inocula, they have been extensively tested and used in different parts of the world with positive results. The increased population of soil microbes in the rhizosphere of the host plant can efficiently colonize its roots and contribute to the fixation and increased uptake of different nutrients. For example, in the symbiotic association between rhizobia and legumes, uptake of nitrogen is significantly increased by the host plant. In the symbiotic association of mycorrhizal fungi with their host plant, uptake of different macro- and micronutrients is significantly enhanced by the plant (Hungria et al., 2006; Lindström et al., 2010; Zhang and Smith, 2002).

For the production of effective inocula the most efficient strain of *Rhizobium* must be selected with respect to the present conditions, which also includes the ability of inoculum bacteria to survive by competing with the native bacteria (Date, 2000). The selection of suitable rhizobia for the production of efficient inocula may be conducted by use of the following steps.

- Collecting the bacteria from root nodules or from the other available collections.
- Purifying and authenticating the rhizobial isolates, by inoculating a similar host plant and isolating the *Rhizobium* bacteria from the nodules. The purity and nodulating capability of the isolates must be tested. Accordingly the effective isolates are selected and used for the following stages (Lupwayi et al., 2000).
- Evaluating the isolated rhizobia for their N-fixing abilities; the isolates must be tested under controlled and sterilized conditions for the inoculation of soybean seedlings. The behavior of inoculated seedlings must be compared with the control treatments, which are fertilized with adequate amounts of N (Date, 2000; Thompson, 1980).
- Finally, the isolates with high N-fixing ability are tested under field conditions and, after passing the required conditions, are used as inocula (Rodriguez-Navarro et al., 2011).

The other important properties that must be considered at the time of producing inocula is the ability of rhizobia to inoculate the host plant under a wide range of field conditions, to grow in the culture medium with high cell densities, and survive during the process of production and storage. Their persistence in the soil, in the presence or absence of the host plant, the genetic stability of strains, and surviving in the presence of agrochemicals are also important factors determining the efficiency of rhizobial inocula. Repeated inoculation may be an effective method when the native bacteria are competitive with the inoculum bacteria (Date, 2000; Herridge et al., 2002; Miransari 2011a, b; Mpepereki et al., 2000).

In the early days of using rhizobial inoculants the containers, which were made of rigid plastic, glass bottles, and metal cans were used for the storage and transfer of microbial inoculants. However, nowadays, the most used container is flexible polyethylene film with a low density (0.038–0.076 mm). The interesting properties of such a container are the ability to hold a high rate of moisture, exchange adequate amounts of gas, and be sealed using heat (Rodriguez-Navarro et al., 2011).

The autoclaving of commercially polyethylene bags containing a peat carrier can be conducted at 121°C for 50 min. The presterilized bags can then be inoculated using a syringe or an automatic dispenser. The inoculated bags are then incubated at a suitable temperature and hence the bacteria would be able to grow further. Figure 5.2. shows the production stages of rhizobial inocula. The most important point about the successful production of microbial inocula is the quality control (Herridge et al., 2002; Thompson, 1980). If the quality of inoculants is not suitable then the use of inocula would not be recommendable. Different surveys have indicated that in most inoculants each gram of carrier has less than 10^8 ml^{-1} of viable rhizobium. The lower the rate of contamination, the higher the number of rhizobia in the inocula (Hungria et al., 2001).

However, because the carriers are sterilized, a lower quantity of rhizobia in the carrier can be due to the production process. Hence, this indicates that the production of rhizobial inocula must be more accurate and have improved supervision. Accordingly, for the production of microbial inoculants, public or private sector agencies, such as those of the USA, England, France, Brazil, Canada, and Australia, must ensure control of the quality of microbial inoculants (Herridge et al., 2002; Lupwayi et al., 2000; Rebah et al., 2002).

With respect to the above mentioned discussion, before the commercial production of rhizobial inoculants, the efficiency of rhizobia for symbiotic association must be determined. It is also important to indicate the acceptable purity of the bacterial culture using Gram-staining and microscopy before using inocula on a large scale. The culture pH is a good indicator of determining the contamination of the bacterial culture. Using the method of plating carrier dilutions it is likely to determine the presence of microbial contaminants in the medium. In addition, the other important factors related to the quality control measures of the bacterial inocula are: (1) the moisture content of the final prepared inocula, (2) the density of the bacteria, (3) the proper sealing of the polyethylene bags, (4) their weights, and (5) the use-by date of the product (Date, 2000; Herridge, 2008; Herrmann and Lesueur, 2013).

The quality of microbial inocula is defined by the conditions after packaging as well as by the conditions during production, transport, and distribution. A longer time for the maturation of rhizobia and a lower storage temperature define the higher efficiency of inocula at the time of inoculating seeds. The presence of native rhizobia in the soil is considered a barrier for the success of the inoculation process. Such native microbes are well adapted to the soil, however, with a low rate of N fixation. As a

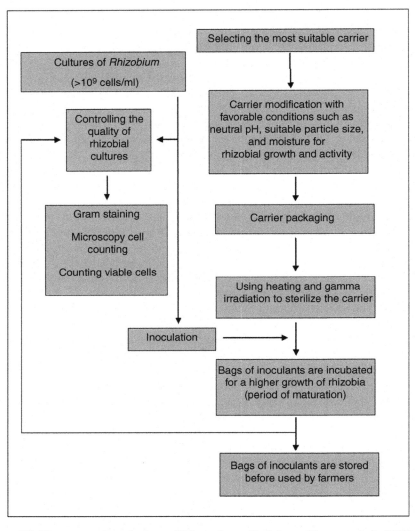

Figure 5.2 The stages of producing solid inoculants (Rodriguez-Navarro et al., 2011).

result, nodule occupancy by the inoculants and hence the rate of N fixation decreases (Herrmann and Lesueur, 2013; Miransari, 2014).

The survival of bacterial inocula on the seed surface is not as great as on the solid carriers due to disadvantageous conditions such as contamination and competition with other microbes, and unfavorable temperature, as well as toxic substances. This is particularly of great significance as the number of viable bacteria is important for the successful inoculation of seeds (Howieson et al., 2008; Hume and Blair, 1992).

Although using liquid inoculants for the inoculation of seeds has some advantages such as being easier and cheaper, their efficiency is not as great as with solid carriers, because the rhizobial cells are not protected. However, broth cultures along with some other compounds may be used to counter such disadvantages under different conditions including stress. There are also oil-based inoculants. The inoculants can be used by pretreating the seeds using adhesives or can be used at seeding in the furrows (Lindström et al., 2010; Melchiorre et al., 2011; Stephens and Rask, 2000).

The success of inoculation is a function of different properties including the rhizobial and plant species, the conditions of soil, such as that of native rhizobia, and the presence of stresses including salinity, acidity, compaction, flooding, heavy metals, mineral N, and suboptimal root zone temperature (Miransari and Smith, 2007, 2008, 2009; Yang et al., 2001; Zhang et al., 1996). Yang et al. (2001) compared the effects of different soybean genotypes affecting the process of symbiotic N fixation. They accordingly indicated that nodule occupancy by rhizobia and the population of native rhizobia in the soil was significantly affected by soybean cultivar as well as by the place of experiment.

With respect to the advantages of using an organic input such as microbial inoculants, from an environmental and economical point of view their use has become more usual across the world. Accordingly, it is a must to use quality control standards so that a greater efficiency from using inoculants will result. For example, such rules and legislation have been defined and used in different countries of the world such as the USA and Brazil (Rebah et al., 2007; Silva et al., 2007).

It is required that biosafety rules are defined and used, especially in the case of using new inoculum strains. For a higher efficiency, use of a mixture of inoculum strains may be tested and used under different conditions including stress. However, it must be mentioned that it is difficult to register and control the quality of such inocula. The use of rhizobium inocula is also useful for the reduction of greenhouse gases (Araujo et al., 2012).

The biodiversity of rhizobium is also an important point affecting the production and quality of inocula achieved by research and evaluation of different rhizobial strains (Daba and Haile, 2000). Rhizobium biodiversity affects the rate and efficiency of biological N fixation, the survival of rhizobia in the soil, and their competition with the other strains of bacteria as well as with the native soil microbes (Aguilar et al., 2001; Doignon-Bourcier et al., 1999; Doyle and Doyle, 1998).

Rhizobium biodiversity is important because it affects the selection of strains which are more suitable for the production of effective inocula under different conditions including stress resulting in a higher rate of legume crop yield (Date, 2000; Lesueur et al., 2001; Martins et al., 2003; Silva et al., 2007). The use of genetics is also of great importance for the production of efficient inocula. It is likely to produce plant species and rhizobium strains which are more efficient under different conditions including stress. The genetically modified plant species and rhizobial strains must be able to establish an efficient symbiotic association and hence fix higher amounts of atmospheric N relative to the nonmodified plant species and rhizobial strains (Chebotar et al., 2001; Delves et al., 1986).

Some examples of investigating efficient strains for use as inoculants under different conditions, including stress, follow. Rhamani et al. (2009) investigated the high temperature tolerance of *Bradyrhizobium* for soybean production in the arid and semi-arid areas of Iran. Fifty-six isolated strains of bradyrhizobia were tested for their thermotolerance ability at 38 and 41°C, under field and greenhouse conditions using solid or liquid yeast extract mannitol medium. There were 19 strains that were able to grow at 38 °C and 10 strains that were able to grow at 41 °C. In the greenhouse, the strains were tested at 28 and 38 °C. Ten isolates indicated a higher index of N-fixing ability related to the N fertilized treatments at 28 °C. There were also some strains that were able to efficiently fix N at 38 °C.

Rhamani et al. (2009) accordingly suggested that if the *Bradyrhizobium* strains are selected from stress conditions they are more likely to tolerate the stress and fix N in association with their host plant related to the strains, which are selected from non-stress conditions. There has also been some related research work in different parts of the world (Araujo et al., 2012; Ruiz Diaz et al., 2009).

Soybean inocula carriers

Liquid inoculants and sterile peat are among the most used and efficient carriers for the production and use of rhizobial inoculants. However, due to problems related to the solid carriers such as the unevenness of composition, method of use in the field, handling, and the sterilization of carrier, new formulations and methods must be tested, assessed, and used under field conditions (Dobbelaere et al., 2003; Rebah et al., 2007).

Similarly, due to the difficulties related to the use of the liquid carriers such as their storage (must be kept at 4 °C) and transportation, homogeneous inoculation of seeds in the field and their use is not likely. As a result the new formulations must be easy to handle, friendly to the environment (biodegradable), and in the case of using mixed inoculants (for example, mixed with PGPR) greater attention must be paid to them (Dobbelaere et al., 2003). The inoculants must be recommendable with respect to the related expenses (Rebah et al., 2007).

The following are the most used solid carriers of rhizobia: (1) peat, soil, lignite, and soil; (2) soybean meal, farmyard manure, cellulose, and different composted plant products; (3) perlite, vermiculite, bentonite, alginate beads, talc, calcium phosphate ground rock phosphate, and polyacrylamide gel; and (4) a combination of the above mentioned products (Smith, 1992). Large numbers of rhizobia must be grown before incorporation into the carriers. The details related to the production of rhizobia including the related specifications, use of fermentor, medium, and culture are well documented (Thompson, 1980; Somasegaran and Halliday, 1982).

Peat is among the most widely used solid carriers, followed by granular (5–15 kg ha^{-1}) and liquid carriers (2–4 ml kg seed^{-1}), for the inoculation of rhizobium to the seed or to the soil (Khavazi et al., 2007; Smith, 1992). The carriers must be sterilized before their use as such a process increases the number of rhizobia and enhance their survival rate during the storage period. For a higher efficiency of the inoculants, use of an adhesive is recommended. The adhesive that renders rhizobia attachable to the seed requires water for use and must be protected from deleterious compounds.

If peat is available, it can be the most suitable carrier for use with inoculants, because it favors the growth and storage of bacteria, and can be easily handled. However, in cases where peat is not available other carriers have been tried such as plant residues and wastewater sludge (Rebah et al., 2002). In such carriers bacteria are adsorbed on the medium surface; however, in formulations with polymer the bacteria are encapsulated and entrapped in the polymer matrix resulting in the solidification of the mixture of bacterial carrier. The polymer mixtures include gels of polyacrylamide, alginate, and xanthan gum (Junior et al., 2012; Rebah et al., 2007). In addition, the carriers must have a high capacity of holding water, with homogeneous chemical and physical characters, and be of neutral pH, nontoxic, and abundant and accessible, and with a reasonable price (Stephens and Rask, 2000). However, the most important properties determining the quality of a carrier are the rhizobial growth and survival.

The inoculum efficiency of different carrier mixtures was compared using diazotrophs under room temperature with peat carrier in Brazil. Polymer mixtures are indicated to be suitable carriers, because they are soluble, biodegradable, nontoxic, etc. The polymer inoculants were able to preincubate the seeds with a high number of bacteria comparable with peat inoculants (Xavier et al., 2010). This product was tested with cowpeas and *Bradyrhizobium* and for sugar cane (*Herbaspirillum, Burkholderia tropica, Azospirillum amazonense, Gluconacetobacter diazotrophicus*) (Silva et al., 2007).

Inocula and soybean N fixation

A great deal of research work has indicated how responsive soybean can be to the inocula and the necessity of using inocula in the field (Herridge, 2008; Vessey, 2004). Accordingly, a threshold of 300 effective rhizobia per gram of soil was defined by the authors to indicate if soybean would be responsive to the use of rhizobial inocula. It is favorable that before the use of inocula, the number of soil rhizobium are determined. If the number of soil rhizobium is less than the threshold number, the probability of soybean response to the use of inocula increases. Research during the 1970s and 1980s indicated that if the soil rhizobia was higher than 100 effective rhizobia per gram of soil, the use of inocula is essential and if the number is higher than 1000 there will be no need for the use of rhizobial inocula (Herridge, 2008).

Among the most important factors affecting the number of rhizobia in the soil is pH. For example, determination of soil rhizobia in 14 sites in Australia indicated that with respect to the environmental factors (pH ranging from 4.6 to 7.9), the number of rhizobia ranged from zero to 4.8 log (10) g^{-1} soil. However, effects of soil pH on the number of rhizobia in the soil are dependent on the strains of rhizobia (Woomer et al., 1988).

Another experiment investigating the number of rhizobium in the soil surface (10 cm, where the number of rhizobia is the highest) of 400 sites indicated that for clover under the pH of 3.7–7.9 the number of rhizobia was less than 300 g^{-1} soil; however, the corresponding pH for Medicago truncatula was in the range of 5.3–8.5. Accordingly, such data indicated that the number of rhizobia was the maximum during autumn, winter, and spring and it was at its minimum during the summer because of the hot climate and less water in the soil (Herridge, 2008).

The other important point related to the use of rhizobial inocula is the selection of the right strains for the highest efficiency of nodulation and N fixation under field conditions. The following determine how a rhizobial strain may be selected and used (Campo et al., 2009; Hungria et al., 2006; Lindström et al., 2010):

- Establish the symbiotic association with the host plant, produce nodules, and fix N.
- Produce the nodules in the presence of native soil rhizobia and fix N.
- Establish a symbiotic association with a wide range of host plant genotypes under different environmental conditions.
- Can be produced in different growing media and grow in the carriers and in the soil.
- Produce nodules and fix N even in the presence of soil nitrate.
- Be resistant on the inoculated seeds, and during the storage of carriers.
- Be tolerant under environmental stresses.
- Have suitable persistency in the soil, especially for legumes, which are planted regularly.
- Be able to approach the host plant roots and colonize them from the initial place of inoculation.
- Be efficiently active in the presence of agrochemicals.
- Grow in the soil and colonize even if the host plant is not present.
- Its genetic combination does not change.

According to Howieson et al. (2000) suitable strains of rhizobia can be isolated using the following: (1) acquisition of rhizobial germplasm and its maintenance as a culture, which is lyophilized, (2) greenhouse experiments for the determination of rhizobial isolates for the ability of producing nodules and fixing N in the presence of selective hosts, (3) selection of strains from the second stage and assessment of their nodulation and N fixing under field conditions for a period of 2–3 years, while colonizing the soil and persistent in the absence of the host plant, and (4) selection of the most suitable strains from stage three and assessing their ability of nodulating and N fixing under a wide range of environmental conditions.

Inocula are used by the farmers using the following: (1) a solid carrier (finely milled) by inoculating the legume seeds or inoculating the water suspension used for the seeds, (2) a solid carrier, with a larger particle size, and (3) using liquid inoculants for the treatment of seeds or the soil. The most widespread medium used in different parts of the world are solid carriers, especially peat used for the treatment of the seeds (Xavier et al., 2004).

An inoculant's efficiency is determined by: (1) the number of rhizobia, as a higher number of rhizobia increases the efficiency of inocula, (2) the process of inoculation, a more effective method resulting in a more efficient inoculation, and (3) the time delay of inoculation between the production and use of inoculants as longer time, including the storage time, decreases the efficiency of inoculants. The advantage of inoculating the soil rather than the seeds is that the number of rhizobia will be higher in the soil, as after seed inoculation during the sowing the number of rhizobia decreases (Miransari, 2014).

Table 5.2 indicates the number of rhizobia in different carriers, their numbers on the seed, and their rate of survival, inoculated to the soil or seed. Table 5.3 represents the effects of storage time on the survival and efficiency of rhizobial symbiosis. As is apparent, the longer the storage time, the less the number of rhizobia and the lower the efficiency of symbiosis (Revellin et al., 2000). The production of efficient and healthy

Table 5.2 Number of soybean rhizobia per seed following inoculation and the rate of survival of rhizobia

Carriers	Inoculated to	Log number rhizobia per seed		Surviving rhizobia (%)
		Theoretical	Actual	
Granular	Soil	7.2	7.2	100
Peat	Seed	6.45	5.37	8
Liquid	Seed	7.38	5.16	<1
Frozen paste	Seed	6	4.93	<1
Freeze-dried	Seed	5.53	4.62	12

Adapted from D.F. Herridge, N. Moore, E. Hartley, and L.G. Gemel, unpublished data.

Table 5.3 Number of rhizobia on the plate, their actual numbers, and the efficiency of rhizobial symbiosis as affected by time of storage using peat carrier (Revellin et al., 2000)

Time of storage (years)	Plate number (log rhizobia g^{-1} peat)	Efficiency (%)	Actual numbers (log rhizobia g^{-1} peat)
1	9.59	36	9.15
4	9.10	42	8.72
6	8.30	11	7.34
7	7.26	3	6.00
8	7.30	5	6.00

inocula is dependent on a regular quality control, which indicates the number and quality of rhizobia and the number of contaminating microbes.

The inoculants must be of good quality to be used efficiently or the use of inoculants will not benefit the host plant. Accordingly, for an efficient inoculation of the host plant, the inoculants must have suitable number of rhizobia of an appropriate strain (Smith, 1992). However, after the production of inoculants, their quality must be tested on a regular basis for the number of rhizobia and the quality of inoculants. The regulations related to the quality of inoculants are different from place to place. The inoculants usually must be registered, be of proven efficiency, and be friendly to the other plants and the environment.

An important point for the production of inoculants with high efficiency is to develop new technologies and products, and update the related regulations. The rhizobia may be used singly or in combination with PGPR. Such strains may be selected from natural environments or be genetically modified to enhance their efficiency. The formulation and storage of inoculants must be improved to enhance their survival and activity under difficult conditions including stress (Date, 2001; Nelson, 2004).

The following points can render the production and use of rhizobial inoculants successful: (1) familiarize farmers with the use of inoculants with the help of international guidelines, (2) gain the help of individuals who have experience of such activities,

(3) observe suitable regulations, standards, and quality control, (4) note research and development related to the production and use of rhizobium strains, carriers, methods of production and use, (5) examine the effects and efficiency of inoculants regarding the growth and yield of legumes under field conditions, and (6) if essential have the inoculants produced by private companies.

The inocula are able to fix large amounts of N and make it available for the use of the host plant. Such a process is of great economic and environmental value. Economically, it is of importance, because the use of N chemical fertilizer significantly decreases. The environmental significance of such a process is that N chemical products are subjected to leaching, volatilization, and denitrification related to the process of biological N fixation.

Economic and environmental significance of soybean inocula

Chianu et al. (2010) investigated the economic significance of biological N fixation by soybean in Africa. They obtained their data from research stations and farms and from the Food and Agriculture Organization of the United Nations from across the African continent. They calculated the amounts of N biologically fixed by different soybean varieties (twice as much as the production of chemical fertilization) and the related financial values ($200 million). With the fertilizer value of $795t^{-1} (2008) the amount would be around $375 million. Accordingly, it is important to find methods that may improve the utilization and efficiency of biological N fixation in Africa. As a result farmers would be able to increase their crop production per hectare.

In Brazil, savings related to the use of commercially produced inoculants, in 13 million hectares of soybean cultivated fields, is about $2.5 billion year^{-1}. In developing nations of the world the amount of biologically fixed N per year is equal to 11.1 million metric tons of N, which is equal to the amounts of $6.7 billion of N chemical fertilizer per year. As a result, to increase the amounts of soybean yield per hectare produced in Africa from 1.1 ton ha^{-1} to 3 ton ha^{-1} at the least, the use of commercially produced inoculants is required (Chianu et al., 2010).

Because the income of farmers is not high in Africa, the process of biological N fixation can be of significance, as the farmers would have to use lower rates of N fertilization and hence this can be to their advantage. However, because of difficulties related to the use of N fertilization and biological N fertilization different possible options were examined in Africa. Accordingly, using biotechnological techniques a new variety of soybean was developed with the ability to grow without using inoculants. As a result, the new soybean variety was able to nodulate efficiently in different places in the absence of inoculants (Mpepereki et al., 2000).

The production of new soybean varieties has significantly increased in Africa; for example, in Nigeria the amount of soybean grain yield produced in 1984 increased from 60,000 tons to 405,000 tons (1999). Developing a variety with the ability to nodulate with the native bacteria was also another idea, which was tested and developed with success. Similarly, in this case also the use of inoculants was not required. There were also other techniques of producing new soybean varieties, which were used for their biomass (improvement of soil) and food and have been successful in Africa (Mpepereki et al., 2000).

The important point about the contribution of biological N fixation to the growth and yield of soybean is determining the proportion of N derived from the soil and that from N chemical fertilization. Accordingly, under a higher rate of N soil, the rate of biological N fixation decreases. Although, given adequate amounts of N soil, the addition of N chemical fertilization may not be required, at seeding and for the proper growth of soybean seedlings a certain amount of N chemical fertilization must be used.

Conclusions and future perspectives

Symbiotic N fixation is an important process affecting the growth and yield of legumes including soybean. There are native rhizobia in the soil which are able to colonize the roots of their host plant; however, such symbiotic association is not of high efficiency. Accordingly, and with respect to the economic and environmental benefits of inoculating soybean seeds with biotechnologically produced inocula, this methodology is widely used in different parts of the world with great benefit. It is important to find methods which increase the efficiency of symbiotic N fixation. For example, the use of more efficient species of legumes and rhizobium strains is one such strategy. Before the production of inocula on a large scale the rhizobial strains must be tested under different conditions, including stress, so that the most tolerant and efficient strains are selected. Commercial rhizobial strains must be produced according to the existing regulations and standards, resulting in the production of efficient and contaminant-free inocula. Future research may focus on the production of more efficient strains which can contribute to a higher N fixation in soybean host plants and hence decrease the amount of N chemical fertilization. The methods of production must be improved so that the number of rhizobia increase in the inocula and the amount of contaminant decreases to a minimum. The handling of commercially produced rhizobia, with respect to inocula properties, is also important and must be conducted with due care so that farmers can use the inocula appropriately for the production of soybean.

References

Afzal, A., Bano, A., Fatima, M., 2010. Higher soybean yield by inoculation with N-fixing and P-solubilizing bacteria. Agron. Sustain. Dev. 30, 487–495.

Aguilar, O.M., López, M.V., Riccillo, P.M., 2001. The diversity of rhizobia nodulating beans in Northwest Argentina as a source of more efficient inoculant strains. J. Biotechnol. 91, 181–188.

Araujo, A., Fernando, L., Leite, C., Iwata, B., Lira, Jr., M., Xavier, G., Figueiredo, M., 2012. Microbiological process in agroforestry systems. A review. Agron. Sustain. Dev. 32, 215–226.

Arzanesh, M.H., Alikhani, H.A., Khavazi, K., Rahimian, H.A., Miransari, M., 2011. Wheat (*Triticum aestivum* L.) growth enhancement by *Azospirillum* spp. under drought stress. World J. Microbiol. Biotechnol. 27, 197–205.

Bacanamwo, M., Harper, J.E., 1997. The feedback mechanism of nitrate inhibition of nitrogenase activity in soybean may involve asparagines and/or products of its metabolism. Physiol. Plant. 100, 371–377.

Barea, J.M., Werner, D., Azcon-Aguilar, C., Azcon, R., 2005. Interactions of arbuscular mycorrhiza and nitrogen fixing symbiosis in sustainable agriculture. In: Werner, D., Newton, W.E. (Eds.), Nitrogen Fixation in Agriculture, Forestry, Ecology and Environment. Springer, Netherlands, pp. 199–222.

Barker, D.W., Sawyer, J.E., 2005. Nitrogen application to soybean at early reproductive development. Agron. J. 97, 615–619.

Bothe, H., 2012. Arbuscular mycorrhiza and salt tolerance of plants. Symbiosis 58, 7–16.

Campo, R., Araujo, R., Hungria, M., 2009. Nitrogen fixation with the soybean crop in Brazil: compatibility between seed treatment with fungicides and bradyrhizobial inoculants. Symbiosis 48, 154–163.

Chebotar, V.K., Asis, C.A., Akao, S., 2001. Production of growth promoting substances and high colonization ability of rhizobacteria enhance the nitrogen fixation of soybean when inoculated with *Bradyrhizobium japonicum*. Biol. Fertil. Soils 34, 427–432.

Chen, W.X., Yan, G.H., Li, J.L., 1988. Numerical taxonomic study of fast-growing soybean rhizobia and a proposal that *Rhizobium fredii* be assigned to *Sinorhizobium* gen. nov. Int. J. System. Bacteriol. 38, 392–397.

Chianu, J., Huising, J., Danso, S., Okoth, P., Chianu, J., Sanginga, N., 2010. Financial value of nitrogen fixation in soybean in Africa: increasing benefits to smallholder farmers. J. Life Sci. 4, 51–59.

Chisholm, S., Coaker Day, B., Staskawicz, B., 2006. Host-microbe interactions: shaping the evolution of the plant immune response. Cell 124, 803–814.

Daba, S., Haile, M., 2000. Effects of rhizobial inoculant and nitrogen fertilizer on yield and nodulation of common bean. J. Plant Nutr. 23, 581–591.

Daei, G., Ardakani, M., Rejali, F., Teimuri, S., Miransari, M., 2009. Alleviation of salinity stress on wheat yield, yield components, and nutrient uptake using arbuscular mycorrhizal fungi under field conditions. J. Plant Physiol. 166, 617–625.

Date, R.A., 2000. Inoculated legumes in cropping systems of the tropics. Field Crops Res. 65, 123–136.

Date, R.A., 2001. Advances in inoculant technology: a brief review. Aust. J. Exp. Agric. 41, 321–325.

De Boer, W., Folman, L.B., Summerbell, R.C., Boddy, L., 2005. Living in a fungal world: impact of fungi on soil bacterial niche development. FEMS Microbiol. Rev. 29, 795–811.

Delves, A.C., Mathews, A., Day, D.A., Carter, A.S., Carroll, B.J., Gresshoff, P., 1986. Regulation of the soybean-*Rhizobium* nodule symbiosis by shoot and root factors. Plant Physiol. 82, 588–590.

Dobbelaere, S., Vanderleyden, J., Okon, Y., 2003. Plant growth promoting effects of diazotrophs in the rhizosphere. CRC Crit. Rev. Plant Sci. 22, 107–149.

Doignon-Bourcier, F., Sy, A., Willems, A., Torck, U., Dreyfus, B., Gillis, M., De Lajudie, P., 1999. Diversity of Bradyrhizobia from 27 tropical Leguminosae species native of Senegal. System. Appl. Microbiol. 22, 647–661.

Doyle, J.J., Doyle, J., 1998. Phylogenetic perspectives on nodulation: evolving views of plants and symbiotic bacteria. Trends Plant Sci. 3, 473–478.

Fuhrmann, J., Wollum, II, A.G., 1989. Nodulation competition among *Bradyrhizobium japonicum* strains as influenced by rhizosphere bacteria and iron availability. Biol. Fertil. Soils 7, 108–112.

Fujikake, H., Yamazaki, A., Ohtake, N., Sueyoshi, K., Matsuhashi, S., Ito, T., Mizuniwa, C., Kume, T., Hashimoto, S., Ishioka, N.S., et al., 2003. Quick and reversible inhibition of soybean root nodule growth by nitrate involves a decrease in sucrose supply to nodules. J. Exp. Bot. 54, 1379–1388.

Glick, B., 2005. Modulation of plant ethylene levels by the bacterial enzyme ACC deaminase. FEMS Microbiol. Lett. 251, 1–7.

Gordon, A.J., Skot, L., James, C.L., Minchin, F.R., 2002. Short-term metabolic response of soybean root nodules to nitrate. J. Exp. Bot. 53, 423–428.

Gremaud, M.F., Harper, J.E., 1989. Selection and initial characterization of partially nitrate tolerant nodulation mutants of soybean. Plant Physiol. 89, 169–173.

Herridge, D., 2008. Inoculation technology for legumes. In: Dilworth, M., James, E., Sprent, J., Newton, W. (Eds.), Nitrogen-Fixing Leguminous Symbioses. Nitrogen Fixation: Origins, Applications, and Research Progress, vol. 7, Springer, pp. 77–115, ISBN: 978-1-4020-3545-6.

Herridge, D., Gemell, G., Hartley, E., 2002. Legume inoculants and quality control. In: Herridge, D. (Ed.), Inoculants and Nitrogen Fixation in Vietnam, ACIAR Proceedings 109e, Canberra, Australia, pp. 105–115.

Herrmann, L., Lesueur, D., 2013. Challenges of formulation and quality of biofertilizers for successful inoculation. Appl. Microbiol. Biotechnol. 97, 8859–8873.

Howieson, J., Malden, J., Yates, R.J., O'Hara, G.W., 2000. Techniques for the selection and development of elite inoculant strains of *Rhizobium leguminosarum* in Southern Australia. Symbiosis 28, 33–48.

Howieson, J., Yates, R., Foster, K., Real, D., Besier, B., 2008. Prospects for the future use of legumes. In: Dilworth, M., James, E., Sprent, J., Newton, W. (Eds.), Nitrogen-Fixing Leguminous Symbioses. Nitrogen Fixation: Origins, Applications, and Research Progress, vol. 7, Springer, pp. 363–394.

Hu, S., Hong, K., Song, Y., Liu, J., Tan, R., 2009. Biotransformation of soybean isoflavones by a marine *Streptomyces* sp. 060524 and cytotoxicity of the products. World J. Microbiol. Biotechnol. 25, 115–121.

Hume, D.J., Blair, D.H., 1992. Effects of numbers on *Bradyrhizobium japonicum* applied in commercial inoculants on soybean yield in Ontario. Can. J. Microbiol. 38, 588–593.

Hungria, M., Campo, R., Chueire, L., Grange, L., Megias, M., 2001. Symbiotic effectiveness of fast-growing rhizobial strains isolated from soybean nodules in Brazil. Biol. Fertil. Soils 33, 387–394.

Hungria, M., Franchini, J.C., Campo, R.J., Crispino, C.C., Moraes, J.Z., Sibaldelli, R.N.R., Mendes, I.C., Arihara, J., 2006. Nitrogen nutrition of soybean in Brazil: contributions of biological N_2 fixation and N fertilizer to grain yield. Can. J. Plant Sci. 86, 927–939.

Jalili, F., Khavazi, K., Pazira, E., Nejati, A., Asadi Rahmani, H., Rasuli Sadaghiani, H., Miransari, M., 2009. Isolation and characterization of ACC deaminase producing fluorescent pseudomonads, to alleviate salinity stress on canola (*Brassica napus* L.) growth. J. Plant Physiol. 166, 667–674.

Jordan, D.C., 1982. Transfer of *Rhizobium japonicum* Buchanan 1980 to *Bradyrhizobium* gen. nov., a genus of slow-growing, root nodule bacteria from leguminous plants. Int. J. System. Bacteriol. 32, 136–139.

Junior, P., Junior, E., Junior, S., Santos, C., de Oliveira, P., Rumjanek, N., Martins, L., Xavier, G., 2012. Performance of polymer compositions as carrier to cowpea rhizobial inoculant formulations: Survival of rhizobia in pre-inoculated seeds and field efficiency. Afr. J. Biotechnol. 11, 2945–2951.

Khavazi, K., Rejali, F., Seguin, P., Miransari, M., 2007. Effects of carrier, sterilization method, and incubation on survival of *Bradyrhizobium japonicum* in soybean (*Glycine max* L.) inoculants. Enz. Microb. Technol. 41, 780–784.

Kuiper, I., Lagendijk, E., Bloemberg, G., Lugtenberg, B., 2004. Rhizoremediation: a beneficial plant-microbe interaction. Mol. Plant Microbe Interact. 17, 6–15.

Kuklinsky-Sobral, J., Araujo, W.L., Mendes, R., Geraldi, I.O., Pizzirani-Kleiner, A.A., Azevedo, J.L., 2004. Isolation and characterization of soybean-associated bacteria and their potential for plant growth promotion. Environ. Microbiol. 6, 1244–1251.

Lesueur, D., Ingleby, K., Odee, D., Chamberlain, J., Wilson, J., Manga, T.T., Sarrailh, J.M., Pottinger, A., 2001. Improvement of forage production in *Calliandra calothyrsus*: methodology for the identification of an effective inoculum containing *Rhizobium* strains and arbuscular mycorrhizal isolates. J. Biotechnol. 91, 269–282.

Lindström, K., Murwira, M., Willems, A., Altier, N., 2010. The biodiversity of beneficial microbe host mutualism: the case of rhizobia. Res. Microbiol. 161, 453–463.

Liu, F., Jensen, C., Andersen, M., 2004. Drought stress effect on carbohydrate concentration in soybean leaves and pods during early reproductive development: its implication in altering pod set. Field Crops Res. 86, 1–13.

Long, S., 2001. Genes and signals in the *Rhizobium*-legume symbiosis. Plant Physiol. 125, 69–72.

Lupwayi, N.Z., Olsen, P.E., Sande, E.S., Keyser, H.H., Collins, M.M., Singleton, P.W., Rice, W.A., 2000. Inoculant quality and its evaluation. Field Crops Res. 65, 259–270.

Martínez-Romero, E., 2009. Coevolution in Rhizobium-legume symbiosis? DNA Cell Biol. 28, 361–370.

Martins, L.M.V., Xavier, G.R., Rangel, F.W., Ribeiro, J.R.A., Neves, M.C.P., Morgado, L.B., Rumjanek, N.G., 2003. Contribution of biological nitrogen fixation to cowpea: a strategy for improving grain yield in the semi-arid region of Brazil. Biol. Fertil. Soils 38, 333–339.

Melchiorre, M., Gonzalez Anta, G., Suarez, P., Lopez, C., Lascano, R., Racca, R., 2011. Evaluation of bradyrhizobia strains isolated from field-grown soybean plants in Argentina as improved inoculants. Biol. Fertil. Soils 47, 81–89.

Miransari, M., 2010. Contribution of arbuscular mycorrhizal symbiosis to plant growth under different types of soil stresses. Plant Biol. 12, 563–569.

Miransari, M., 2011a. Interactions between arbuscular mycorrhizal fungi and soil bacteria. Appl. Microbiol. Biotechnol. 89, 917–930.

Miransari, M., 2011b. Soil microbes and plant fertilization. Appl. Microbiol. Biotechnol. 92, 875–885.

Miransari, M., 2014. Microbial inoculums. In: Miransari, M. (Ed.), Use of Microbes for the Alleviation of Soil Stresses. Alleviation of Soil Stress by PGPR and Mycorrhizal Fungi, vol. 2. Springer, New York, pp. 175–184.

Miransari, M., Mackenzie, A.F., 2011. Development of a soil N test for fertilizer requirements for wheat. J. Plant Nutr. 34, 762–777.

Miransari, M., Mackenzie, A.F., 2014. Optimal N fertilization, using total and mineral N, affecting corn (*Zea mays* L.) grain N uptake. J. Plant Nutr. 37, 232–243.

Miransari, M., Smith, D.L., 2007. Overcoming the stressful effects of salinity and acidity on soybean [*Glycine max* (L.) Merr.] nodulation and yields using signal molecule genistein under field conditions. J. Plant Nutr. 30, 1967–1992.

Miransari, M., Smith, D.L., 2008. Using signal molecule genistein to alleviate the stress of suboptimal root zone temperature on soybean-*Bradyrhizobium* symbiosis under different soil textures. J. Plant Interact. 3, 287–295.

Miransari, M., Smith, D., 2009. Alleviating salt stress on soybean (*Glycine max* (L.) Merr.) – *Bradyrhizobium japonicum* symbiosis, using signal molecule genistein. Eur. J. Soil Biol. 45, 146–152.

Miransari, M., Bahrami, H.A., Rejali, F., Malakouti, M.J., 2008. Using arbuscular mycorrhiza to reduce the stressful effects of soil compaction on wheat (*Triticum aestivum* L.) growth. Soil Biol. Biochem. 40, 1197–1206.

Miransari, M., Riahi, H., Eftekhar, F., Minaie, A., Smith, D.L., 2013. Improving soybean (*Glycine max* L.) N2 fixation under stress. J. Plant Growth Regul. 32, 909–921.

Mpepereki, S., Javaheri, F., Davis, P., Giller, K.E., 2000. Soybeans and sustainable agriculture: "Promiscuous" soybeans in southern Africa. Field Crops Res. 65, 137–149.

Nelson, L.M., 2004. Plant growth promoting rhizobacteria (PGPR): prospects for new inoculants. Available from: http://www.plantmanagementnetwork.org/pub/cm/review/2004/inoculant/.

Neo, H.H., Layzell, D.B., 1997. Phloem glutamine and the regulation of O$_2$ diffusion in legume nodules. Plant Physiol. 113, 259–267.

Njira, K., Nalivata, P., Kanyama-Phiri, G., Lowole, M., 2013. An assessment for the need of soybean inoculation with *Bradyrhizobium japonicum* in some sites of Kasungu district, Central Malawi. Int. J. Curr. Microbiol. Appl. Sci. 2, 60–72.

Ohyama, T., Nicholas, J.C., Harper, J.E., 1993. Assimilation of ^{15}N$_2$ and ^{15}NO$_3$ by partially nitrate-tolerant nodulation mutants of soybean. J. Exp. Bot. 44, 1739–1747.

Oldroyd, G., Downie, A., 2008. Coordinating nodule morphogenesis with rhizobial infection in legumes. Annu. Rev. Plant Biol. 59, 519–546.

Pankhurst, C., 1977. Symbiotic effectiveness of antibiotic-resistant mutants of fast- and slow-growing strains of *Rhizobium* nodulating *Lotus* species. Can. J. Microbiol. 23, 1026–1033.

Pozo, M., Van Loon, L., Pieterse, C., 2005. Jasmonates – signals in plant-microbe interactions. J. Plant Growth Regul. 23, 211–222.

Rebah, F.B., Tyagi, R.D., Prévost, D., 2002. Wastewater sludge as a substrate for growth and carrier for rhizobia: the effect of storage conditions on survival of *Sinorhizobium meliloti*. Bioresour. Technol. 83, 145–151.

Rebah, F.B, Prevost, D., Yezza, A., Tyagi, R.D., 2007. Agro-industrial waste materials and wastewater sludge for rhizobial inoculant production: a review. Bioresour. Technol. 98, 3535–3546.

Reid, D.E., Ferguson, B.J., Gresshoff, P.M., 2011. Inoculation- and nitrate-induced CLE peptides of soybean control NARK-dependent nodule formation. Molecular mechanisms controlling legume autoregulation of nodulation. Mol. Plant Microbe Interact. 24, 606–618.

Revellin, C., Meunier, G., Giraud, J.J., Sommer, G., Wadoux, P., Catroux, G., 2000. Changes in the physiological and agricultural characteristics of peat-based *Bradyrhizobium japonicum* inoculants after long-term storage. Appl. Microbiol. Biotechnol. 54, 206–221.

Rhamani, H.A., Saleh-Rastin, N., Khavazi, K., Asgharzadeh, A., Fewer, D., Kiani, S., Lindstrom, K., 2009. Selection of thermotolerant bradyrhizobial strains for nodulation of soybean (*Glycine max* L.) in semi-arid regions of Iran. World J. Microbiol. Biotechnol. 25, 591–600.

Rodriguez-Navarro, D.N., Oliver, I.M., Contreras, M.A., Ruiz-Sainz, J.E., 2011. Soybean interactions with soil microbes, agronomical and molecular aspects. Agron. Sustain. Dev. 31, 73–190.

Ruiz Diaz, D., Pedersen, P., Sawyer, J., 2009. Soybean response to inoculation and nitrogen application following long-term grass pasture. Crop Sci. 49, 1058–1062.

Saito, A., Tanabata, S., Tanabata, T., Tajima, S., Ueno, M., Ishikawa, S., Ohtake, N., Sueyoshi, K., Ohyama, T., 2014. Effect of nitrate on nodule and root growth of soybean (*Glycine max* (L.) Merr.). Int. J. Mol. Sci. 15, 4464–4480.

Salvagiotti, F., Cassman, K.G., Specht, J.E., Walters, D.T., Weiss, A., Dobermann, A., 2008. Nitrogen uptake, fixation and response to fertilizer N in soybeans: a review. Field Crops Res. 108, 1–13.

Schmitt, M.A., Lamb, J.A., Randall, G.W., Orf, J.H., Rehm, G.W., 2001. In-season fertilizer nitrogen applications for soybean in Minnesota. Agron. J. 93, 983–988.

Schuller, K.A., Minchin, F.R., Gresshoff, P., 1988. Nitrogenase activity and oxygen diffusion in nodules of soybean cv. Bragg and a supernodulating mutant: effect of nitrate. J. Exp. Bot. 39, 865–877.

Seneviratne, G., Van Holm, L., Ekanayake, E., 2000. Agronomic benefits of rhizobial inoculant use over nitrogen fertilizer application in tropical soybean. Field Crops Res. 68, 199–203.

Silva, V.N., Silva, L.E.S.F., Figueiredo, M.V.B., Carvalho, F.G., Silva, M.L.R.B., Silva, A.J.N., 2007. Caracterização e seleção de populações natives de rizóbios de solo da região semi-árida de Pernambuco. Pesquisa Agropecuaria Trop. 37, 16–21.

Sinclair, T., 2004. Improved carbon and nitrogen assimilation for increased yield. In: Boerma, H.R., Specht, J.E. (Eds.), Soybeans: Improvement, Production and Uses. ASA, CSSA, SSSA, Madison, WI, pp. 537–568.

Smith, R.S., 1992. Legume inoculant formulation and application. Can. J. Microbiol. 38, 485–492.

Somasegaran, P., Halliday, J., 1982. Dilution of liquid *Rhizobium* cultures to increase production capacity of inoculant plants. Appl. Environ. Microbiol. 44, 330–333.

Specht, J.E., Hume, D.J., Kumudini, S.V., 1999. Soybean yield potential – a genetic and physiological perspective. Crop Sci. 39, 1560–1570.

Stephens, J.H.G., Rask, H.M., 2000. Inoculant production and formulation. Field Crops Res. 65, 249–258.

Streeter, J.G., 1988. Inhibition of legume nodule formation and N_2 fixation by nitrate. CRC Crit. Rev. Plant Sci. 7, 1–23.

Sun, Y., Cheng, Z., Glick, B.R., 2009. The presence of a 1-aminocyclopropane-1-carboxylate (ACC) deaminase deletion mutation alters the physiology of the endophytic plant growth promoting bacterium *Burkholderia phytofirmans* PsJN. FEMS Microbiol. Lett. 296, 131–136.

Thompson, J.A., 1980. Production and quality control of legume inoculants. In: Bergersen, F.J. (Ed.), Methods for Evaluating Biological Nitrogen Fixation. Wiley, New York, pp. 489–533.

Thorpe, M.R., Walsh, K.B., Minchin, P.E.H., 1998. Photoassimilate partitioning in nodulated soybean. I. ^{11}C methodology. J. Exp. Bot. 49, 1805–1815.

Trabelsi, D., Mhamdi, R., 2013. Microbial inoculants and their impact on soil microbial communities: a review. BioMed Res. Int. 2013:863240.

Vessey, J.K., 2004. Benefits of inoculating legume crops with rhizobia in the northern Great Plains. Available from http://www.plantmanagementnetwork.org/pub/cm/review/2004/inoculant/.

Vessey, J.K., Walsh, K.B., Layzell, D.B., 1988. Can a limitation in phloem supply to nodules account for the inhibitory effect of nitrate on nitrogenase activity in soybean? Physiol. Plant. 74, 137–146.

Wang, X., Pan, Q., Chen, F., Yan, X., Liao, H., 2011. Effects of co-inoculation with arbuscular mycorrhizal fungi and rhizobia on soybean growth as related to root architecture and availability of N and P. Mycorrhiza 21, 173–181.

Wang, D., Yang, S., Tang, F., Zhu, H., 2012. Symbiosis specificity in the legume– rhizobial mutualism. Cell. Microbiol. 14, 334–342.

Wilcox, J.R., 2004. World distribution and trade of soybean. In: Boerma, H.R., Specht, J.E. (Eds.), Soybeans: Improvement, Production and Uses. ASA, CSSA, ASSA, Madison, WI, pp. 1–13.

Woomer, P., Singleton, P.W., Bohlool, B.B., 1988. Ecological indicators of native rhizobia in tropical soils. Appl. Environ. Microbiol. 54, 1112–1116.

Xavier, I.J., Holloway, G., Leggett, M., 2004. Development of rhizobial inoculant formulations. Available from http://www.plantmanagementnetwork.org/pub/cm/review/2004/inoculant/.

Xavier, G.R., Correia, M.E.F., Aquino, A.M., Zilli, J.É., Rumjanek, N.G., 2010. The structural and functional biodiversity of soil: an interdisciplinary vision for conservation agriculture in Brazil. In: Dion, P. (Ed.), Soil Biology and Tropical Agriculture. Springer, Heidelberg, pp. 65–80.

Yang, S.S., Bellogin, R., Buendia-Claveria, A.M., Camacho, M., Chen, M., Cubo, T., Daza, A., Diaz, C.L., Espuny, M.R., Gutierrez, R., Harteveld, M., Li, X.H., Lyra, M.C.C.P., Madina-beitia, N., Medina, C., Miao, L., Ollero, F.J., Olsthoorn, M.M.A., Rodriguez, D.N., San-tamaria, C., Schlaman, H.P., Spaink, H.P., Temprano, F., Thomas-Oates, J.E., van Brussel, A.A.N., Vinardell, J.M., Xie, F., Yang, J., Zhang, H.Y., Zhen, J., Zhou, J., Ruiz-Sainz, J.E., 2001. Effect of pH and soybean cultivars on the quantitative analyses of soybean rhizobia populations. J. Biotechnol. 91, 243–255.

Young, I.M., Crawford, J.W., 2004. Interactions and self-organization in the soil-microbe complex. Science 304, 1634–1637.

Zabihi, H.R., Savaghebi, G.R., Khavazi, K., Ganjali, A., Miransari, M., 2010. *Pseudomonas* bacteria and phosphorus fertilization, affecting wheat (*Triticum aestivum* L.) yield and P uptake under greenhouse and field conditions. Acta Physiol. Plant. 33, 145–152.

Zhang, F., Smith, D.L., 1995. Preincubation of *Bradyrhizobium japonicum* with genistein accel-erates nodule development of soybean at suboptimal root zone temperatures. Plant Physiol. 108, 961–968.

Zhang, F., Smith, D.L., 2002. Interorganismal signaling in suboptimum environments: the legume-rhizobia symbiosis. Adv. Agron. 76, 125–161.

Zhang, F., Dashti, N., Hynes, R.K., Smith, D.L., 1996. Plant-growth promoting rhizobacteria and soybean (*Glycine max* [L.] Merr.) nodulation and nitrogen fixation at suboptimal root zone temperatures. Ann. Bot. 77, 453–459.

Zhou, X., Liang, Y., Chen, H., Shen, S., Jing, Y., 2006. Effects of rhizobia inoculation and nitro-gen fertilization on photosynthetic physiology of soybean. Photosynthetica 44, 530–535.

Plant growth promoting rhizobacteria to alleviate soybean growth under abiotic and biotic stresses

Sakshi Tewari*, Naveen Kumar Arora*[1], Mohammad Miransari**[1]
*Department of Environmental Microbiology, Babasaheb Bhimrao Ambedkar University, Lucknow, Uttar Pradesh, India; **Department of Book & Article, AbtinBerkeh Ltd. Company, Isfahan, Iran

Introduction

Agriculture is considered to be the sector most vulnerable to the global impact of climate change. Abrupt change in climatic conditions increases the incidence of abiotic and biotic stresses that have become major causes for stagnation of productivity in important crops (Grover et al., 2010). Followed by the USA, China, and Brazil, the Indian vegetable oil economy is the fourth largest in the world, accounting for an area of about 14% of the world's oilseed and about 7% of global production of oilseed. It also significantly contributes to the global production of oil meal (about 6%) and exports (about 4%) (Mehta, 2015).

Oils and oilseeds are two of the most sensitive and essential commodities. India is one of the largest producer as well as consumers of oilseeds and edible oils in the world, and this sector occupies an important position in the agricultural economy. India, as mentioned, contributes 7.4% of world oilseeds production, being the fourth largest edible oil economy in the world (http://www.sunvingroup.com/oil.php). India is fortunate to have a wide range of oilseed crops grown in its different agro - climatic zones.

Soybean (*Glycine max* (L.) Merr.) acquires the lion's share of the overall oilseed production in India. The total oilseed production globally for the years 2013–2014 was estimated at 504.43 metric tons (MT), and of this total, soybean accounts for 284.5 MT. The estimated production of world vegetable oil for the year 2013–2014 was 169.53 million metric tons (MMT), and of this total, soybean oil accounts for 44.60 MMT (http://www.sunvingroup.com/oil.php).

The soybean plant is well adapted to a number of environmental conditions but certain biotic and abiotic stresses cause significant reductions in growth and yield. These environmental stresses result in a large negative effect on production of the crop. Among abiotic stresses salinity, temperature, drought, and pH are the main constraints

[1] Dr Naveen Kumar Arora is the first corresponding author, Dr Mohammad Miransari is the second corresponding author.

Abiotic and Biotic Stresses in Soybean Production. http://dx.doi.org/10.1016/B978-0-12-801536-0.00006-2

that limit plant productivity and cause loss of arable land whereas among biotic stresses attack of diverse fungal pathogens is the main issue that needs to be resolved.

When subjected to such environmental stressors, a large number of plant processes are affected, decreasing plant growth and development. Under stress conditions, the plant reduces its vegetative growth to conserve and redistribute resources that are essential for plant survival under severe conditions (Skirycz and Inze, 2010). These stressors affect the plant growth and development in one way or another. For example, in water-limited conditions, reduction in initial plant growth decreases the size and number of cells in plant leaf (Aguirrezabal et al., 2006; Skirycz et al., 2010).

Under saline conditions plant growth is affected by a number of negative impacts including ion toxicity, osmotic effect, and specific ion effect (Ashraf, 1994; Conde et al., 2011; Habib and Ashraf, 2014; Munns, 1993; Sosa et al., 2005). In the context of climate change and mounting population pressure, increasing crop yields under such stresses is essential. A number of techniques have been used to understand the mechanisms and provide tools to enhance plant tolerance under environmental stresses. Some of these techniques are based on the usage of chemicals that may be toxic for the environment.

One of the alternatives and emerging technologies used to solve this problem is naturally occurring plant growth promoting rhizobacteria (PGPR). There are number of reports that show the effectiveness of PGPR for enhancing plant growth and development (Glick, 2012; Paul and Nair, 2008; Zahir et al., 2004). This environmentally friendly bacterial population is equally effective in promoting crop productivity and disease management under normal and stress conditions (Nadeem et al., 2010a; Saharan and Nehra, 2011; Tewari and Arora, 2013, 2014; Zahir et al., 2004). Several studies have revealed the mechanism of growth promotion by PGPR that may be either direct or indirect (Ahemad and Kibret, 2014; Glick, 2012; Nadeem et al., 2010b). The use of this environmentally friendly approach could be among the most efficient methods for minimizing the use of chemicals, which can adversely impact human health directly and indirectly.

The use of this effective microbial technology could also be useful for enhancing yields of soybean crops that otherwise looks difficult to achieve due to the limited availability of agricultural inputs, as well as stress environments. This chapter highlights the versatility of soybean crops around the globe along with the deleterious impact of various abiotic and biotic stressors that hampers growth and production of this very important oilseed plant. This chapter also highlights the significant contributions of diverse PGPR in mitigating the harsh effect of abiotic and biotic stressors to protect the plant.

Soybean: a versatile crop around the globe

Soybean is recognized as the "golden bean" or the "miracle bean." The western world provided suitable conditions that resulted in its enhanced growth during the early part of the century. Soybean, which is originally from China, contains over 40% protein and 20% oil. It was named as one of the five sacred grains as early as 2853 BC by the Emperor Sheng-Nung of China, indicating the cultivation of soybean in China for more than 4000 years (Hymowitz, 1970).

It is believed that with the development of sea and land trade, soybean moved out of China to nearby countries such as Burma (Myanmar), Japan, India, Indonesia, Malaysia, Nepal, the Philippines, Thailand, and Vietnam between the first century AD and AD 1100. At present, more than 50 countries grow soybean as the leading oilseed crop, being consumed worldwide (Wilcox, 2004). Soybean is now the largest source of protein and vegetable oil in the world and its large-scale cultivation is concentrated in a few countries such as Argentina, Brazil, Canada, China, India, Paraguay, and the USA, which together produce about 96% of the world's 189 million tons annual soybean production (http://shodhganga.inflibnet.ac.in).

With a huge potential for the production of food and large number of industrial products, soybean has had great effects on the agricultural economy of the USA. At present, the USA and Brazil are the leaders in soybean production and the USA is enjoying hegemony (Wrather et al., 2010). The USA now has over 50% of the total soybean area in the world, producing over 50% of the world's soybeans. For a long time, soybean has been used as a nutritious source of food just like bread, meat, milk, cheese, and oil to the people of different nations including China, Japan, Korea, Manchuria, the Philippines, and Indonesia. This indicates why soybean is described by expressions such as "cow of the field," or "gold from soil" by such nations.

Soybean is now recognized as one of the premier agricultural crops. In brief, soybean is a major source of vegetable oil, protein, and animal feed. Soybean, with over 40% protein and 20% oil, has now been recognized all over the world as a potential supplementary source of edible oil and nutritious food (Krishnamurthy and Shivashankar, 1975). The protein of soybean is called a complete protein, because it supplies sufficient amounts of the kinds of amino acids required by the body to build and repair tissues. Its food value in heart-related diseases and diabetes is well known (Bhise and Kaur, 2013).

It is significant that Chinese infants using soybean milk in place of cow's milk are practically free from rickets. Soybean is a rich source of edible oil containing no cholesterol and almost none of the saturated fats. The oil is superior to all other oils because it is an ideal food for heart patients as well as for avoiding coronary heart diseases (Anderson et al., 1999). The oil, with a large amount of lecithin and a fair amount of fat-soluble vitamins, is an important constituent of all organs of the human body and especially of the nervous tissue, heart, and liver. Besides its nutritional qualities, functional properties of soy protein have made it a suitable product for producing new foodstuff and improve the quality of existing standard food products (Krishnamurthy and Shivashankar, 1975).

A chain of soy-based industries have emerged in the USA. Oil is extracted for human consumption and industrial uses, and defatted soy meal is converted into various protein rich foods and feed products. In industry, soybean is used in the manufacture of edible lard, margarine, vegetable ghee milk, and pastries, as well as for manufacturing paints, varnishes, adhesives etc. (Craig, 1997).

The concentrate of the soybean protein, the isolate of protein, and textured protein are being used extensively for production by multifarious commercial food. As a versatile crop, soybean can be used for production in many agro-based industries. Soybean is used not only for the production of food for humans and animals, but also can be used for improving the properties of soil through its ability to fix atmospheric nitrogen. As a legume, it can be ideally used as a component of a sound agricultural system. Due to all these advantages of soybean and its adaptability and productivity across tropical, subtropical, and temperate environments, significant strides have indicated its potential for innovation. In fact, the significantly increased production of soybean across the world has been characterized as one of the striking developments of recent decades (Kim et al., 2005).

India is one of the largest producers as well as the consumers of oilseeds and edible oil in the world, significantly affecting its agricultural economy. India ranks fifth in the production of soybean in the world (Wrather et al., 2010). The progress of soybean cultivation in India is a case of success in diversification. In 1970, the introduction of new varieties for commercial purposes indicated the future of soybean as a pulse (*dal*) for the production of ample amounts of soybean for the growing demand and under protein deficient conditions. Other uses of soybean include soya milk, nuggets, etc. In fact, as it has turned out, soybean production in India has been greatly affected by the demand for edible oil in the domestic market and as oilcake in overseas markets (Chand, 2007).

In India, which is predominantly a vegetarian society, fats and proteins of vegetable origin are of great significance. Since soybean is used as an oilseed and a pulse crop, and India has been trying hard to fill the oil and protein gap, fresh attempts were started in the 1960s to produce soybean on a larger scale and as a commercial crop (Krishnamurthy and Shivashankar, 1975). It was indicated that production of soybean would increase farm income and that it could be utilized as a cheap and high quality protein suitable for the use by humans as well as a source of essential edible oil.

Many factors have resulted in motivating India to be one of the main soybean developing nations since the beginning of the 1970s. Even though India has contributed to the production of about 1% of the world's soybean at present, its developing rate of expansion during the last 15 years is rated as one of the most striking occurrences of agricultural development. In just 10 years, the area of soybean production in India grew from about 32 thousand hectares in 1971–1972 to about 814 thousand hectares in 1983–1984, resulting in a 25-fold increase (Bisaliah, 1985). The prospects of using soybean as a major crop in India are good. It appears that the importance of soybean is increasing while the availability of pulses, the nation's cheapest source of protein, is decreasing.

In spite of its versatility and tremendous utility, soybean crops are vulnerable to infestation by biotic (fungi) and abiotic (salinity) factors, and are highly susceptible to mechanical injury and damage occurring during post-harvest handling. Among abiotic factors salinity, drought, temperature, and pH are the principal factors that cause stagnation in crop productivity. Among biotic stressors, diverse phytopathogens cause reductions in yield and production of soybean crops (discussed in next sections in detail). Evidence indicates that biotic and abiotic stresses can cause 26.3% and 69.3% loss of soybeans production, respectively. A detailed list of phytopathogens of soybean and their biocontrol agents are mentioned in Table 6.1. (Wang et al., 2013).

Table 6.1 Deleterious effects of different phytopathogens on soybean crop and their biocontrol using PGPR

Phytopathogens	Diseases	Symptoms	Biocontrols	References
M. phaseolina	Soybean charcoal rot disease, root rot disease.	Spindle-shaped lesions with light gray centers and brown margins with scattered pycnidial bodies are the early symptoms. Spindle-shaped lesions on the stem to extended lesions that result in the wilting of the plant.	*R. japonicum, Pseudomonas, Bacillus,* etc.	Zambolium et al. (1983); Ramezani et al. (2007); Rakib et al. (2012)
F. solani	Root rot.	Lesions developing on the taproot range from a nondescript brown to a dark purple brown or black. These lesions may increase in size until they girdle the taproot.	*Pseudomonas, Bacillus, Rhizobium,* etc.	Rakib et al. (2012); Wrather et al. (2010)
Pythium spp.	Seed decay, pre-emergence damping-off, or early post emergence damping-off.	Seed may decay before germinating. Infected seed becomes soft and rotted. Areas of brown discoloration and soft, watery rot develop on infected hypocotyls and cotyledons. Infected seedlings wilt, collapse and shrivel up.	*Pseudomonas, Bacillus*	Laura et al. (2008); Wrather et al. (2010)
Phytophthora sojae	Seed decay, pre-emergence, or postemergence damping-off and seedling blight of soybeans.	Infected seed has a soft, mushy, and fuzzy appearance. Dark, reddish brown to black, water soaked lesions are evident on the hypocotyls. Cotyledons and hypocotyls turn brown to black and have a wet, rotted appearance. In the seedling blight phase of the disease young seedlings that appear to be established turn off-color to yellow, wilt, and die.	*Pseudomonas; Bacillus, Rhizobium*	Laura et al. (2008)
R. solani	Seed decay and pre-emergence damping-off of soybean seedlings.	The infected seed or seedling root is discolored and decayed. Localized red to reddish brown lesions develop on the hypocotyl near the soil line. The cankers eventually turn brown to tan or even a bleached white with reddish brown borders.	*Pseudomonas, Paenibacillus, Bacillus, Pseudomonas, Rhizobium,* etc.	Zambolium et al. (1983); Laura et al. (2008)

(Continued)

Table 6.1 Deleterious effects of different phytopathogens on soybean crop and their biocontrol using PGPRs (*cont.*)

Phytopathogens	Diseases	Symptoms	Biocontrols	References
Sclerotium rolfsii	Southern blight or white mold	A yellowing or wilting of scattered plants in the field. Light-brown to brown lesions may be present on the lower stem close to the soil line.	*Pseudomonas, Trichoderma, Gliocladium*	Laura et al. (2008); Angelique et al. (2012); Wrather et al. (2010)
S. sclerotiorum	Stem rot	Leaves may have a gray-green or off-color and wilted appearance. Cankers may be evident on stems at the nodes.	*Pseudomonas, Trichoderma, Gliocladium*	Laura et al. (2008); Angelique et al. (2012)
P. pachyrhizi	Soybean rust	Foliar lesions develop on the upper leaf surface; initial symptoms may be small, yellow flecks or specks in the leaf tissue. These lesions darken and may range from dark brown or reddish brown to tan or gray-green in color.	*Pseudomonas, Bacillus, Rhizobium*, etc.	Laura et al. (2008); Wrather et al. (2010)
Phomopsis longi-colla	Seedling blight and pod and stem blight	Infected seed may fail to germinate or it may germinate slowly.	*Pseudomonas, Rhizobium, Bradyrhizobium, Bacillus*, etc.	Laura et al. (2008); Wrather et al. (2010)
Septoria glycines	Septoria brown spot	Causes small, angular to somewhat circular, red to brown spots on the unifoliolate and lower trifoliolate leaves.	*Pseudomonas, Bacillus, Rhizobium*	Laura et al. (2008)
Cercospora sojina	Frogeye leaf spot	Lesions are small, circular to somewhat irregular spots that develop on the upper leaf surfaces. Initially the spots are dark and water soaked in appearance. As the lesions age, the centers become light brown.	*Pseudomonas, Bacillus, Rhizobium*	Laura et al. (2008); Wrather et al. (2010)
P. manshurica	Downy mildew	Pale green to light yellow spots or blotches on the upper surface of young leaves. These areas enlarge into pale to bright yellow lesions of indefinite size and shape.	*Pseudomonas, Bacillus, Rhizobium*	Laura et al. (2008); Koenning and Wrather (2010)
Colletotrichum truncatum	Anthracnose of soybean	Irregularly shaped brown areas might develop on stems, petioles, and pods. The anthracnose fungus also produces small black fruiting bodies in infected tissue.	*Pseudomonas, Bacillus, Rhizobium*	Laura et al. (2008); Wrather and Koenning (2009)

Impact of abiotic stressors on growth of soybean

Soybean production globally has increased from 26 MT to 223 MT due to increase in harvest area and yield (Bisaliah, 1985). From 1961 onwards, soybean yield increased at an average rate of 22.76 kg ha^{-1} year^{-1}, from 1129 kg ha^{-1} in 1961 to 2243 kg ha^{-1} in 2009 (Kobraei et al., 2011). The total soybean production of the world during 2006 was 220.4 MT (USDA, 2006). The top eight soybean-producing countries were USA (83.4 MT), Brazil (57.0 MT), Argentina (40.5 MT), China (16.4 MT), India (7.0 MT), Paraguay (3.6 MT), Canada (3.16 MT), and Bolivia (2.0 MT) (Wrather et al., 2010).

These countries produced about 97% of the world supply for 2006. High yields are essential to soybean producers' profit margins, even at the time of high soybean prices. Unfortunately, yields in major soybean producing countries have been suppressed by numerous biotic and abiotic factors in the past, and as a result income derived from this crop has been less than expected (Wrather et al., 2001). This financial reduction is important for the economies of rural and related industries.

Therefore, knowledge about plant responses to these stresses is very important for acquisition of cultivars with high yield potential. Drought, salinity, and temperature stresses decrease yield components and hence yield of soybean (Wang et al., 2013). Two important parameters determining the impact of stress on the growth and final yield of soybean are: (1) plant growth stage and (2) stress duration. Water deficit and high temperature adversely affect the stage of flowering to maturity, shortening seed filling period and reducing grain weight. Hence, research must focus on management of these abiotic stressors that cause extensive losses.

Salinity

Soil salinity derives from the soluble salts extracted from soil and water (Richards, 1954). The negative effects of soil salinity also decrease the caloric and nutritional value of agricultural production. Salinity is among the most important major abiotic stresses limiting plant growth and productivity (Khan and Panda, 2008), affecting an estimated land rate of 900 million ha, which is 6% of the total global land area (Flowers, 2004).

According to the Food and Agricultural Organization (FAO), if proper methods are not undertaken the rate of land loss, due to salinity, the loss will be at a level of 30% in the next year and up to 50% by the year 2050 (Munns, 2002). Salt stress reduces crop growth and yield of soybean crops in different ways. However, NaCl, which is the salt most commonly found in nature, has two major effects on soybean plants by reason of its osmotic potential and ionic toxicity.

Under normal conditions the osmotic potential in plant cells is higher than that in soil solution. Plant cells use this higher osmotic potential to take up water and essential minerals through root cells from the soil solution. However, under salt stress the higher osmotic potential in the soil solution related to the osmotic potential of plant cells (higher salt concentration) deceases the plant's ability to absorb water and other essential nutrients (Munns, 1993).

On the other hand, prevalent Na^+ and Cl^- ions can enter into cells and exert direct toxic effects on cell membranes, as well as on metabolic activities in the cytosol (Hasegawa et al., 2000). These major effects of salinity stress adversely influence plant cellular activities by decreasing: (1) the expansion of cells, (2) production of assimilate, (3) membrane functioning, and (4) cytosolic metabolism and also cause the production of reactive oxygen species (ROS). As a result, under highly salt stressed conditions, plants may stop functioning and growing. Soybean is generally a salt-sensitive crop and under saline conditions, symptoms of leaf chlorosis and reduced growth and biomass resulting from chloride-induced toxicity appear (Lauchli, 1984).

More visible symptoms of salinity stress are caused by chloride accumulation in the leaf (Yang and Blanchar, 1993), which include decreased photosynthesis and formation of superoxide radicals, which cause membrane damage (Marschner, 1995). In soybean, germination of seeds, growth of seedlings, nodulation, biomass accumulation, and yield are adversely affected by salinity (Essa, 2002). These effects constitute the interference resulting from osmotic potential on the uptake of water and nutrients (Brady and Weill, 2002). Singleton and Bohlool (1984) reported that the process of nodule initiation in soybean is extremely sensitive to NaCl. They stated that 26.6 mM NaCl in the rooting medium caused a 50% reduction in nodulation rate. Soybean grown in saline environments exhibits compressed photosynthetic activity, reduced nodule number and weight, and diminished growth and yields.

Drought

Agricultural drought refers to the shortage of precipitation that reduces soil water and ground water or reservoir levels, negatively affecting farming and crop production. The effects of drought on the growth and yield of soybean have been extensively investigated, significantly decreasing seed yield (24–50%) of soybean crop (Sadeghipour and Abbasi, 2012). Drought alters morphological changes in the vegetative growth of plant, and reduces seed quantity and quality, as well as pod number and dry weight (Ku et al., 2013). It retards the growth and metabolic activity of soybean genotypes.

Numerous experiments have examined the effects of drought on various vegetative stages of soybean production. A 2-year field experiment accompanied by Brown et al. (1985) on different soybean cultivars demonstrated the significant reduction of seed yield. An in-depth analysis of drought effects at various growth stages on seed yield of soybean was reported by Eck et al. (1987). A field experiment was conducted under drought stress conditions by Eck et al. (1987) that showed 45% and 88% reduction of seed yield, respectively, in two consecutive years.

Desclaux et al. (2000) directed a comprehensive analysis of yield components when soybean genotypes were subjected to drought stress at different developmental stages. In this experiment, the stress condition was established by temporarilly withholding watering for 4–5 days until the plant's available water decreased to 50% or 30% of that found in normal conditions. It has been clearly demonstrated that although water deficit will affect seed yield, depending on a variety of properties the growth stages that are most sensitive to drought stress differ among different varieties. Germination rate is an important measure for assessing seed quality.

A 2-year field study exhibited on soybean cultivars in the southern USA indicated that drought stress resulted in an 80% reduction in seed germination (Heatherly, 1993). This observation is supported by a greenhouse experiment reporting that the germination rate was decreased in medium seeds from plants under salinity stress during the seed filling period (Samarah et al., 2009). Hence, it could be suggested that drought stress is one of the major abiotic-stress factors that limits productivity of soybean crop.

Temperature

High temperature is a major environmental stress limiting plant growth, metabolism, and productivity worldwide. Plant growth and development rely on numerous biochemical reactions that are sensitive to temperature. Soybean is a temperate legume crop, with an ideal daytime temperature of 85°F, and at higher air temperature the negative impact of stress may appear on plant growth regardless of reproductive stage. The adverse effects of heat stress on soybeans growth may result in yield reductions, especially under limiting conditions of soil moisture (Gibson and Mullen, 1996).

During flowering, heat stress can result in pollen sterility and reduced seed set. Temperatures higher than 85°F may decrease the number of pods, while temperatures above 99°F severely limits the formation of pods. Heat stress at the growth stage of seed production has the greatest impact on soybean yield. During the period of seed fill, daytime temperatures of 91–96°F decrease the number of seeds per plant (Lindemannand and Ham, 1979). During seed fill, daytime temperatures greater than 85°F may reduce the weight of soybean yield.

The nodulation of soybean is influenced by temperature. Greatest nodule weight and nitrogen fixation (the conversion of gaseous nitrogen to plant-available nitrogen) has been found to be in the heat range of up to 85°F. At soil heat higher than 86°F, initiation and growth of nodules decreases. A soybean canopy may decrease soil temperatures. However, in fields without canopy closure, high soil temperatures may result in decreased nodulation and nitrogen fixation. Applying nitrogen fertilizer can increase soybean yield under stress; however, it may have some environmental and economical consequences. Applying ureal forms of nitrogen to warm and damp soils results in nitrogen loss as a gas (volatilization). Additionally, under dry weather conditions, the nitrogen may not move down into the soil and be accessible to the soybean roots. Chennupati et al. (2011) reported that soybean has a range of compounds including tocopherols and isoflavones, with putative health benefits. The concentrations of isoflavone and tocopherol, when soy plants are subjected to high temperatures, are adversely affected. High temperatures also affect seed formation, germination, and other growth stages such as pre and post emergence of seedlings. Hasanuzzaman et al. (2013) reported that temperature stress significantly reduces total chlorophyll content, and hence the abortion and abscission of flowers, young pods, and developing seeds result thereby decreasing the seed numbers in soybean.

pH

One external factor that can adversely affect nodulation in soybean crops is unfavorable soil pH. Soybeans are adversely affected by acidity when the pH falls beneath 4.7 (Mubarik and Sunatmo, 2014). Soil pH decreases as the acidity increases, pH being the rate of acidity as the negative logarithm of H^+ ion concentration. Soybeans generally grow in soil at pH 5.5–7 while the optimum pH is 6.8. The decrease in production is due to the chemical properties of acidic soil with high levels of aluminum, iron, and manganese. The estimated land surface of the world subjected to acidity (pH < 5.5) is at about 30%, which includes 40% of arable land (Von Uexkull and Mutert, 1995).

Low soil pH reduces the availability of nutrients, increases the toxicity of Al^{3+}, and is generally unfavorable to crop yields. It has been estimated that Al^{3+} toxicity is among the most limiting factors on plant productivity in 67% of the world's acidic soil regions (Eswaran et al., 1997). These poor growth conditions reduce root development, nodulation, and compromise nutrient transport (Horst, 1995; Marschner, 1991).

The high rate of N fertilizer produced worldwide, and used for soybean production (75% of 3.6 million tons), especially under acidic conditions, indicates the importance of nodulation loss and nitrogen fixation resulting from low soil pH (Lin et al., 2012). The reduced nodulation observed in acidic soil is one of the reasons for hindered plant growth and development (Bhagwat and Apte, 1989; Graham et al., 1994). Since acidic soils have both mineral deficiency and toxicity, acid-tolerant strains of soybean should be developed to overcome the constraint of reduced crop yield.

Rainfall

Nearly all soybean production is under rain-fed conditions. The rate of rainfall during the growing season, essential for soybean growth and yield production, is 60 mm (Peters and Johnson, 1960). Water is often a major limiting factor in soybean production, and hence must be managed efficiently. Except in critical periods such as germination, flowering, and pod formation, soybean is generally considered to be tolerant to shortages of moisture. Moisture stress or rain deficiency, during vegetative growth, may be very detrimental, causing the plants to be too small for high yields.

The water stress during seed development often reduces seed yield. However, excessive and continuous rainfall may delay planting or adversely affect seed germination, increasing pathogen activity and anaerobic conditions. Shanmugasundaram (1980) reported a 28% reduction in soybean plant establishment due to flooding that occurs immediately after sowing. Field observations and laboratory studies suggest that soybean related to the other legumes is relatively tolerant to noncontinuous waterlogging. In the absence of disease, it quickly regrows after waterlogging ceases (Stanley et al., 1980). If the intensity of precipitation is significantly higher than the infiltration rate, then surface runoff will be considerable. Soils differ in their capacity to store and carry over readily available moisture in the crop root zone. The amount of water available to the crop from soil moisture depends upon the crop rooting characteristics under prevailing conditions and the capacity of soil to store moisture and make water available within that depth.

Alleviation of abiotic stresses with PGPR

As discussed in the previous section, soybean plants face a variety of abiotic stressors which are responsible for low productivity and diminished growth. Researchers are focusing on utilizing stress-tolerating microbes that alleviate stresses so that plants may survive under such conditions (Arora et al., 2012). Kasotia et al. (2012) reported that *Pseudomonas* sp. strain VS1 had *in vitro* plant growth-promoting characteristics and promoted soybean seed emergence under salt stress (200 mM NaCl). Strain VS1 produced indole 3-acetic acid under salinity stress, which resulted in high numbers of lateral roots compared to control.

Ian and Francisco (2012) reported that inoculation of soybean with mycorrhizae, *Glomus etunicatum*, enhanced the root and shoot growth of plants grown under the concentrations of 50 and 100 mM NaCl. *Pseudomonas* strains *P. trivialis* 3Re27 and *P. extremorientalis* TSAU20, having excellent root-colonizing capability and plant growth-promoting activity, also tolerated 4% NaCl and displayed the ability to alleviate salt stress in soybean plants (Egamberdieva et al., 2010).

Egamberdieva et al. (2013) observed that coinoculation of salt-stressed soybean with *Bradyrhizobium japonicum* USDA110 and *Pseudomonas putida* TSAU1 improved root and shoot length, dry weight, and nodulation relative to those plants inoculated with *B. japonicum* alone. Coinoculation of *Pseudomonas* and *Rhizobium* enhanced nodulation, N fixation, plant biomass, and grain yield in different leguminous plant species including soybean under stress conditions (Dashti et al., 1998).

Miransari and Smith (2009) reported that under saline conditions soybean production is hampered due to the inhibitory effects on the signaling symbiosis between the two symbionts (soybean and *B. japonicum*). Hence, to alleviate the effect of salinity, genistein (a nod gene inducer) was added to enhance soybean nodulation and growth. The effects of genistein on the alleviation of salinity stress became greater with time under high salinity levels. The direct role of genistein (5 μM) in overcoming the stressful effects of salinity on the symbiosis between *B. japonicum* and soybean is a novel finding that may be useful to increase soybean yields in salty croplands.

Soybean plants inoculated with drought-tolerant strains of *B. japonicum* displayed enhanced growth attributes and nodulation in comparison with control seeds (Milošević et al., 2012). The mycorrhizal symbiosis established between mycorrhizal fungi and the host plants can improve drought tolerance in soybean, which might be closely related to the fungal mitogen-activated protein kinase (MAPK) response and the molecular dialogue between fungal and soybean MAPK cascades (Liu et al., 2005).

Some strains of root nodulating bacteria like *Rhizobium, Sinorhizobium*, and *Bradyrhizobium* are tolerant to acidic soil conditions (Eaglesham and Ayanaba, 1984). Strains of *B. japonicum* with slimy colonies, are generally more tolerant to acid–Al stress conditions and are able to provide fixed nitrogen to soybean plants and enhance growth and development even at pH 4.0–4.5 (Mubarik and Sunatmo, 2014).

Jawson et al. (1989) indicated that *B. japonicum* entrapped in a polyacrylamide gel could be used for the inoculation of soybeans. The other inoculant carriers used by the authors were wet blocks and wet crushes of the gel, air-dried gel blocks, and gel powder to offer protection to soybean seedlings against environmental stressors. Inoculation of soybean with gel-based carriers of *Bradyrhizobium* resulted in enhancement in plant length, nodule number, and nodule weight under stress conditions.

Impact of biotic stressors on growth of soybean

Large number of pathogens attacking soybean cause varying degrees of loss and this is one of the major hurdles in increasing the productivity of this important crop (Srivastava and Agarwal, 1989). Soybean is one of the crops most sensitive to pathogens; it may be affected simultaneously by 30 different pathogens including fungi, bacteria, nematodes, and viruses. The range of diseases affecting soybean crops is rather extensive, and those caused by fungi are considered of major importance; not only by their higher number, but also by the loss caused to the quality of the seeds and yield (Diniz et al., 2013) (Table 6.1). Fungal losses are reported to be 16% every year (Oerke, 2006).

Soybean is prone to attack by root-infecting fungi like *Alternaria, Cladosporium, Bipolaris, Fusarium oxysporum, Fusarium solani, Macrophomina phaseolina, Pythium* spp., *Rhizoctonia solani,* and *Sclerotinium* (Inam-Ul-Haq et al., 2012). Fungi from the genera *Peronospora manshurica, Aspergillus, Penicillium, Diaporthe/Phomopsis,* etc. are also found abundantly in seed samples of soybean, causing seed-borne diseases in plants (Medić-Pap et al., 2007). Among all the phytopathogens, *Sclerotinia sclerotiorum, Macrophomina phaeolina, Fusarium,* and *Phakopsora pachyrhizi* cause major losses in soy production (Figure 6.1).

The number of diseases that have been so far identified in soybean in India is about 35, among which 14 are important with respect to their magnitude of yield loss. The

    (A)   (B)   (C)   (D)

Figure 6.1 Diverse phytopathogens causing diseases in soybean crop. (A) Sclerotia of S. sclerotiorum inside of a soybean stem (photo: D.S. Mueller; Angelique et al., 2012). (B) Soybean stem with microsclerotia of M. phaseolina (photo: Jack Walters, University of Arkansas). (C) Lower part of stem showing dark discoloration and decay from Fusarium root rot (Laura et al., 2008; http://ipm.missouri.edu/ipm_pubs/ipm1002.pdf). (D) Soybean rust on a mature and a senescing leaf (photo:Steve Koenning; http://www.ces.ncsu.edu/depts/pp/notes/ Soybean/soy008/soy008.htm).

highest losses of yield in 2006 were caused by anthracnose, rust, sclerotium blight, charcoal rot, rhizoctonia aerial blight, and viruses. A few diseases were distributed throughout the country while others were restricted to specific regions. Myrothecium, leaf spot, frogeye leaf spot, and brown spot are among the other diseases that reduced yield, however not on a regular basis and only in small pockets (Wrather et al., 2010).

Phakopsora pachyrhizi

P. pachyrhizi originated in Asia–Australia and is a plant pathogenic basidiomycete fungi causing rust disease in soybean plants (Goellner et al., 2010). In 2002–2003, the disease spread through Brazil and caused losses estimated at US\$2 billion for 2003 (Yorinori et al., 2005). Soybean rust causes major economic losses in practically all the locations where it is present. The damage is caused by rapid deterioration of the leaf tissue, resulting in the drying and prematurely falling of plant leaf and hence precluding the formation of full grain. The earlier the leaves fall, the smaller will be the grain size and, consequently, the greater the loss in yield and quality.

The extensive areas of soybean rust distribution are the Eastern and Western Hemispheres. In the Eastern Hemisphere, it is distributed from Japan to Australia and westward to India and in the People's Republic of China, Hawaii, and central and southern Africa; in the Western Hemisphere it is found generally in Latin America and regions of the Caribbean. As of 2006, it has also been found in North America in the south and south-east reaching upward to Illinois. Significant losses are found to the Eastern Hemisphere: 10–90% in India, 40% in Japan, 10–50% in southern China, 23–90% in Taiwan, and 10–40% in Thailand (http://soydiseases.illinois.edu/index.cfm?category=diseases&disease=98).

Among the worst soybean diseases in India is soybean rust, which may result in reduction of yield depending on the time of infection, genotype planted, and climate (Wrather et al., 2010). Initially, and prior to 1977, soybean rust was found in the northeastern states of India, the hills of Uttar Pradesh and West Bengal, resulting in little loss of soybean yield.

This pathogen is able to survive on winter-sown and volunteer soybean in southern India and then during the rainy season spreads to soybean in more northern areas (Wrather et al., 2010). During the fall of 2001, soybean rust was first confirmed in Paraguay, with yield losses as high as 60% in some fields. Farmers have planted rust resistant genotypes but due to pathogen variability, the resistance was not durable. Yield losses were greater in late-planted compared to early planted soybeans because the late-planted crop reached the seed development stage when air temperatures and humidity were more suitable for rust.

Soybean rust was then detected in 2004 in the United States during late fall, but it did not result in yield reduction (Wrather and Koenning, 2006). However, during 2005, soybean rust did reduce yield in Georgia and South Carolina, and the yield reduction in Alabama was 7347t, Georgia 2721t, Louisiana 7891t, and North Carolina 1633 t in 2006 (Koenning, 2007). Relative to the other diseases, soybean rust caused more yield losses in Brazil in 2006. The whole country was infected with the disease and its severity was related to the frequency of rain. Yield losses due to soybean rust were greater than 60% in some fields being treated with fungicides, and using geno-types with different tolerance, and different cultural practices. The constant use of in-ocula at the time of planting soybean nearly every month of the year and the presence of other hosts renders rust management difficult (Wrather et al., 2001).

Sclerotinia sclerotiorum

Soybean *Sclerotinia* stem rot caused by *S. sclerotiorum* (Lib.) de Bary, was first found in the United States in 1946 and reported in 1951 (Angelique et al., 2012). The fun-gus has more than 370 plant species as its host and causes diseases on a wide range of crops including soybean. *Sclerotinia* stem rot can result in significant yield losses under temperate climates worldwide when conditions are favorable to the develop-ment of disease.

It is estimated that yield losses from 1996 to 2009 by *Sclerotinia* stem rot equalled up to 10 million bushels (270 million kg) in seven of the 14 years (Koenning and Wrather, 2010; Wrather and Koenning, 2009). In particular, *Sclerotinia* stem rot re-sulted in large yield losses in 1997, 2004, and 2009, with 35, 60, and 59 million bushels (953 million, 1.63, and 1.61 billion kg) lost, respectively. The market value of soybean in each of those years indicated that producers lost the equivalent of 227, 344, and 560 million dollars, respectively (USDA/NASS, 2011).

Decreased seed number and weight, and hence yield losses result from *Sclerotinia* stem rot (Danielson et al., 2004; Hoffman et al., 1998). During the growing season, potential yield loss can be estimated on the basis of the percentage of diseased plants or disease incidence. Soybean yield is reported to be reduced by 2–5 bushels per acre (approximately 133-333 kg/ha) with every 10% increment in Sclerotina stem rot dis-ease (Chun et al., 1987; Danielson et al., 2004; Hoffman et al., 1998; Yang et al., 1999).

In addition to causing yield loss, *Sclerotinia* stem rot can reduce seed quality. Angelique et al. (2012) reported that the pathogen spreads from field to field by air-borne spores (sclerotia), which can be mixed with seeds and possibly within the infested seed. Sclerotia produce apothecia, which release ascospores to infect soy-bean stems. The fungal mycelium is able to grow in and outside stems inhibiting the growth of the infected plants in late growth stages (Boland and Hall, 1987). Severe diseases can occur during an extended period of cool and moist weather (Grau and Hartman, 1999).

The disease was initially a minor production problem, causing localized epidem-ics occurring in some areas of Michigan, Minnesota, and Wisconsin where mature soybeans are grown. It was ranked twelfth before 1991 in disease losses, in the area of soybean production in US north-central region (Doupnik, 1993). The disease has emerged as a major production problem in the north-central region (USA), with major outbreaks in mature soybean crops.

S. sclerotiorum can also infect soybean seeds and act as an important source of in-ocula if they are planted in fields which have not been previously infected with *Sclero-tinia* stem rot (Hartman et al., 1998; Mueller et al., 1999; Yang et al., 1999). Seed germination as well as protein and oil concentrations are decreased in infected seeds (Danielson et al., 2004). Grau and Hartman (1999) also observed a linear relationship in yield and severity of stem rot on soybean due to *Sclerotinia* spp., which is one of the important diseases of soybean in India, where it has resulted in yield reduction of up to 40% in some areas. Fifteen days after its emergence, seedling death due to girdling near the hypocotyl may appear. Maximum yield loss due to this pathogen was reported in year 2006 (Wrather et al., 2010).

Macrophomina phaseolina

M. phaseolina (Tassi) Goidanich, is among the most important soil-borne pathogens, with a host range of more than 500 plant species in more than 100 plant families around the world (Smith and Wyllie, 1999). Charcoal rot, caused by the ascomycete fungus *M. phaseolina*, can significantly decrease the yield in oilseed crops including soybean. *M. phaseolina* has been a problem for soybean farmers in the United States for many years.

The pathogen has resulted in significant yield losses estimated at 8.54×10^5 ton year^{-1} from 1974 to 1994 in nonirrigated fields in the 16 southern states (Wrather and Koenning, 2009). Charcoal rot ranked fifth among diseases in the USA that decreased yield and was most prevalent in Arkansas, Illinois, Indiana, Kansas, Kentucky, Missouri, Mississippi, and Tennessee (Wrather et al., 2001). Hussain et al. (1989) reported that *M. phaseolina* was among the six most common contaminants of seed affecting soybean growth in the Sindh province of Pakistan.

Charcoal rot infection may alter soybean seed composition and influence nitrogen fixation activity. In the United States, estimated yield loss due to charcoal rot in soybean (2 million metric tons) can reach 1.56% (Wrather and Koenning, 2006). Charcoal rot is also referred to as dry weather wilt or summer wilt, because the plant symptoms appear under heat and drought stresses (Smith and Wyllie, 1999). Under irrigated soybeans these stresses can also appear resulting in losses from 6% to 33% in experimental plots (Mengistu et al., 2011); however the presence of *M. phaseolina* and the combination of stresses caused higher yield reduction in soybeans.

The pathogen assails the plant throughout the season, often causing progressive weakness of the host. The symptoms appear after flowering in the epidermal and subepidermal tissue of the taproot, and the lower part of the stem with a light gray or shiny staining color. Peeling away the epidermis from the stem can be the best diagnostic symptom showing numerous small, black bodies of microsclerotia, frequently produced in the xylem and pith of the stem, and these may block water flow (Mengistu et al., 2013). Charcoal rot has resulted in more yield loss in India since 2004 due to irregular rainfall and greater drought periods. The maximum damage to soybean has occurred in the major soybean-producing states of Maharashtra, Rajasthan, Karnataka, Chattisgarh and Andhra Pradesh. Yield losses due to *M. phaseolina* have been as high as 77% in some fields (Wrather et al., 2010).

Fusarium

Fusariosis of soybean was first found in 1917 in the USA and has ever since been found in many parts of the world (Sinclair and Backman, 1989). In India, the disease was first observed and described by Aćimović (1988) in 1964, and later by Tošiă et al. (1986) as well (Jasnić et al., 2005). Although *Fusarium* root rot is common and widespread, its effects on the rate of yield has not been easy to quantify as it occurs in combination with other pathogens or plant health problems, and the symptoms related to below-ground might be not be differentiable from those of other root rots (Arias, 2012).

According to Wrather et al. (2001) average soybean yield losses due to *Fusarium* diseases in the United States from 1994 to 2010 were estimated at approximately 36.2 million bushels year^{-1}. The yield losses related to *Fusarium* root rot on average were estimated at 6.63 million bushels year^{-1}. However, this estimate is not related to the role of *Fusarium* species in seedling diseases, resulting in average losses of 34.3 million bushels year^{-1} over the same time period.

At any stage of development, *Fusarium* wilt or blight can affect soybeans, as is apparent in the south-eastern USA. The cause of the disease is the soil-borne fungus *F. oxysporum* (Nelson et al., 1981). In the affected plants the stem tips wilt and the upper leaves become scorched. The leaves in the middle and lower part of the plant may turn yellow or pale yellow spots may appear. Under severe conditions the leaves will dry up and drop off prematurely.

When subjected to deficit moisture and hot conditions, the symptoms are more noticeable. In infected plants, brown vascular tissue also appears in the root and stems (Naito et al., 1993). The *Fusarium* wilt resulting in necrosis of root and lower parts of soybean is an important disease in many countries. It can cause great damage, as it may reduce the average yield of soybean by up to 59% (Sinclair and Backman, 1989).

Alleviation of biotic stresses with PGPR

Diverse phytopathogens, which cause diseases in soybean crop, can be controlled by utilizing PGPR (Table 6.1). Biological control can also be part of an integrated system to manage *Sclerotinia* stem rot. The fungus *Coniothyrium minitans* was identified as a pathogen of *S. sclerotiorum* in 1947 (Campbell, 1947) and as a biological control can be efficiently used for managing *Sclerotinia* stem rot. It is commercially available under different names such as Contans (PROPHYTA Biologischer Pflanzenschutz GmbH; Malchow/Poel, Germany) or KONI (Belchim Crop Protection; Londerzeel, Belgium). *C. minitans* should be incorporated and mixed into soil as thoroughly as possible to a depth of 2 in. (5 cm).

Application of *C. minitans* should occur at least 3 months before *Sclerotinia* stem rot is likely to develop (Crop Data Management Systems Inc., 2011). Hence, the fungus has adequate time to colonize and degrade sclerotia. Degraded sclerotia is not able to produce apothecia and because of not producing ascospores will not be able to infect soybean. Additional tillage is not essential because it can bring uncolonized sclerotia to the soil surface.

Other biological control microbial species such as the bacterium *Streptomyces lydicus* (Actinovate AG; SipcamAdvan, Inc., Durham, NC) and the fungus *Trichoderma harzianum* (Plant ShieldHC; BioWorks, Inc., Victor, NY) also have been promising in the management of *Sclerotinia* stem rot in limited field experiments and under growth chamber conditions (Zeng et al., 2012). Research work on the usage of bacterial biocontrol treatments for the management of *S. sclerotiorum* indicates that *Bacillus amyloliquefaciens*, *Pseudomonas chlororaphis*, and *Pantoea agglomerans* are able to suppress carpogenic germination and mycelial growth by performing a set of actions including the production of volatile and nonvolatile antimicrobial antibiotics (Fernando et al., 2004).

The mycoparasitic fungi and bacteria associated with parasitized sclerotia include *C. minitans*, *Gliocladium* spp., *Sporidesmium sclerotivorum*, *Trichoderma* spp., *Fusarium*, *Hormodendrum*, *Aspergillus*, *Mucor*, *Penicillium*, *Stachybotrys*, and *Verticillium* (Budge et al., 1995). Among them, *Gliocladium virens* and *C. minitans* have shown practical potential for biological control of *S. sclerotiorum* (Fernando et al., 2004).

Further research is needed to check the efficacy of biological control products and their potential to alleviate *Sclerotinia* stem rot of soybean, especially in fields where the native populations of biological control fungi are present. Rakib et al. (2012) evaluated the antagonistic activity of *R. japonicum* against *F. solani* and *M. phaseolina*, causative pathogens of root rot disease in soybean. Ebtehag et al. (2009) also reported the impact of certain PGPR like *Azospirillum brasilense*, *Azotobacter chroococcum*, *Bacillus megaterium*, *Bacillus cereus*, *B. japonicum* and *Pseudomonas* for their biocontrol efficacy against *M. phaseolina* for soybean plants. Tewari and Arora (2014) reported the role of fluorescent pseudomonads in suppressing *M. phaseolina* even under saline stress conditions in fields.

Vasebi et al. (2012) reported the antagonistic effect of *T. harzianum* T100 as a potential biocontrol agent against soybean charcoal rot caused by *M. phaseolina* under greenhouse experiments. Treatment of infested fields with T100 brought 53.6% enhancement in dry weight of soybean in comparison with the control. Additionally, controlling and reducing microsclerotial formation on soybean root and stem confirmed the antagonist capability of T100.

Chakraborty and Purkayastha (1984) reported that soybean was protected by some rhizobitoxine producing strains of *Bradyrhizobium* from infection by *M. phaseolina*. Vyas (1994) described treatment of soybean seeds with plant growth promoting and biocontrol strains of *Trichoderma viride*, *Bacillus subtilis*, *T. harzianum*, and *Pseudomonas fluorescens* being quite effective for controlling root-rot fungus *M. phaseolina*. Ari et al. (2012) isolated diverse strains of fluorescent *Pseudomonas* from the rhizosphere of soybean plants that displayed strong biocontrol potential against *T. viride*, *Fusarium*, and *T. harzianum* and these have been indicated to inhibit the mycelial growth of all root rot fungi (Ahmad et al., 2013).

The biocontrol agent *Clonostachys rosea* strain ACM941 was identified as an antagonist against a number of soil-borne and seed-borne pathogens, including *F. solani*, *F. oxysporum*, and *F. graminearum* and its potential was checked in field experiments taking soybean as a test crop (Xue et al., 2011). Biological control of soybean rust with *Trichothecium roseum* was also described by More and Kamble (2009).

Shamarao (2014) indicated the role of bioformulations produced by the use of *T. harzianum* in combination with cow urine to significantly reduce disease incidence of Asian soybean rust. These biointensive strategies worked together in suppression occurrence of disease in soybean. Ward et al. (2012) documented the disease-suppressive and mycophilic nature of *Simplicillium lanosoniveum* (as a biocontrol treatment) controlling the phytopathogen *P. pachyrhizi*, which causes soybean rust.

Conclusions

Soybean is an Asiatic leguminous plant, planted in large acreages of land worldwide for its oil and protein. However, due to climatic fluctuations and extreme weather patterns the yield and production of the crop is declining. Several abiotic and biotic factors affect yield of soybean very badly. Various researchers aim to develop stress-tolerant soybean crops to overcome the losses, but development of such genetic breeds of soybean is not an easy, recommendable, and economical approach for sustainable agriculture. Hence, the use of microbial inocula to alleviate the effect of diverse abiotic and biotic stresses is a better and a most promising approach to enhance production and yield in stress-affected regions. There is a need to discover and utilize diverse strains of PGPR displaying the triple activities of plant growth promotion, stress amelioration, and disease management working under natural environmental stresses for improved future security of soybean crops. Bioformulations or biointensive preparations developed from PGPR with biocontrol potential should be taken into consideration for the growth promotion and disease management of soybean crops. Enhancements in the yield of such a commercially important oil seed crop will serve as a boon in boosting the economies of developing countries. In the future, the potentialities of soybean will be exploited globally by agro-industries, due to its multifarious uses.

Acknowledgments

Thanks are due to DBT, DST, New Delhi and CST, Lucknow, India for support. The authors also thank Professor R.C. Sobti, Vice Chancellor, BBA University, for providing ceaseless support.

References

Aćimović, M., 1988. Prouzrokovaåi bolesti soje i njihovo suzbijanje, Nauåna knjiga. 1–260, Beograd.

Aguirrezabal, L., Sandrine, B.C., Amandine, R.J., Myriam, D., Sarah, J.C., Christine, G., 2006. Plasticity to soil water deficit in *Arabidopsis thaliana*: dissection of leaf development into underlying growth dynamic and cellular variables reveals invisible phenotypes. Plant Cell Environ. 29, 2216–2227.

Ahemad, M., Kibret, M., 2014. Mechanisms and applications of plant growth promoting rhizo-bacteria: current perspective. J. King Saud Uni. Sci. 26, 1–20.

Ahmad, M., Lee, S.S., Oh, S.E., Mohan, D., Moon, D.H., Lee, Y.H., Ok, Y.S., 2013. Modeling adsorption kinetics of trichloroethylene onto biochars derived from soybean stover and peanut shell wastes. Environ. Sci. Pollut. Res. Int. 20, 8364–8373.

Anderson, J.W., Smith, B.M., Washnok, C.S., 1999. Cardiovascular and renal benefits of dry bean and soybean intake. Am. J. Clin. Nutr. 70, 464–474.

Angelique, J.P., Carl, A.B., Martin, I.C., Dean, K.M., Daren, S.M., Kiersten, A.W., Paul, D.E., 2012. Biology yield loss and control of sclerotinia stem rot of soybean. JIPM 3, 2012.

Ari, M.M., Ayanwale, B.A., Adama, A.Z., Olatunji, E.A., 2012. Effects of different fermentation methods on the proximate composition, amino acids profile and some antinutritional factors in soybean (*Glycine max*). J. Ferment. Bioeng. 2, 6–13.

Arias, M.M.D., 2012. Fusarium species infecting soybean roots: frequency, aggressiveness, yield impact and interaction with the soybean cyst nematode. Graduate theses and dissertations. Paper 12314.

Arora, N.K., Tewari, S., Singh, S., Lal, N., 2012. PGPR for protection of plant health under saline conditions. In: Maheshwari D.K. (Eds.). Bacteria in Agrobiology Springer, The Netherlands, pp. 239–258.

Ashraf, M., 1994. Salt tolerance of pigeon pea (*Cajanus cajan* (L.) Millsp.) at three growth stages. Ann. Appl. Biol. 124, 153–164.

Bhagwat, A.A., Apte, S.K., 1989. Comparative analysis of proteins induced by heat shock, salinity, and osmotic stress in the nitrogen-fixing cyanobacterium *Anabaena* sp. strain L-31. J. Bacteriol. 171, 5187–5189.

Bhise, S., Kaur, A., 2013. Development of functional chapatti from texturized deoiled cake of sunflower, soybean and flaxseed. Int. J. Eng. Res. Appl. 3, 1581–1587.

Bisaliah, S., 1985. The Study on Soybean Development in India: a methodological frame. Paper presented at the workshop on "towards recommendations for research, policy and extension: methodological issues in the socio-economic analysis of food legumes and coarse grains," Organized by the ESCAP CGPRT Centre, November 18–23, 1985, Bandung, Indonesia.

Boland, G.J., Hall, R., 1987. Evaluating soybean cultivars for resistance to *Sclerotinia sclerotiorum* under field conditions. Plant Dis. 71, 934–936.

Brady, N.C., Weill, R.R., 2002. The nature and property of soils, thirteenth ed. Prentice Hall, Upper Saddle River. p. 960.

Brown, E., Brown, D., Caviness, C., 1985. Response of selected soybean cultivars to soil moisture deficit. Agro. J. 77, 274–278.

Budge, S.P., McQuilken, M.P., Fenlon, J.S., Whipps, J.M., 1995. Use of *Coniothyrium minitans* and *Gliocladium virens* for biocontrol of *Sclerotinia sclerotiorum* in glasshouse lettuce. Biol. Control 5, 513–522.

Campbell, W.A., 1947. A new species of *Coniothyrium* parasitic on sclerotia. Mycologia 39, 190–195.

Chakraborty, U., Purkayastha, 1984. Role of rhizobiotoxine in protection soybean roots from *Macrophomina phaseolina* infection. Can. J. Microbiol. 30, 285–289.

Chand, R., 2007. Agro-industries characterization and appraisal: soybeans in India. National Centre for Agricultural Economics and Policy Research (NCAP), New Delhi, India. 2007. Working Document of the Food and Agriculture Association's (FAO) Agricultural Management, Marketing and Finance Service (AGSF). http://www.fao.org/Ag/AGS/publications/docs/AGSF_WorkingDocuments/agsfwd20.pdf.

Chennupati, P., Seguin, P., Liu, W., 2011. Effects of high temperature stress at different development stages on soybean isoflavone and tocopherol concentrations. J. Agric. Food Chem. 59, 13081–13088.

Chun, D., Kao, L.B., Lockwood, J.L., Isleib, T.G., 1987. Laboratory and field assessment of resistance in soybean to stem rot caused by *Sclerotinia sclerotiorum*. Plant Dis. 71, 811–815.

Conde, A., Chaves, M.M., Geros, H., 2011. Membrane transport, sensing and signaling in plant adaptation to environmental stress. Plant Cell Physiol. 52, 1583–1602.

Craig, W.J., 1997. Phytochemicals: guardians of our health. J. Am. Diet Assoc. 97, S199–S204.

Crop Data Management Systems, Inc., 2011. ContansWG specimen label. Prophyta Biologischer P flanzenschutz GmbH; Malchow/Poel, Germany. EPA registration number: 72444-1 (http://www.cdms.net/).

Danielson, G.A., Nelson, B.D., Helms, T.C., 2004. Effect of sclerotinia stem rot on yield of soybean inoculated at different growth stages. Plant Dis. 88, 297–300.

Dashti, N., Zhang, F., Hynes, R., Smith, D.L., 1998. Plant growth promoting rhizobacteria accelerate nodulation and increase nitrogen fixation activity by field grown soybean [*Glycine max* (L.) Merr.] under short growing seasons. Plant Soil 200, 205–213.

Desclaux, D., Huynh, T.T., Roumet, P., 2000. Identification of soybean plant characteristics that indicate the timing of drought stress. Crop Sci. 40, 716–722.

Diniz, F.O., Reis, M.S., Araújo, E.F., dos Santos Dias, L.A., Sediyama, T., Sediyama-Bhering, C.A.Z., 2013. Incidence of pathogens and field emergence of soybean seeds subjected to harvest delay. J. Seed Sci. 35, 478–484.

Doupnik, B., 1993. Soybean production and disease loss estimates for North Central United States from 1989 to 1991. The Amer. Phytopath. Soc. 77, 1170–1172.

Eaglesham, A.R.J., Ayanaba, A., 1984. Tropical stress ecology of rhizobia, root-nodulation and legume fixation. In: Subba Rao, N.S. (Ed.), Current Developments in Biological Nitrogen Fixation. Edward Arnold, London, pp. 1–35.

Ebtehag, El-B., Nemat, M.A., Azza, S.T., Hoda, A.H., 2009. Antagonistic activity of selected strains of rhizobacteria against *Macrophomina phaseolina* of soybean plants. Am. Eurasian J. Agric. Environ. Sci. 5, 337–347.

Eck, H.V., Mathers, A.C., Musick, J.T., 1987. Plant water stress at various growth stages and growth and yield of soybeans. Field Crops Res. 17, 1–16.

Egamberdieva, D., Berg, G., Lindstrom, K., Rasanen, L.A., 2010. Coinoculation of *Pseudomonas* spp. with *Rhizobium* improves growth and symbiotic performance of fodder galega (*Galega orientalis* Lam.). Eur. J. Soil Biol. 46, 269–272.

Egamberdieva, D., Berg, G., Lindström, K., Räsänen, L.A., 2013. Alleviation of salt stress of symbiotic *Galega officinalis* L. (Goat's Rue) by coinoculation of *Rhizobium* with root colonizing *Pseudomonas*. Plant Soil 369, 453–465.

Essa, T.A., 2002. Effect of salinity stress on growth and nutrient composition of three soybean (Glycine max (L.) Merrill) cultivars. J. Agro. Crop Sci. 188, 86–93.

Eswaran, H., Reich, P., Beinroth, F., 1997. Global distribution of soils with acidity. In: Moniz, A.C. et al. (Eds.), Plant-Soil Interactions at Low pH. Brazil. Soil Sci. Soc., pp. 159–164.

Fernando, W.G.D., Nakkeeran, S., Zhang, Y., 2004. Ecofriendly methods in combating *Sclerotinia sclerotiorum* (Lib.) de Bary. Dev. Toxicol. Environ. Sci. 1, 329–347.

Flowers, T., 2004. Improving crop salt tolerance. J. Exp. Bot. 55, 307–319.

Gibson, L.R., Mullen, R.E., 1996. Soybean seed quality reductions by high day and night temperature. Crop Sci. 36, 1615–1619.

Glick, B.R., 2012. Plant growth-promoting bacteria: mechanisms and applications. Scientifica 2012:963401, 1–15.

Goellner, K., Loehrer, M., Langenbach, C., Conrath, U., Koch, E., Schaffrath, U., 2010. *Phakopsora pachyrhizi*, the causal agent of Asian soybean rust. Mol. Plant Pathol. 11, 169–177.

Graham, P., Draeger, K., Ferrey, M., Conroy, M., Hammer, B., Martinez, E., Aarons, S., Quinto, C., 1994. Acid pH tolerance in strains of *Rhizobium* and *Bradyrhizobium*, and initial studies on the basis for acid tolerance of *Rhizobium tropici* UMR 1899. Can. J. Microbiol. 40, 198–207.

Grau, C.R., Hartman, G.L., 1999. Sclerotinia stem rot. Hartman, G.L., Sinclair, J.B., Rupe, J.C. (Eds.), Compendium of Soybean Diseases, 4, APS Press, St. Paul, MN, pp. 46–48.

Grover, M., Ali, S.Z., Sandhya, V., Rasul, A., Venkateswarlu, B., 2010. Role of microorganisms in adaptation of agriculture crops to abiotic stress. World J. Microbiol. Biotechnol. 27, 1231–1240.

Habib, N., Ashraf, M., 2014. Effect of exogenously applied nitric oxide on water relations and ionic composition of rice (Oryza sativa L.) plants under salt stress. Pak. J. Bot. 46, 111–116.

Hartman, G.L., Kull, L., Yuang, Y.H., 1998. Occurrence of Sclerotinia sclerotiorumin soybean fields in east-central Illinois and enumeration of inocula in soybean seed lots. Plant Dis. 82, 560–564.

Hasegawa, P.M., Bressan, R.A., Zhu, J.K., Bohnert, H.J., 2000. Plant cellular and molecular responses to high salinity. Annu. Rev. Plant Mol. Plant Physiol. 51, 463–499.

Hasanuzzaman, M., Nahar, K., Alam, M., Roychowdhury, R., Fujita, M., 2013. Physiological, biochemical, and molecular mechanisms of heat stress tolerance in plants. Int. J. Mol. Sci. 14, 9643–9684.

Heatherly, L.G., 1993. Drought stress and irrigation effects on germination of harvested soybean seed. Crop Sci. 33, 777–781.

Hoffman, D.D., Hartman, G.L., Mueller, D.S., Leitz, R.A., Nickell, C.D., Pedersen, W.L., 1998. Yield and seed quality of soybean cultivars infected with Sclerotinia sclerotiorum. Plant Dis. 82, 826–829.

Horst, W.J., 1995. The role of the apoplast in aluminum toxicity and resistance of higher plants: a review. Z. Pflanzenernahr. Bodenk. 158, 419–428.

Hussain, S., Hassan, S., Khan, B.A., 1989. Seedborne mycoflora of soybean in the North West Frontier Province of Pakistan. Sarhad J. Agric. 5, 421–424.

Hymowitz, T., 1970. On the domestication of the soybean. Econ. Bot. 24, 408–421.

Ian, C.D., Francisco, P.A., 2012. Microbial amelioration of crop salinity stress. J. Exp. Bot. 63, 3415–3428.

Inam-Ul-Haq, M., Mehmood, S., Rehman, H.M., Ali, Z., Tahir, M.I., 2012. Incidence of root rot diseases of soybean in Multan Pakistan and its management by the use of plant growth promoting rhizobacteria. Pak. J. Bot. 44, 2077–2080.

Jasnić, S.M., Vidić, M.B., Bagi, F.F., Đorđević, V.B., 2005. Pathogenicity of Fusarium species in soybean. Zbornik Matice srpske za prirodne nauke, 109, 113–121.

Jawson, M.G., Franzluebbers, A.J., Berg, R.K., 1989. Bradyrhizobium japonicum survival in and soybean inoculation with fluid gels. Appl. Env. Microbiol. 55, 617–622.

Kasotia, A., Jain, S., Vaishnav, A., Kumari, S., Gaur, R.K., Choudhary, D.K., 2012. Soybean growth-promotion by Pseudomonas sp. strain VS1 under salt stress. Pak. J. Biol. Sci. 15, 698–701.

Khan, M.H., Panda, S.K., 2008. Alterations in root lipid peroxidation and antioxidative responses in two rice cultivars under NaCl-salinity stress. Acta Physiol. Plant 30, 89–91.

Kim, S.W., Mateo, R.D., Ji., F., 2005. Fermented soybean meal as a protein source in nursery diets replacing dried skim milk. J. Anim. Sci. 83, 116.

Kobraei, S., Etminan, A., Mohammadi, R., Kobraee, S., 2011. Effects of drought stress on yield and yield components of soybean. Ann. Biol. Res. 2, 504–509.

Koenning, S.R., 2007. Southern United States soybean disease loss estimates for 2006. Proc. South. Soybean Dis. Work. 34, 1–6.

Koenning, S., Wrather, J., 2010. Suppression of soybean yield potential in the continental United States by plant disease from 2006 to 2009. Plant Health Prog., doi:10. 1094/PHP-2010-1122-01-RS.

Krishnamurthy, K., Shivashankar, K., 1975. Soybean Production in Karnataka. University of Agricultural Sciences, UAS Technical Series, Bangalore, India, Publication No. 12.

Ku, Y.S., Yeung, W.K.A., Yung, Y.L., Li, M.W., Wen, C.Q., Liu, X., Lam, H.M., 2013. Drought stress and tolerance in soybean. In: Board, J.E., (Ed.), A comprehensive survey of international soybean research - genetics, physiology, agronomy and nitrogen relationships. pp 209–238.

Laura, E.S., Allen, W., Simeon, W., 2008. Integrated Pest Management Soybean Diseases. University of Missouri Extension, Extension Publications 2800, Maguire Blvd., Columbia, MO65211; IPM 1002.

Lauchli, A., 1984. Salt exclusion: an adaptation of legume crops and pastures under saline conditions. In: Staples, R.C., Toeniessen, G.H., (Eds.), Salinity tolerance in plants: strategies for crop improvement. New York: John Wiley and Sons. pp. 171–187.

Lin, Y.H., Zhang, Z., Docherty, K.S., Zhang, H., Budisulistiorini, S.H., Rubitschun, C.L., Shaw, S.L., Knipping, E.M., Edgerton, E.S., Kleindienst, T.E., Gold, A., Surratt, J.D., 2012. Isoprene epoxydiols as precursors to secondary organic aerosol formation: Acid-catalyzed reactive uptake studies with authentic compounds, Environ. Sci. Technol. 46, 250–258.

Lindemann, W.C., Ham, G.E., 1979. Soybean plant growth, nodulation, and nitrogen fixation as affected by root temperature. Soil Sci. Soc. Am. J. 43, 1134–1137.

Liu, Y., Zhu, Y.G., Chen, B.D., Christie, P., Li, X.L., 2005. Influence of the arbuscular mycorrhizal fungus *Glomus mosseae* on uptake of arsenate by the As hyperaccumulator fern *Pteris vittata* L. Mycorrhiza 15, 187–192.

Marschner, H., 1991. Mechanisms of adaptation of plants to acid soils. Plant Soil 134, 1–24.

Marschner, H., 1995. Mineral nutrition of higher plants. Academic Press London, UK.

Medić-Pap, S., Milošević, M., Jasnić, S., 2007. Soybean Seed-Borne Fungi in the Vojvodina Province. The Polish Phytopathological Society, Poznań, ISSN 1230.

Mehta, B.V., 2015. India as a strategic partner in vegetable oil market. Presentation in 26th Annual palm and lauric oils conference and exhibition POC, Kuala Lumpur, Malaysia. pp. 1–42.

Mengistu, A., Smith, J.R., Ray, J.D., 2011. Seasonal progress of charcoal rot and its impact on soybean productivity. Plant Dis. 95, 1159–1166.

Mengistu, A., Reddy, K.N., Bellaloui, N., Walker, E.R., Kelly, H.M., 2013. Effect of glyphosate on *Macrophomina phaseolina in vitro* and its effect on disease severity of soybean in the field. Crop Protect. 54, 23–28.

Milošević, N.A., Marinković, J.B., Branislava, B.T., 2012. Mitigating abiotic stress in crop plants by microorganisms. 123, 17–26.

Miransari, M., Smith, D.L., 2009. Alleviating salt stress on soybean (*Glycine max* (L.) Merr.) – *Bradyrhizobium japonicum* symbiosis, using signal molecule genistein. Eur. J. Soil Biol. 45, 146–152.

More, S.B., Kamble, S.S., 2009. Biological control of soybean rust with *Trichothecium roseum*. Bioinfolet 6, 342–343.

Mubarik, N.R., Sunatmo, T.I., 2014. Symbiotic of nitrogen fixation between acid aluminium tolerant *Brayrhizobium japonicum* and soybean. In: Ohyama, T. (Ed.), Advances in Biology and Ecology of Nitrogen Fixation. Rijeka (HR) In Tech, pp. 259–274.

Mueller, D.S., Hartman, G.L., Pedersen, W.L., 1999. Development of sclerotia and apothecia of *Sclerotinia sclerotiorum* from infected soybean seed and its control by fungicide seed treatment. Plant Dis. 83, 1113–1115.

Munns, R., 1993. Physiological processes limiting plant growth in saline soils: some dogmas and hypotheses. Plant Cell Environ. 16, 15–24.

Munns, R., 2002. Comparative physiology of salt and water stress. Plant Cell Environ. 25, 239–250.

Nadeem, S.M., Zahir, Z.A., Naveed, M., Asghar, H.N., Arshad, M., 2010a. Rhizobacteria capable of producing ACC–deaminase may mitigate salt stress in wheat. Soil Sci. Soc. Am. J. 74, 533–542.

Nadeem, S.M., Zahir, Z.A., Naveed, M., Ashraf, M., 2010b. Microbial ACC deaminase: prospects and applications for inducing salt tolerance in plants. Crit. Rev. Plant Sci. 29, 360–393.

Naito, S., Mohamad, D., Nasution, A., Purwanti, H., 1993. Soil-borne diseases and ecology of pathogens on soybean roots in Indonesia. JARQ 26, 247–253.

Nelson, P.E., Toussoun, T.A., Cook, R.J., 1981. *Fusarium*: Diseases, Biology and Taxonomy. Pennsylvania State University Press, University Park, PA, USA.

Oerke, E.C., 2006. Centenary Review: crop losses to pests. J. Agric. Sci. 144, 31–43.

Paul, D., Nair, S., 2008. Stress adaptations in a plant growth promoting rhizobacterium (PGPR) with increasing salinity in the coastal agricultural soils. J. Basic Microbiol. 48, 378–384.

Peters, D.B., Johnson, L.C., 1960. Soil moisture use by soybeans. Agron. J. 52, 687–689.

Rakib, Al-Ani, A., Mustafa, A.A., Majda, H.M., Hadi, M.A., 2012. *Rhizobium japonicum* as a biocontrol agent of soybean root rot disease caused by *Fusarium solani* and *Macrophomina phaseolina*. Plant Protect. Sci. 4, 149–155.

Ramezani, J., Schmitz, M.D., Davydov, V.I., Bowring, S.A., Snyder, W.S., Northrup, C.J., 2007. High-precision U-Pb zircon age constraints on the Carboniferous-Permian boundary in the southern Urals stratotype. Earth Planet. Sci. Lett. 256, 244–257.

Richards, L.A., 1954. Diagnosis and improvement of saline and alkali soils. USDA Agriculture Handbook 60, Washington, D.C.

Sadeghipour, O., Abbasi, S., 2012. Soybean response to drought and seed inoculation. World Appl. Sci. J. 17, 55–60.

Saharan, B.S., Nehra, V., 2011. Plant growth promoting rhizobacteria: a critical review. Life Sci. Med. Res. 2011, 1–30.

Samarah, N.H., Mullen, R.E., Anderson, I., 2009. Soluble sugar contents, germination, and vigor of soybean seeds in response to drought stress. J. New Seeds. 10, 63–73.

Shamarao, J., 2014. Bioformulations and indigenous plant protection measures in enhancing the vitalities of bio-control agents for induced systemic resistance suppressing Asian soybean rust in India. International Conference on Biological, Civil and Environmental Engineering (BCEE-2014).

Shanmugasundaram, S., 1980. The role of AVRDC I the improvement of soybean and mungbean for the developing tropical countries. In: Grain Legume Production in Asia. Asian Productivity Organization, Tokyo, Japan. pp. 137–166.

Sinclair, J.B., Backman, P.A., 1989. Compendium of Soybean Diseases, 3rd edn. The American Phytopathological Society, St. Paul, MN, pp. 106.

Singleton, P.W., Bohlool, B.B., 1984. Effect of salinity on nodule formation by soybean. Plant Physiol. 74, 72–76.

Skirycz, A., Inze, D., 2010. More from less: plant growth under limited water. Curr. Opin. Biotechnol. 21, 197–203.

Skirycz, A., De Bodt, S., Obata, T., De Clercq, I., Claeys, H., De Rycke, R., Andriankaja, M., Van Aken, O., Van Breusegem, F., Fernie, A.R., Inzé, D., 2010. Developmental stage specificity and the role of mitochondrial metabolism in the response of *Arabidopsis* leaves to prolonged mild osmotic stress. Plant Physiol. 152, 226–244.

Smith, G.S., Wyllie, T.D., 1999. Charcoal rot. In: Hartman, G.L., Sinclair, J.B., Rupe, J.C. (Eds.), Compendium of Soybean Disease, vol. 4. American Phytopathological Society, St. Paul, MN, pp. 29–31.

Sosa, L., Llanes, A., Reinoso, H., Reginato, M., Luna, V., 2005. Osmotic and specific ion effect on the germination of *Prospis strombulifera*. Ann. Bot. 96, 261–267.

Srivastava, S.K., Agarwal, S.C., 1989. Roga Niyantran (In Hindi). In: Singh, O.P., Srivastava, S.K. (Eds.), Soybean. Agro Botanical Publishers (India), Bikaner, India, 133–167.

Stanley, C.D., Kaspar, T.C., Taylor, H.M., 1980. Soybean top and root response to temporary water tables imposed at three different stages of growth. Agron. J. 72, 341–346.

Tewari, S., Arora, N.K., 2013. Transactions amongst microorganisms and plant in the composite rhizosphere habitat. In: Arora, N.K. (Ed.), Plant Microbe Symbiosis: Fundamentals and Advances. Springer, India, pp. 411–449.

Tewari, S., Arora, N.K., 2014. Talc based exopolysaccharides formulation enhancing growth and production of *Helianthus annus* under saline conditions. Cell. Mol. Biol. 60, 73–81.

Teawri, S., Arora, N.K., 2015. Multifunctional exopolysachharides from *Pseudomonas aeruginosa* PF 23 involved in plant growth stimulation, biocontrol and stress amelioration in sunflower under saline conditions. Curr Microbiol. 69(4), 484–494.

Tošiã, M., Paviã, M., Stojanoviã, T., Antonijeviã, D., 1986. Bolesti soje na području S.R. Srbije u 1985 godini. R.O. Industrija biljnih ulja i proteina, Zbornik radova Republiãkog savetovanja o unapreÐenju soje, suncokreta Iuljane repice, str. Beograd pp. 1–21.

USDA., 2006. World Agriculture Production: Crop Production Tables. Online Production, Supply and Distribution. USDA-Foreign Agricultural Service, Washington, DC.

United States Department of Agriculture/National Agricultural Statistics Service, USDA/NASS. 2011. United States soybean prices. U.S. Department of Agriculture/NASS, Washington, DC.

Vasebi, Y., Alizadeh, A., Safaee, N., 2012. Biological control of soybean charcoal rot caused by *Macrophomina phaseolina* using *Trichoderma harzianum*. J. Agric. Sci. 22, 41–54.

Von Uexkull, H.R., Mutert, E., 1995. Global extent, development and economic impact of acid soils. Plant Soil 171, 1–15.

Vyas, S.C., 1994. Integrated biological and chemical control of dry root rot on soybean. Ind. J. Mycol. Plant Pathol. 24, 132–134.

Wang, Q., Ge, X., Tian, X., Zhang, Y., Zhang, J., Zhang, P., 2013. Soy isoflavones: the multipurpose phytochemical (review). Biomed. Rep. 1, 697–701.

Ward, N.A., Schneider, R.W., 2012. Effects of Simplicillium lanosoniveum on *Phakopsora pachyrhizi*, the soybean rust pathogen, and its use as a biological control agent. Phytopathol. 102, 749–760.

Wilcox, J.R., 2004. World distribution and trade of soybean. In: Boerma, H.R., Specht, J.E. (Eds.), Soybeans Chemistry, Production, Processing, and Utilization, vol. 3. Agron. Monogr. 16. American Society for Agronomy, Madison, WI, pp. 1–14.

Wrather, J.A., Koenning, S.R., 2006. Estimates of disease effects on soybean yields in the United States, 2003 to 2005. J. Nematol. 38, 173–180.

Wrather, J.A., Koenning, S.R., 2009. Effects of diseases on soybean yields in the United States, 1996 to 2007. Plant Health Prog., doi: 10.1094/PHP-2009-0401-01-RS.

Wrather, J.A., Anderson, T.R., Arsyad, D.M., Tan, Y., Ploper, L.D., Porta-Puglia, A., Ram, H.H., Yorinori, J.T., 2001. Soybean disease loss estimates for the top 10 soybean producing countries in 1998. Can. J. Plant Pathol. 23, 115–121.

Wrather, A., Shannon, G., Balardin, R., Carregal, L., Escobar, R., Gupta, G.K., Ma, Z., Morel, W., Ploper, D., Tenuta, A., 2010. Effect of diseases on soybean yield in the top eight producing countries in 2006. Plant Health Prog. doi:10.1094/PHP-2010-0125-01-RS.

Xue, S., Yao, X., Luo, W., Jha, D., Tester, M., Horie, T., Schroeder, J.I., 2011. AtHKT1; 1 mediates nernstian sodium channel transport properties in *Arabidopsis* root stellar cells. PLoS One 6, e24725.

Yang, J., Blanchar, R.W., 1993. Differentiating chloride susceptibility in soybean cultivars. Agron. J. 85, 880–885.

Yang, X.B., Lundeen, P., Uphoff, M.D., 1999. Soybean varietal response and yield loss caused by *Sclerotinia sclerotiorum*. Plant Dis. 83, 456–461.

Yorinori, J.T., Paiva, W.M., Frederick, R.D., Costamilan, L.M., Bertagnolli, P.F., Hartman, G.E., Godoy, C.V., Nunes, J., 2005. Epidemics of soybean rust (*Phakopsora pachyrhizi*) in Brazil and Paraguay from 2001 to 2003. Plant Dis. 89, 675–677.

Zahir, Z.A., Arshad, M., Frankenberger, W.T.J., 2004. Plant growth promoting rhizobacteria: application and perspectives in agriculture. Adv. Agron. 81, 97–168.

Zambolium, L., Schenck, N.C., Mitchell, D.J., 1983. Inoculum density, pathogenicity and interactions of soybean root infecting fungi. Phytopathology 73, 1398–1402.

Zeng, W., Wang, D., Kirk, W., Hao, J., 2012. Use of Coniothyrium minitans and other microorganisms for reducing *Sclerotinia sclerotiorum*. Biol. Control. 60, 225–232.

Soybean production and salinity stress

7

Mohammad Miransari
Department of Book & Article, AbtinBerkeh Ltd. Company, Isfahan, Iran

Introduction

Plants are subjected to different kinds of stresses, including salinity, adversely affecting plant growth and crop production. The deleterious effects of salinity on the growth and yield of crop plants consist of decreasing water efficiency inside the plant cells and the unfavorable effects of sodium (Na^+) and chloride (Cl^-) negatively affecting plant physiology and morphology. Although these are among the most important effects of salinity adversely affecting plant growth and crop production, under saline conditions the availability and uptake of nitrogen (N) also decreases. Hence, planting crop plants which are tolerant of salinity and able to fix atmospheric N may partly mitigate the unfavorable effects of this stress because an important part of N in such plants is supplied by the symbiotic rhizobia (El Idrissi and Abdelmoumen, 2008).

Different techniques have been used to alleviate the stress of salinity on crop production and plant growth including the use of tolerant plant varieties and microbial species, as well as using biotechnological and molecular methods. Plant-tolerant varieties are able to resist the unfavorable effects of stress and hence ensure growth using the following mechanisms: (1) decreased uptake of Na^+ and Cl^- by plant roots, (2) extrusion of Na^+ and Cl^- from the plant leaf, (3) controlled activities of plant stomata, (4) higher uptake of ions such as K^+, (5) adjusting the rate of K^+/Na^+ in the plant, and (6) production of different signaling molecules including plant hormones, mitogen-activated protein kinase (MAPK), and antioxidant enzymes and products (Finet et al., 2010; Lau et al., 2011; Van Ha et al., 2013).

Soybean (*Glycine max* L.) is one of the most important legume crops, and it is a suitable and inexpensive source of protein and oil for human consumption. Containing 40% protein, soybean has the highest level of protein among various crop plants, and with 20% oil it is second among these plants (Toorchi et al., 2009). According to the Food and Agriculture Organization of the United Nations (FAO, 2010), the production of soybean worldwide is about 260 million tons, produced over more than 100 million hectares. More than 50% of these soybean fields are in the USA and Brazil.

Soybean is a crop sensitive-to-moderately tolerant to salinity (Kao et al., 2006). Drought stress significantly decreases soybean yield, by 24–50% (Frederick et al., 2001), as the most sensitive stage is seed filling. The plant obtains its required N mostly by the process of N-symbiotic fixation and the remaining N needed is supplied by the use of chemical fertilization. The process of biological-N fixation

Abiotic and Biotic Stresses in Soybean Production. http://dx.doi.org/10.1016/B978-0-12-801536-0.00007-4

is of environmental and economical importance significantly affecting plant growth and crop production. The specific rhizobial bacteria of soybean are of the genus *Bradyrhizobium japonicum* (Phang et al., 2008).

It is important to assess the drought tolerance of different soybean varieties. Different parameters are used to indicate soybean tolerance under stress using the following steps: (1) the direct method, which is according to the plant yield with respect to the average yield, water use efficiency, and environmental index, which is the yield at the region related to the yield at the other part of the area and (2) the use of drought tolerance coefficient, which is soybean yield at the water deficit year related to the soybean yield in a sufficient water year (Liu, 2009).

Under drought stress, soybean is able to perform some morphological adjustments to avoid the adverse effects of stress (Liu et al., 2005). Under drought, root distribution changes at the surface, and the deeper parts of the root with a low root density appear on the dry soil surface and a high root density at a deeper depth (Benjamin and Nielsen, 2006). Interestingly, it was indicated that there is a positive correlation between drought tolerance and plant weight including root weight in drought-tolerant soybeans (Liu et al., 2005). Under stresses such as salinity and drought, plants devote more carbohydrates to their roots and as a result the root weight increases. There are also some alterations of root cell properties including the cell wall structure and growth. Soybean plants can adjust the rate of lignin content in the cell wall, probably by the activity of cell wall peroxidase (Wang et al., 2012).

The tolerant varieties of soybean adjust the conductance of their stomata under drought stress. Maintaining cell turgidity in soybean plants is also another important tolerating process under drought stress. Soybean varieties tolerant to drought decrease the solute potential, increase water potential, and water use efficiency (Benjamin and Nielsen, 2006).

In the process of N symbiotic fixation the following stages occur: (1) the plant roots produce signaling molecules, which are able to induce morphological and physiological alterations in rhizobia, (2) in response bacterial *nod* genes are activated and produce biochemicals called lipochitooligosaccharides (LCOs) with the ability to alter the morphology and physiology of root hairs, (3) root bulging and curling, plant responses to LCO molecules, (4) the formation of infection thread results in the entrance of bacteria into the root cortical cells and eventual production of nodules, and (5) rhizobia fix atmospheric N_2 into ammonium, organic N, and proteins used by the host plant (Maj et al., 2010; Miransari et al., 2013).

Although the process of N fixation can provide the plant with its required N, similarly to the other plant activities, under stress its efficiency decreases. Under salinity stress both plant and symbiotic rhizobia are adversely affected, so depending on the tolerance rate of bacteria and the host plant, they can survive the stress. The bacteria are not tolerant under stress, although some tolerant strains have been produced (Marinkovic et al., 2013).

With respect to the salinity tolerance of soybean, it is classified as a moderately tolerant crop and under a salinity higher than 5 dS m^{-1}, soybean yield decreases. Different researchers have investigated the effects of salinity on the growth and yield of soybean (Abel and Mackenzie, 1964; Miransari and Smith, 2007; Miransari et al., 2013;

Singleton and Bohlool, 1984). Chang et al. (1994) examined the salt tolerance of 20 soybean varieties under the saline conditions of 14–15 and 18–20 dS m^{-1}. Related to control treatment, the yield of soybean decreased by 52.5 and 61.1%, respectively. Berstein and Ogata (1966) found that soybeans inoculated with *B. japonicum* are more sensitive to salinity stress than soybean fertilized with chemical N fertilization. They measured the growth and N-fixing related factors under a salinity of 0–5.6 atm of NaCl, using gravel medium. At the highest level of salinity, the nodulating ability of soybean was thoroughly inhibited.

Different growth stages of soybean are adversely affected by salt stress. Soybean has a different salt tolerance at different growth stages. For example, soybean seed germination was delayed under the salt stress of 0.05–0.1%. High salt concentration decreased the rate of seed germination in soybean (Abel and Mackenzie, 1964). Salt-tolerant varieties indicated higher rate of seed germination relative to the salt-sensitive ones (Abel, 1969). Different stages of seed germination are affected according to the following factors: imbibitions > radical emergence > radical growth > lateral root growth (Shao et al., 1994).

The stage of soybean seedling growth is a much more sensitive to salinity stress than the stage of seed germination (Hosseini et al., 2002). Accordingly, a 5% decrease in the seedling growth was resulted under a salinity of 220 mM l^{-1} relative to a salinity of 300 mM l^{-1}, where the seedling growth was completely suppressed. Forty percent of the seeds were able to germinate when the concentration of sodium in the embryonic axis was at 9.3 mg g^{-1} (fresh weight), while the seedling growth was completely suppressed at a sodium concentration of 6.1 mg g^{-1} (fresh weight).

The growth of different soybean plant parts is adversely affected by salinity stress including plant height, plant biomass, leaf size, number of branches, internodes, pods and weight of 100 seeds, and plant weight (Abel and Mackenzie, 1964; Chang et al., 1994; Miransari and Smith 2007, 2009). However, salt-tolerant varieties can grow more efficiently under salinity stress compared with the salt-sensitive varieties. Soybean seed quality is also affected under salinity stress as seed protein content decreases under saline conditions (Chang et al., 1994; Wan et al., 2002). There are some indications that the oil content of soybean seed is also adversely affected by salinity stress, so it is variety-dependent (Chang et al., 1994; Ghassemi-Golezani et al., 2009; Wan et al., 2002).

With respect to the aforementioned text some of the most important properties, signaling pathways, processes, and mechanisms related to the effects of salinity on the growth and yield of soybean, including its effects on the symbiotic association with *B. japonicum* are now presented and analyzed. Such details can be used for the production of soybean species and rhizobial strains which are more tolerant under salinity stress, can establish a symbiotic association, and can fix N in reasonable amounts.

Soybean and *B. japonicum* under salinity

The diazotrophic Gram-negative bacteria, rhizobia, are able to fix atmospheric N using the nitrogenase-enzyme complex. These bacteria belong to the Rhizobiaceae family (α-proteobacteria) and are able to develop a symbiotic association with leguminous plants and reduce the atmospheric N for the use of host plant. As a result, an important part of N for plant use is supplied by this process, so decreasing the need for and use of chemical fertilization (Chang et al., 2009).

The bacteria approach the host plant roots in response to certain signaling communications between the two symbionts. As a result the NodD protein and the nodulation (*nod*) gene of rhizobia are activated and create products called lipochitooligosaccarides (Nod factors). Nod factors are able to initiate cellular division in the root cortex (Denarie and Cullimore, 1993; Peck et al., 2006). The response of plant host to the signals produced by the bacterial symbiont increases the level of calcium in the host-plant roots and hence the root cytoskeleton is changed (Sieberer et al., 2005).

The production of Nod factors by bacteria results in the curling and bulging of plant root hairs and hence the bacteria are trapped, and as a result of cortical cell division in the root hairs, nodules are produced. Small amounts of oxygen exist in the environment around the bacteria in the nodules, so that bacteria can reduce atmospheric N using the nitrogenase complex using the process of N-symbiotic fixation (Fischer, 1994). Leghemoglobin, an oxygen protein binding protein produced by the host plant in the nodules, is able to adjust the level of oxygen (Ott et al., 2005).

Respiration of bacteria during the process of N fixation provides the nitrogenase enzyme with 16 ATP and eight electrons, resulting in the reduction of one N_2 to two NH_4^+. The ammonium ions produced by the bacteria are assimilated by the host plant using the enzymes asparagine and glutamine synthetases. Dicarboxylic acid such as malate produced by the host plant and is used by the bacteria as metabolite for the differentiation of bacteria and the process of N fixation (Jones et al., 2007).

The adverse effects of salinity on the process of N fixation by soybean and the symbiotic rhizobia are dependent on the following: (1) the growth and survival of rhizobia, (2) the number of rhizobia in nodule, (3) the protein and lipochitooligosaccharide content of rhizobial cells, (4) the infection process, and (5) the functioning of root nodules (Soussi et al., 1999, 2001). The rhizobia are able to alleviate adverse effects of salinity by producing osmolites such as glutamate, ectoine, betaine, and sugars (Boncompagni et al., 1999; Gouffi et al., 1998, 1999; Gouffi and Blanco, 2000).

Under drought, betaine is not readily available, however carbohydrates and sugars are available in the soil and in the rhizosphere (El Idrissi and Abdelmoumen, 2008). These authors examined the effects of salinity on the utilization of different carbon products under salinity stress. Under nonsaline conditions a large number of carbohydrates were used by rhizobia as a source of activity. Esculin was the most suitable carbohydrate under saline conditions.

However, under milder saline conditions the preferable sources of carbohydrates such as xylose, mannose, galactose, fructose, maltose, and sucrose were dominant. El Idrissi and Abdelmoumen (2008) speculated that the use of such organic products by rhizobia may be activated under salt stress. Use of a large amount of carbohydrate by rhizobia may be a suitable method for the bacteria to survive and be active in the nodules under stress (Jensen et al., 2002). There was a diverse pattern of bacterial growth seen under salinity stress. Relative to the sensitive strains, the tolerant ones were able to use a greater quantity of carbohydrate, although their growth rate was affected to some extent. Certain bacterial genes regulate the utilization of carbon sources by bacteria (El Idrissi and Abdelmoumen, 2008).

Gil-Quintana et al. (2013) investigated the effects of drought on the N fixation of soybeans nodules. One part of soybean root was irrigated at field capacity, while the other part was subjected to the stress. Although drought decreased N fixation, the level of nitrogenase was at a normal level in the control treatments. Under water stress the rate of amino acid and ureide increased in the stressed part of the root regardless of transpiration. Ureide increased because of its decreased degradation rate and not because of its increased biosynthesis. Proteomic analysis indicated that protein synthesis, cell growth, and metabolism of amino acids and carbons are among the processes that are affected the most in soybean nodules under drought stress (Marino et al., 2007).

The technique of proteomics has been used for the investigation of activities in different soybean parts including root (Brechenmacher et al., 2009), leaf, root hairs (Wan et al., 2005), mitochondria (Hoa et al., 2004), nodule cytosol, and the peribacteriod membrane (Panter et al., 2000). However, research using proteomics for soybean performance under drought stress is sparse.

The main reasons for the accumulation of ureide in soybean nodules are: (1) decreased rate of transpiration, (2) decreased demand for N by the vegetative part, and (3) the changed metabolism of ureide (Valentine et al., 2011). Under drought stress the activities of urate oxidase and allantoate amidohydrolase in nodules decreased even during the first stages of nodule activity, when the water potential of the nodules had not been adversely affected. Drought also adversely affected the activity of glutamine synthetase, an N assimilation enzyme in nodules (similarly to the results for *Medicago truncatula*), and hence the level of N fixation (Gil-Quintana et al., 2013).

Soybean tolerating mechanisms under salinity

Soybean use different mechanisms to alleviate the stress of salinity. Wei et al. (2014) examined the effects of melatonin (*N*-acetyl-5-methoxytryptamine) on the growth of soybean plants under abiotic stress. Seeds were coated with melatonin, which resulted in increased soybean growth and yield under different stresses including salt and drought. Using transcriptome analysis, it was indicated that salt deactivated the genes associated with oxidoreductase activity, binding, and metabolic processes. However, melatonin upregulated the activity of the genes, which were deactivated by salt stress and so alleviated the adverse effects of stress. Such genes are associated with the following functions: (1) carbohydrate production, (2) cell division, (3) photosynthesis, and (4) production of ascorbate.

Control of ion homeostasis

The concentration of ions in plant cell is an important factor determining the salt tolerance of soybean. Sodium is among the ions most deleterious to the growth and yield of plants under salinity stress (Gao et al., 2007). If the plant is able to adjust and control the concentration of unfavorable ions inside its cellar environment, it will be able to tolerate the stress of salinity, efficiently. In plant health the properties of the cellular membrane are among the most important parameters affecting the cellular concentration of ions in plant cells. Under high salt concentrations, tolerant plants are able to excrete extra amounts of salt ions from their cells and hence keep the concentration constant (Phang et al., 2009; Miransari, 2011; Serraj et al., 1998).

It has also been indicated that the concentration of chloride inside the plant cells can determine plant response to stress (Abel, 1969; Abel and Mackenzie, 1964). Elevated levels of chloride in plants can result in leaf chlorosis (Pantalone et al., 1997). Interestingly, salt-tolerant soybean was able to accumulate lower amounts of sodium and chloride in its leaves (Essa, 2002; Li et al., 2006). The other important factor that enables soybean plants to survive stress is the accumulation of sodium and chloride in the roots (Luo et al., 2005).

Both inter and intracompartmentalization of sodium regulates sodium homeostasis in soybean plants. The role of active transfer of ions across tonoplasts is an important factor affecting the homeostasis of sodium and chloride in soybean plants. Such ability has been indicated by the high activities of H^+–ATPase and H^+–PPiase in the vesicles of tonoplasts isolated from the roots of salt-tolerant soybean under salinity stress (Yu et al., 2005).

The other important parameter which may affect the salt-tolerating ability of soybean is the interaction between phosphorus and sodium. Although these results may differ from situation to situation, it has been indicated in most cases that increased levels of phosphorus may adversely affect soybean salt tolerance. Under salt stress, NaCl may prevent the activity of the Pi transporter in plants and hence plants may become phosphorus deficient. Higher P concentration may intensify the adverse effects of salinity on the growth of plants because there would be a synergetic interaction between phosphorus and chloride. As a result under such conditions the plant may be less tolerant to the stress (Grattan and Maas, 1984; Phang et al., 2009).

However, while these authors found that this type of interaction does not exist between phosphorus and sodium, other research has indicated the toxic effects of sodium on the growth of plants (Munns and Tester, 2008). Although the sensitive varieties of soybean plants indicated unfavorable effects of high phosphorus uptake by plants, the tolerant varieties did not show such effects under salinity stress (Phang et al., 2009). Accordingly, these authors found that phosphorus can intensify the salinity effect on the growth of plants by adversely affecting the induction of *GmSOS1* and *GmCNGC*.

Use of osmoprotectants for osmotic adjustments

As previously mentioned, high uptake of ions such as sodium and chloride under salinity stress interrupts different plant activities. Under such conditions, plant morphological and physiological properties are affected. As a result, an other important mechanism used by soybean under salinity stress is the use of osmoprotectants. Such small neutral metabolites are hydrophilic compounds with the ability to decrease the cell-osmotic potential, while plant activities remain unchanged. In addition to adjusting plant osmotic potential, such solutes can also, for example, stabilize the structure of proteins and cellular membrane by glycine betaine (Papageorgiou and Murata, 1995) and precipitate in the presence of reactive oxygen species via mannitol (Shen et al., 1997a, b).

The adjustment of osmotic potential is a method used by plants to regulate cell turgidity. As a result the concentration of solutes in the cell increases. Under drought stress the related soybean gene *P5CS* is upregulated resulting in the activation of enzyme Δ1-pyrroline-5-carboxylate synthase, as an important factor in the biosynthesis of proline (Manavalan et al., 2009). If such a gene is inactivated the plant is not able to tolerate the stress (de Ronde et al., 2000). However, the role of proline under drought stress has yet to be further investigated.

Soybean does not accumulate high amounts of glycine betaine, with an average of less than 5 μM g^{-1} DW. However, foliar use of glycine betaine may increase the amount of glycine betaine up to 60 μM g^{-1} DW. As a result the photosynthetic activity of soybean plants, the rate of N fixation, leaf area growth, and seed yield under saline and nonsaline conditions increase. Hence, under both water and salt stress conditions glycine betaine can be used for the alleviation of stress (Agboma et al., 1997).

The other important osmoprotectant is trigonelline (TRG). Under stress the solute is able to adjust plant osmotic potential and hence alleviate the stress of drought and salinity (Chen and Wood, 2004; Malencic et al., 2003; Manavalan et al., 2009). Although the concentration rate of TRG under nonstressed conditions is in the range of 63.8–162.4 μg g^{-1} DW, it can increase up to 75.4–218.7 μg g^{-1} DW in a plant leaf under stressed conditions. However, it is a cultivar-related trait (Phang et al., 2008). Proline may sometimes accumulate under salt stress in soybean plants and in some cases it may not alleviate salinity stress in them. Different results regarding the role of proline under salt stress may be related to the conditions of the experiment (Yoon et al., 2009).

The role of certain proteins such as late embryogenesis abundant (LEA) proteins (hydrophilic and thermostable) under salt stress has also been investigated in soybean plants. Such proteins can alleviate stress by the following mechanisms: (1) their antioxidant activities; (2) stabilizing the structure of proteins and cellular membrane; and (3) maintaining the cellular structure (Tunnacliffe and Wise, 2007). Around 20 LEA proteins have been so far indicated in soybean including glycine, glutamate, and glutamine (Esperlund et al., 1992; Lan et al., 2005).

Using the proteomic method, Toorchi et al. (2009) investigated the proteins that were induced under osmotic stress in soybean roots. The role of roots is important under stress, helping the plant to survive. The presence of osmolites in roots regulates plant water potential and the expansion of leaf, and hence is an important factor under water stress. Roots are also suitable for use in the proteomic method because relative to the plant leaf, there is little ribulose 1,5-bisphosphate carboxylase/oxygenase (RuBisCO) present that can adversely affect the method of proteomic. Proteomic data related to the response of soybean to osmotic stress is scarce.

Using proteomics, Ma et al. (2014) analyzed soybean seedlings to determine the responsive proteins under salt stress. They used two different varieties including Lee68 (salt tolerant) and Jackson (salt sensitive). Sixty-eight proteins were determined, which were associated 13 metabolic and cellular processes. Under salt stress the proteins of brassinosteroid and gibberellin signaling pathways were only enhanced in Lee68. The role of abscisic acid was also evident in this variety under salt stress.

Proteins related to the calcium signaling pathway were also induced in seedling roots by salt stress more significantly in Lee68, relative to Jackson. The tolerant variety was also able to scavenge reactive oxygen species more efficiently and maintain cellular K^+/Na^+ homeostasis more than the Jackson variety. These results indicated that such differences are the main reasons for the tolerance in the variety Lee68 (Ma et al., 2014).

Reactive oxygen species are continuously produced in plants as a result of metabolic activities and in response to different stresses. The main production sites of reactive oxygen species are the chloroplasts and mitochondria. The production of reactive oxygen species can adversely affect plant growth and activity by the unfavorable effects on the activity and structure of enzymes and chlorophyll, as well as lipid peroxidation. The plant response to the production of reactive oxygen species is the production of antioxidant enzymes and products (Phang et al., 2008; Sajedi et al., 2010, 2011).

The antioxidant enzymes include superoxide dismutase (SOD), catalase (CAT), and peroxidase, and antioxidant products include ascorbic acid, carotenoid, and glutathione. Chen et al. (1997) determined the activity of SOD in soybean seedlings and found that 9%, 11%, and 80% of the enzyme activity was related to chloroplasts, mitochondria, and cytosol, respectively. *GmPAP3* activates the acid phosphatase in the mitochondria. The mitochondrion is among the most important production sites of reactive oxygen species. Stresses, such as salinity, induce the activity of *GmPAP3* (Li et al., 2008b; Liao et al., 2003).

Zhang et al. (2014) examined the effects of the polyamine putrescine on the growth of soybean under salt stress. Soybean seedlings were grown in a nutrient solution with 100 mM salt concentration and were treated with putrescine 10 (mM) and its biosynthetic inhibitor D-arginine (0.5 mM). According to the results, while putrescine was able to alleviate the stress of salinity, D-arginine intensified the effects of salinity stress on the growth of soybean plants. The alleviating effect of putrescine on the growth of soybean under salt stress was the induction of the activity of antioxidant enzymes.

The important point about the tolerance of soybean under salinity stress is maintaining its K^+/Na^+ cellular homeostasis. Different varieties of soybean differ in tolerance to potassium deficiency and hence salt stress. Accordingly, to avoid or alleviate the adverse effects of salinity on the growth of soybean, the selection of a variety that is more tolerant to potassium deficiency or providing the plant with adequate amounts of potassium is of great significance. Wang et al. (2014) investigated the effects of potassium deficiency on the growth and photosynthesis of two different soybean varieties (tolerant and sensitive) with respect to tolerance to potassium deficiency. Potassium deficiency decreased plant growth, photosynthesis rate, and stomatal conductance in the sensitive variety and not in the tolerant variety.

Alteration of cellular structure

Production of gland-like tissues and glandular hairs under salinity stress is one of the mechanisms used by soybean plants for the alleviation of salt stress. A typical such gland usually resembles a ball with a diameter of, for example, around 20 μm and a length of 1.2 μm, found on the epidermis of soybean leaf and stem with the ability to excrete salt. It usually has large vacuole and some crystals. The glandular hairs are produced from the leaf veins with a height of 35–45 μm and a diameter of 10–18.5 μm (Wang et al., 1999).

The other mechanism associated with the alteration of cellular structure under salt stress is the modification of the cell wall and cell membrane. These structures are two important cell components determining the size and shape of the cell. The strength of the cell membrane is an important parameter affecting cellular response under salinity stress. The cellular membrane is composed of glycolipids, phospholipids, and steroids. Under salinity stress it may be affected resulting in membrane leakage of organic products and electrolytes from cells (Shao et al., 1993).

For example, salinity stress resulted in a decreased phospholipid content as well as in an altered ratio of saturated to unsaturated fatty acids (Huang, 1996). The plasma membrane of soybean root cells was investigated under salinity stress. Although salinity did not affect the protein content, it decreased the amount of unsaturated fatty acid in the plasma membrane. Such alterations may enhance plant growth under salinity stress (Huang, 1996). Interestingly, the amount of phospholipids and galactolipids in the plasma membrane may also alter under salinity stress. For example, in salt-tolerant varieties, there was an increase while in salt-sensitive varieties a decrease occurred (Yu et al., 2005).

Genetic pathway under stress

The induction of an important protein called NAC (NAM, ATAF1/2, CUC1), acting as a transcription family, increases soybean tolerance under stress. The proteins in this family have a domain of N-terminal DNA binding called the NAC domain. The related C-terminus in NAC proteins has different sequences, with the ability to regulate transcriptional activities. Soybean has 54 NAC genes using the Expressed Sequence Tags database and 101 NAC proteins using the whole gene sequence of soybean (Hao et al., 2011).

This protein is able to regulate different plant activities including morphogenesis, senescence, stress signal transduction, and development (Olsen et al., 2005a, b). NAC proteins are able to regulate the auxin signaling pathway and root formation. Xie et al. (2002) indicated that auxin induces the activity of NAC protein and the gene is able to regulate the auxin signaling pathway and the formation of *Arabidopsis* lateral root.

There is also another NAC gene, *AtNAC2*, which is activated by different hormones such as auxin, abscisic acid, ethylene, and salinity stress. If the gene is overexpressed, the lateral roots of transgenic plants are formed (Hao et al., 2011). The NAC proteins can also affect plant response to biotic and abiotic stresses. Overexpression of *SNAC1* and *SNAC2* in rice transgenic plants increased plant tolerance to stresses such as drought, salinity, and cold (Hu et al., 2006, 2008). The expression of the 54 NAC proteins in soybean was analyzed under stresses such as salinity, drought, cold, and abscisic acid treatments and it was found that 15 proteins were induced under one of these treatments (Hao et al., 2011).

Hao et al. (2011) also indicated the role of two NAC proteins, GmNAC11 and Gm-NAC20, for plant stress conditions. Different abiotic stresses and plant hormones induced the activity of the two proteins and their transcripts were found in high amounts in roots and cotyledons. In the protoplast, GmNAC11 acted as a transcriptional activator and GmNAC20 acted as a mild repressor. If overexpressed, GmNAC20 improved the plant response under both salinity and cold stress, while GmNAC11 improved the plant response only to salinity stress. GmNAC20 affected the formation of lateral roots by regulating the auxin signaling pathway. Accordingly, the authors concluded that the growth and production of important plants under stress is achievable by regulating the acidity of NAC proteins.

The response of soybean under salinity stress has been investigated using different techniques including the use of proteomics. The expression levels of stress proteins are adjusted under stress and hence enable the plant to survive the stress. By using proteomics, it is possible to indicate which proteins are overexpressed and which are underexpressed when plants are subjected to stress. Using the proteomic method, the plant response to salinity stress has been examined by different authors (Nouri and Komatsu, 2010; Toorchi et al., 2009; Xu et al., 2011).

Using two different soybean varieties, tolerant (Lee68) and sensitive (N2899), Xu et al. (2011) investigated how plant seeds may respond to salinity stress during germination. The response was seen at the physiological and proteomic level. Although the germination percentage was not affected by salt stress, the time of germination was delayed by 0.3 and 1.0 d in Lee68 and N2899, respectively, relative to the control. Salinity increased abscisic acid content and decreased gibberellin and isopentenyladenosine content.

In both varieties, salinity stress resulted in the upregulation of ferritin and 20S proteasome subunit b-6. The downregulation of glutathione S-transferase (GST9), GST10, glyceraldehyde 3-phosphate dehydrogenase, and the seed protein PM36 was resulted in Lee68 under salinity stress. However, in the control treatment, such proteins were present in N2899, and salt stress resulted in their upregulation. Xu et al. (2011), accordingly, suggested that these proteins can help the soybean seeds germinate under salinity and hence alleviate the stress experienced (Xu et al., 2011).

Sodium and chloride concentrations in plants determine soybean tolerance under salinity stress. Research into soybean seedlings has indicated that the transport of ions in plants is a function of some mechanisms in the xylem–symplast confirming that sodium and chloride concentrations in plant are associated with plant tolerance under salinity. Soybean plants have the ability to resist a high concentration of sodium relative to chloride, which is due to the ability of plant in restricting sodium transport to the leaf (Sobhanian et al., 2011).

Under stresses such as heat and drought the activity of antioxidant enzymes including CAT, ascorbate peroxidase (APX), and glutathione reductase (GR) increases. Among the mechanisms by which soybean plants are able to tolerate such stress are the activities of APX and GR, which can scavenge the reactive oxygen species produced in the plastids and probably in the plant cell extraplastidic compartments (Sajedi et al., 2010, 2011).

As previously mentioned, the other plant response to salinity stress is the increased concentration of proline. It has also been indicated that if soybean plants are pretreated with salt their tolerance under salinity stress increases, which is due to the compartmentalization of sodium into the vacuole of root cells and hence its subsequent decreased concentration in plant leaf. Accordingly, if some strategies and techniques are used to adjust the concentration of sodium and chloride in soybean plant roots, it is possible to increase plant tolerance under salinity stress (Sobhanian et al., 2011).

It has also been indicated that the use of AM fungi is a suitable method to alleviate the stress of salinity on the growth and yield of soybean. Among the most important mechanisms of alleviating salinity stress by AM fungi is the increased uptake of nutrients by the host plant. However, more research must be conducted to elucidate the details related to the mechanisms of salinity stress alleviation by soybean (Miransari, 2010).

The use of the proteomic technique is essential for the determination of proteins which are induced under salinity stress. Accordingly, their functions and properties can also be determined (Li et al., 2008a; Parker et al., 2006; Sobhanian et al., 2010). Proteome analysis of hypocotyls and roots indicated that when soybean plants are subjected to salt stress, a large number of abundant proteins are upregulated during the late embryogenesis period, while lectin is downregulated. The authors, accordingly suggested that the expression of such proteins under salinity stress confers tolerance to survive under stress. The level of such proteins also increases in the vegetative part when the plants are under stress (Chandler and Robertson, 1994; Ndong et al., 2002).

The function of salt responsive proteins and their importance was determined in soybean plants under salinity stress by analyzing the proteomes of hypocotyls, leaf, and roots. Salt stress resulted in the downregulation of proteins related to soybean photosynthesis proteins, protein biosynthesis, and calreticulin (Sobhanian et al., 2010). Sodium chloride adversely affects the process of photosynthesis and protein biosynthesis, and the calcium signaling pathway under salinity stress. NADP dehydrogenase is also downregulated under salinity, as a result of which the production of ATP and plant growth decreases.

There are a number of other proteins which are upregulated under salinity stress and enable the plant to survive under stress including alcohol dehydrogenase (which produces ATP and consumes glycolytic products under salinity stress), trypsin inhibitor (which has a role in prevention of protein proteolysis, and detoxifying H_2O_2 under saline conditions), annexin, and kinesin motor protein (which regulates the processes of cell cycle and cell integrity under stress) (Sobhanian et al., 2011).

However, some proteins are downregulated under salinity stress including isoflavone reductase (which regulates the metabolism of flavonoid related compounds), caffeoyl-CoA-O-methyltransferase (which regulates the lignification of cell wall), and quinone oxidoreductase (which reduces quinine), indicating that such proteins may not have a role in plant tolerance under salinity stress. But it is a different situation for ripening-related protein as it is upregulated under stress due to the increased concentration of ethylene in soybean seedlings under salinity stress (Sobhanian et al., 2010).

These authors also found that glyceraldehyde-3-phosphate dehydrogenase is downregulated at the protein and mRNA level in soybean leaf and hypocotyls, when the plant is subjected to salinity stress. As a result the production of ATP and plant growth decreases. The authors, accordingly, suggested that such a gene can be used to increase soybean tolerance under salinity stress. They also suggested that improving the calcium signaling and NADH dehydrogenase pathway may also be a suitable method of improving soybean tolerance under stress. However, the improvement of the signaling pathway, which may result in the production of upregulated proteins during salinity stress, may not improve soybean tolerance under stress.

Signaling pathways under salinity stress

When a plant is subjected to different stresses such as salinity, different signaling pathways are activated to help the plant survive the stress. The stress is usually perceived by the receptors in the cellular membrane and as a result the gene and related signaling pathways are activated resulting in the plant response to stress. The activation of plant hormonal signaling and MAPK signaling, and production of antioxidant enzymes and products are among such responses (Miransari et al., 2013).

The response of plants under stress is mediated by stimuli (osmosensors), which have yet to be indicated, results in the activation of the associated signaling pathway and hence plant tolerance. The mechanism of tolerance is the combined activity of kinases, receptors, transcription factors, and effectors as well as the production of metabolites, which results in the induction of the related signaling pathway. Drought tolerance is a multidimensional process including some molecular and genetic responses. There is a great overlap between the signaling pathways of drought and salinity. Drought stress is perceived with sensors, which have yet to be fully understood, and it is speculated that such sensors change the porosity and integrity of the membrane as well as the turgor pressure (Mahajan and Tuteja, 2005).

The potential sensors of osmotic stress include the receptors of the cell wall, proteins of the cell membrane, and cytosolic enzymes (Kader and Lindberg, 2010). Two-component histidine kinases, calcium channels, calcium binding proteins, and receptor-like protein kinases are also among the potential osmosensors (Grene et al., 2011; Kacperska, 2004; Xiong and Zhu, 2002). Some receptor-like protein kinases including GmCLV1 and GmRLK, and two-component histidine kinases including GmHK have been determined in soybean (Yamamoto et al., 2000; Yamamoto and Knap, 2001).

Biological N fixation as affected by salinity stress

Salinity also has adverse effects on the process of nodule production and hence biological N fixation. The suppressing effect of salinity on the process of biological N fixation is by decreasing the rate of aerobic respiration by the N-fixing bacteria, reducing the rate of leghemoglobin (in the root nodules) and metabolites for bacterial activities (Delgado et al., 1994). The production of signaling molecules by soybean plants, depending on the variety, can also affect the plant response under salinity stress.

The efficiency of biological N fixation, including the number of nodules, is affected by both the properties of the host plant and the symbiotic N-fixing rhizobia

(Clement et al., 2008; Wang et al., 2011). It is speculated that the adverse effects of salt stress on *B. japonicum* is more related to an ion effect rather than an osmotic effect. Salt tolerance of *B. japonicum* is determined by carbon, temperature, pH, and osmoprotectants in the solute and not by their ecological origin.

It has been indicated that under stresses such as salinity, acidity, and suboptimal root zone temperature, the initial stages of biological N fixation by soybean and *B. japonicum* including the process of signaling exchange between the two symbionts are adversely affected. Accordingly, Miransari and Smith (2007, 2008, 2009) hypothesized and proved that addition of the bacterium signal, genistein, to the bacterial inocula can overcome the negative effects of stress on the process of N fixation by the two symbionts.

The salt-tolerant species of soybean are able to handle the stress more efficiently related to the sensitive species, resulting in a higher rate of N fixation by the two symbionts. The growth of the salt-sensitive variety "411" on a medium with a salinity of 150 mM l^{-1} NaCl significantly decreased the leghemoglobin content, and at the time of N fixation, the rate of malondialdehyde increased while the rate of SOD, APX, GR, and CAT decreased.

However, it was a different case for the salt-tolerant variety "377" as the process of N fixation, as well as leghemoglobin and MDA contents, was not significantly affected. There was an increase in the content of SOD, APX, GR, and CAT. Interestingly, when the soybean plants were subjected to a mild salinity stress: (1) the increased level of antioxidant enzymes and decreased level of glutathione content acted as protection against the reactive oxygen species; (2) lipid and protein were not peroxidized; and (3) the structure of leghemoglobin remained unchanged (Phang et al., 2008).

The salt tolerance of 1716 soybean varieties from different Chinese provinces and different parts of the world were evaluated by Shao et al. (1986) using a saline field and a mixture of seawater and fresh water. The salt tolerance of different soybean varieties was determined by the evaluation of salt stress appearance on the plant leaf with respect to their growth stage. Seven varieties indicated salt tolerance at different growth stages. However, 242 varieties were tolerant at the growth stage of young seedlings and 85 were salt tolerant during the reproductive stage.

In another experiment, Shao et al. (1993) examined the salt tolerance of 10,128 Chinese soybean varieties at different growth stages among which 9.1% were indicated to be tolerant at the stage of seed germination, 4.5% at the stage of seedling growth, and 2.8% at both growth stages. Accordingly, 83 soybean varieties were indicated to have a high salt tolerance under stress. The ability of soybean species under stress is determined by the structure of their stress gene and the growth conditions. As a result, the production of tolerant varieties is likely with respect to the stress properties, plant growth stage, and growth conditions.

Alleviation of symbiotic N fixation under salinity stress

Silvente et al. (2012) investigated the effects of drought on the growth of two soybean varieties including the drought tolerant NA5009RG and the drought sensitive DM50048. Accordingly, they examined the adverse effects of drought on the growth, biological N fixation, and production of ureide and proline in stressed soybean varieties. There were significant differences in the physiological responses of the two soybean varieties and the related signaling pathway under stress. The production of metabolites by soybean varieties under drought stress was the most important highlight of this research.

The adverse effects of drought on the growth of soybean affect: (1) phosphate supply, (2) oxygen delay into the soybean nodules, and (3) extra amounts of nitrogenous compounds in nodules (Marino et al., 2007; Neo and Layzell, 1997; Serraj et al., 2001). The response of plants to stress is usually an increased level of minerals and organic metabolites or solutes. As a result the stressed cell will be able to keep its water and the maintain integrity of the cellular membrane (Hare et al., 1998; Silvente et al., 2012; Yamaguchi-Shinozaki and Shinozaki, 2006). The metabolites that are produced under stress differ greatly in different soybean varieties. In legumes, as a result, osmotic adjustment is the plant response to the stress.

Accordingly, under stress the following may be used by soybean and rhizobia to alleviate the adverse effects of salinity: (1) use of tolerant soybean species and *B. japonicum* strains, (2) physiological and morphological alteration by plant and bacteria, (3) production of different organic products including osmolytes, (4) pretreatment of rhizobia with signal products, (5) improving the tolerance of bacteria and soybean using the related stress gene, and (6) use of tripartite symbiosis with soybean, mycorrhizal fungi, and rhizobia (Kao et al., 2006; Miransari and Smith, 2007, 2008, 2009; Phang et al., 2008; Zhang et al., 2014).

Conclusions and future perspectives

Salinity is among the most important stresses affecting the growth and yield of plants and their associated microbes. Due to the presence of large numbers of saline fields across the globe, a significant amount of research work has been conducted in different parts of the world to find the most suitable techniques and strategies which may alleviate the stress of salinity on the growth and activity of microbes and their host plants. For soybean and the symbiotic bacteria, *B. japonicum*, research has indicated that during the process of N symbiotic fixation, the initial stages of symbiotic association including the production of signals by the host plant and rhizobia are among the most sensitive. Accordingly, research has indicated that pretreatment of rhizobia with symbiosis signals can alleviate the stress of salinity on the process of N fixation by soybean and *B. japonicum*. Some of the most recent findings related to the effects of salinity on the growth and activity of *B. japonicum* and soybean have been presented and analyzed. Future research work may focus on the production of *B. japonicum* strains and soybean species, which are the most tolerant under salinity stress and which can demonstrate the highest efficiency during the process of biological symbiotic association. The associated signaling pathways must be determined with as much detail as possible and, accordingly, the optimum technique and strategy which can be used for the alleviation of salinity stress must be suggested and tested.

References

Abel, G.H., 1969. Inheritance of the capacity for chloride inclusion and chloride exclusion by soybeans. Crop Sci. 9, 697–698.

Abel, G., Mackenzie, A., 1964. Salt tolerance of soybean varieties (*Glycine max* L. Merrill) during germination and later growth. Crop Sci. 4, 157–161.

Agboma, P.C., Sinclair, T.R., Jokinen, K., Peltonen-Sainio, P., Pehu, E., 1997. An evaluation of the effect of exogenous glycinebetaine on the growth and yield of soybean: timing of application, watering regimes and cultivars. Field Crops Res. 54, 51–64.

Benjamin, J.G., Nielsen, D.C., 2006. Water deficit effects on root distribution of soybean, field pea and chickpea. Field Crops Res. 97, 248–253.

Berstein, L., Ogata, G., 1966. Effects of salinity on nodulation, nitrogen fixation, and growth of soybeans and alfalfa. Agron. J. 58, 201–203.

Boncompagni, E., Osteras, M., Poggl, M.C., Le Rudulier, D., 1999. Occurrence of choline and glycine betaine uptake and metabolism in the family of Rhizobiaceae and their role in osmoprotection. Appl. Environ. Microbiol. 65, 2072–2077.

Brechenmacher, L., Lee, J., Sachdev, S., Song, Z., Nguyen, T.H., Joshi, T., Oehrle, N., Libault, M., Mooney, B., Xu, D., Cooper, B., Stacey, G., 2009. Establishment of a protein reference map for soybean root hair cells. Plant Physiol. 149, 670–682.

Chandler, P.M., Robertson, M., 1994. Gene expression regulated by abscisic acid and its relation to stress tolerance. Annu. Rev. Plant Physiol. 45, 113–141.

Chang, R.Z., Chen, Y.W., Shao, G.H., Wan, C.W., 1994. Effect of salt stress on agronomic characters and chemical quality of seeds in soybean. Soybean Sci. 13, 101–105.

Chang, C., Damiani, I., Puppo, A., Frendo, P., 2009. Redox changes during the legume – rhizobium symbiosis. Mol. Plant 2, 370–377.

Chen, Y.W., Shao, G.H., Chang, R.Z., 1997. The effect of salt stress on superoxide dismutase in various organelles from cotyledon of soybean seedling. Acta Agron. Sinica 23, 214–219.

Chen, X., Wood, A.J., 2004. Purification and characterization of *S*-adenosyl-L-methionine nicotinic acid-*N*-methyltransferase from leaves of *Glycine max*. Biol. Plant. 48, 531–535.

Clement, M., Lambert, A., Herouart, D., Boncompagni, E., 2008. Identification of new upregulated genes under drought stress in soybean nodules. Gene 426, 15–22.

Delgado, M.J., Ligero, F., Liuch, C., 1994. Effects of salt stress on growth and nitrogen fixation by pea, faba-bean, common bean and soybean plant. Soil Biol. Biochem. 26, 371–376.

Denarie, J., Cullimore, J., 1993. Lipo-oligosaccharide nodulation factors: a new class of signalling molecules mediating recognition and morphogenesis. Cell 74, 951–954.

de Ronde, J.A., Spreeth, M.H., Cress, W.A., 2000. Effect of antisense L-Δ1-pyrroline-5-carboxylate reductase transgenic soybean plants subjected to osmotic and drought stress. J. Plant Growth Regul. 32, 13–26.

El Idrissi, M., Abdelmoumen, H., 2008. Carbohydrates as carbon sources in rhizobia under salt stress. Symbiosis 46, 33–44.

Esperlund, M., Saeboe-Larssen, S., Hughes, D.W., Galau, G., Larsen, F., Jakobsen, K., 1992. Late embryogenesis abundant genes encoding proteins with different numbers of hydrophobic repeats are regulated differentially by abscisic acid and osmotic stress. Plant J. 2, 241–252.

Essa, T.A., 2002. Effect of salinity stress on growth and nutrient composition of three soybean (*Glycine max* L. Merrill) cultivars. J. Agron. Crop Sci. 188, 86–93.

FAO., 2010. FAOSTAT. http://faostat.fao.org.

Finet, C., Fourquin, C., Vinauger, M., Berne-Dedieu, A., Chambrier, P., Paindavoine, S., Scutt, C.P., 2010. Parallel structural evolution of auxin response factors in the angiosperms. Plant J. 63, 952–959.

Fischer, H.M., 1994. Genetic regulation of nitrogen fixation in rhizobia. Microbiol. Rev. 58, 352–386.

Frederick, J.R., Camp, C.R., Bauer, P.J., 2001. Drought-stress effects on branch and main stem seed yield and yield components of determinate soybean. Crop Sci. 41, 759–763.

Gao, J.P., Chao, D.Y., Lin, H.X., 2007. Understanding abiotic stress tolerance mechanisms: recent studies on stress response in rice. J. Integr. Plant Biol. 49, 742–750.

Ghassemi-Golezani, K., Taifeh-Noori, M., Oustan, S., Moghaddam, M., 2009. Response of soybean cultivars to salinity stress. J. Food Agric. Environ. 7, 401–404.

Gil-Quintana, E., Larrainzar, E., Seminari, A., Diaz-Leal, J., Alamillo, J., Pineda, M., Arrese-Igor, C., Wienkoop, S., Gonzalez, E., 2013. Local inhibition of nitrogen fixation and nodule metabolism in drought-stressed soybean. J. Exp. Bot. 64, 2171–2182.

Gouffi, K., Pichereau, V., Roland, J.P., Thomas, D., Bernard, T., Blanco, C., 1998. Sucrose is a non accumulated osmoprotectant in *Sinorhizobium meliloti*. J. Bacteriol. 180, 5044–5051.

Gouffi, K., Pica, N., Pichereau, V., Blanco, C., 1999. Disaccharides as new class of non-accumulating osmoprotectants for *Sinorhizobium meliloti*. Appl. Environ. Microbiol. 65, 1491–1500.

Gouffi, K., Blanco, C., 2000. Is the accumulation of osmoprotectant the unique mechanism involved in bacterial osmoprotection? Int. J. Food Microbiol. 55, 171–174.

Grattan, S.R., Maas, E.V., 1984. Interactive effects of salinity and substrate phosphate on soybean. Agron. J. 76, 668–676.

Grene, R., Vasquez-Robinet, C., Bohnert, H.J., 2011. Molecular biology and physiological genomics of dehydration stress. In: Luttge, U., Beck, E., Bartels, D. (Eds.), Plant Desiccation Tolerance, Ecological Studies vol. 215, Springer, Berlin, Heidelberg, Germany, pp. 255–287.

Hao, Y., Wei, W., Song, Q., Chen, H., Zhang, Y., Wang, F., Zou, H., Lei, G., Tian, A., Zhang, W., Ma, B., Zhang, J., Chen, S., 2011. Soybean NAC transcription factors promote abiotic stress tolerance and lateral root formation in transgenic plants. Plant J. 68, 302–313.

Hare, P.D., Cress, W.A., Van Staden, J., 1998. Dissecting the roles of osmolyte accumulation during stress. Plant Cell Environ. 21, 535–553.

Hoa, L.T.P., Nomura, M., Kajiwara, H., Day, D.A., Tajima, S., 2004. Proteomic analysis on symbiotic differentiation of mitochondria in soybean nodules. Plant Cell Physiol. 45, 300–308.

Hosseini, M.K., Powell, A.A., Bingham, I.J., 2002. Comparison of the seed germination and early seedling growth of soybean in saline conditions. Seed Sci. Res. 12, 165–172.

Hu, H.H., Dai, M.Q., Yao, J.L., Xiao, B.Z., Li, X.H., Zhang, Q.F., Xiong, L.Z., 2006. Overexpressing a NAM, ATAF, and CUC (NAC) transcription factor enhances drought resistance and salt tolerance in rice. Proc. Natl Acad. Sci. USA 103, 12987–12992.

Hu, H.H., You, J., Fang, Y.J., Zhu, X.Y., Qi, Z.Y., Xiong, L.Z., 2008. Characterization of transcription factor gene SNAC2 conferring cold and salt tolerance in rice. Plant Mol. Biol. 67, 169–181.

Huang, C-.Y., 1996. Salt-stress induces lipid degradation and lipid phase transition in plasma membrane of soybean plants. Taiwania 41, 96–104.

Jensen, J.B., Peters, N.K., Bhuvaneswaril, T.V., 2002. Redundancy in periplasmic binding protein-dependent transport systems for trehalose, sucrose, and maltose in *Sinorhizobium meliloti*. J. Bacteriol. 184, 2978–2986.

Jones, K.M., Kobayashi, H., Davies, B.W., Taga, M.E., Walker, G.C., 2007. How rhizobial symbionts invade plants: the *Sinorhizobium–Medicago* model. Nat. Rev. Microbiol. 5, 619–633.

Kacperska, A., 2004. Sensor types in signal transduction pathways in plant cells responding to abiotic stressors: do they depend on stress intensity? Physiol. Plant. 122, 159–168.

Kader, M.A., Lindberg, S., 2010. Cytosolic calcium and pH signaling in plants under salinity stress. Plant Signal. Behav. 5, 233.

Kao, W.Y., Tsai, T.T., Tsai, H.C., Shih, C.N., 2006. Response of three Glycine species to salt stress. Environ. Exp. Bot. 56, 120–125.

Lan, Y., Cai, D., Zheng, Y., 2005. Expression in *Escherichia coli* of three different soybean *Late Embryogenesis Abundant* (*LEA*) genes to investigate enhanced stress tolerance. J. Integr. Plant Biol. 47, 613–621.

Lau, S., De Smet, I., Kolb, M., Meinhardt, H., Jurgens, G., 2011. Auxin triggers a genetic switch. Nat. Cell Biol. 13, 611–615.

Li, X., An, P., Inanaga, S., Eneji, E., Tanabe, K., 2006. Salinity and defoliation effects on soybean growth. J. Plant Nutr. 29, 1499–1508.

Li, W.Y.F., Shao, G., Lam, H.M., 2008b. Ectopic expression of *GmPAP3* alleviates oxidative damage caused by salinity and osmotic stresses. New Phytol. 178, 80–91.

Li, D., Inoue, H., Takahashi, M., Kojima, T., Shiraiwa, M., Takahara, H., 2008a. Molecular characterization of a novel salt inducible gene for an OSBP (oxysterol-binding protein)-homologue from soybean. Gene 407, 12–20.

Liao, H., Wong, F.L., Phang, T.H., Cheung, M.Y., Li, W.Y.F., Shao, G.H., 2003. *GmPAP3*, a novel purple acid phosphatase-like gene in soybean induced by NaCl stress but not phosphorus deficiency. Gene 318, 103–111.

Liu, Y., Gai, J.Y., Lu, H.N., Wang, Y.J., Chen, S.Y., 2005. Identification of drought tolerant germplasm and inheritance and QTL mapping of related root traits in soybean (*Glycine max* (L.) Merr.). Yi Chuan Xue Bao. 32 (8), 855–863.

Liu, X., 2009. Drought. In: Lam, H.M., Chang, R., Shao, G., Liu, Z. (Eds.), Research on Tolerance to Stresses in Chinese Soybean. China Agricultural Press, Beijing.

Luo, Q., Yu, B., Liu, Y., 2005. Differential sensitivity to chloride and sodium ions in seedlings of *Glycine max* and *G. soja* under NaCl stress. J. Plant Physiol. 162, 1003–1012.

Ma, H., Song, L., Huang, Z., Yang, Y., Wang, S., Wang, Z., Tong, J., Gu, W., Ma, H., Xiao, L., 2014. Comparative proteomic analysis reveals molecular mechanism of seedling roots of different salt tolerant soybean genotypes in responses to salinity stress. EuPA Open Proteom. 4, 40–57.

Mahajan, S., Tuteja, N., 2005. Cold, salinity and drought stresses: an overview. Arch. Biochem. Biophys. 444, 139–158.

Maj, D., Wielboa, J., Marek-Kozaczuka, M., Skorupska, A., 2010. Response to flavonoids as a factor influencing competitiveness and symbiotic activity of *Rhizobium leguminosarum*. Microbiol. Res. 165, 50–60.

Malencic, D., Popovic, M., Miladinovic, J., 2003. Stress tolerance parameters in different genotypes of soybean. Biol. Plant. 46, 141–143.

Manavalan, L.P., Guttikonda, S.K., Tran, L.S., Nguyen, H.T., 2009. Physiological and molecular approaches to improve drought resistance in soybean. Plant Cell Physiol. 50, 1260–1276.

Marinkovic, J., Djordjevic, V., Baleševic-Tubic, S., Bjelic, D., Vucelic-Radovic, B., Josic, D., 2013. Osmotic stress tolerance, PGP traits and RAPD analysis of *Bradyrhizobium japonicum* strains. Genetika 45, 75–86.

Marino, D., Frendo, P., Ladrera, R., Zabalza, A., Puppo, A., Arrese-Igor, C., González, E.M., 2007. Nitrogen fixation control under drought stress. Localized or systemic? Plant Physiol. 143, 1968–1974.

Miransari, M., 2010. Contribution of arbuscular mycorrhizal symbiosis to plant growth under different types of soil stresses. Plant Biol. 12, 563–569.

Miransari, M., 2011. Hyperaccumulators, arbuscular mycorrhizal fungi and stress of heavy metals. Biotechnol. Adv. 29, 645–653.

Miransari, M., Smith, D.L., 2007. Overcoming the stressful effects of salinity and acidity on soybean [*Glycine max* (L.) Merr.] nodulation and yields using signal molecule genistein under field conditions. J. Plant Nutr. 30, 1967–1992.

Miransari, M., Smith, D.L., 2008. Using signal molecule genistein to alleviate the stress of suboptimal root zone temperature on soybean-*Bradyrhizobium* symbiosis under different soil textures. J. Plant Interact. 3, 287–295.

Miransari, M., Smith, D., 2009. Alleviating salt stress on soybean (*Glycine max* (L.) Merr.) – *Bradyrhizobium japonicum* symbiosis, using signal molecule genistein. Eur. J. Soil Biol. 45, 146–152.

Miransari, M., Riahi, H., Eftekhar, F., Minaie, A., Smith, D.L., 2013. Improving soybean (*Glycine max* L.) N_2-fixation under stress. J. Plant Growth Regul. 32, 909–921.

Munns, R., Tester, M., 2008. Mechanisms of salinity tolerance. Annu. Rev. Plant Biol. 59, 651–681.

Ndong, C., Danyluk, J., Wilson, K.E., Pocock, T., Huner, N.P., Sarhan, F., 2002. Cold-regulated cereal chloroplast late embryogenesis abundant-like proteins: molecular characterization and functional analyses. Plant Physiol. 129, 1368–1381.

Neo, H.H., Layzell, D.B., 1997. Phloem glutamine and the regulation of O_2 diffusion in legume nodules. Plant Physiol. 113, 259–267.

Nouri, M.Z., Komatsu, S., 2010. Comparative analysis of soybean plasma membrane proteins under osmotic stress using gel-based and LC MS/MS-based proteomics approaches. Proteomics 10, 1930–1945.

Olsen, A.N., Ernst, H.A., Leggio, L.L., Skriver, K., 2005a. DNA-binding specificity and molecular functions of NAC transcription factors. Plant Sci. 169, 785–797.

Olsen, A.N., Ernst, H.A., Leggio, L.L., Skriver, K., 2005b. NAC transcription factors: structurally distinct, functionally diverse. Trends Plant Sci. 10, 79–87.

Ott, T., van Dongen, J.T., Gunther, C., Krusell, L., Desbrosses, G., Vigeolas, H., Bock, V., Czechowski, T., Geigenberger, P., Udvardi, M.K., 2005. Symbiotic leghemoglobins are crucial for nitrogen fixation in legume root nodules but not for general plant growth and development. Current Biol. 15, 531–535.

Pantalone, V.R., Kenworthy, W.J., Salughter, L.H., James, B.R., 1997. Chloride tolerance in soybean and perennial *Glycine* accessions. Emphytica 97, 235–239.

Panter, S., Thomson, R., de Bruxelles, G., Laver, D., Trevaskis, B., Udvardi, M., 2000. Identification with proteomics of novel proteins associated with the peribacteroid membrane of soybean root nodules. Mol. Plant Microbe. Interact. 13, 325–333.

Papageorgiou, G., Murata, N., 1995. The unusually strong stabilizing effects of glycine betaine on the structure and function of the oxygen-evolving photosystem II complex. Photosynth. Res. 44, 243–252.

Parker, R., Flowers, T.J., Moorem, A.L., Harpham, N.V.J., 2006. An accurate and reproducible method for proteome profiling of the effects of salt stress in the rice leaf lamina. J. Exp. Bot. 57, 1109–1118.

Peck, M.C., Fisher, R.F., Long, S.R., 2006. Diverse flavonoids stimulate NodD1 binding to nod gene promoters in *Sinorhizobium meliloti*. J. Bacteriol. 188, 5417–5427.

Phang, T., Shao, G., Lam, H., 2008. Salt tolerance in soybean. J. Integr. Plant Biol. 50, 1196–1212.

Phang, T., Shao, G., Liao, H., Yan, X., Lam, H., 2009. High external phosphate (Pi) increases sodium ion uptake and reduces salt tolerance of "Pi-tolerant" soybean. Physiol. Plant. 135, 412–425.

Sajedi, N.A., Ardakani, M.R., Rejali, F., Mohabbati, F., Miransari, M., 2010. Yield and yield components of hybrid corn (*Zea mays* L.) as affected by mycorrhizal symbiosis and zinc sulfate under drought stress. Physiol. Mol. Biol. Plants 16, 343–351.

Sajedi, N.A., Ardakani, M.R., Madani, H., Naderi, A., Miransari, M., 2011. The effects of selenium and other micronutrients on the antioxidant activities and yield of corn (*Zea mays* L.) under drought stress. Physiol. Mol. Biol. Plants 17, 215–222.

Serraj, R., Vadez, V., Sinclair, T.R., 2001. Feedback regulation of symbiotic N_2 fixation under drought stress. Agronomie 21, 621–626.

Serraj, R., Vasquez-Diaz, H., Drevon, J.J., 1998. Effects of salt stress on nitrogen fixation, oxygen diffusion, and ion distribution in soybean, common bean, and alfalfa. J. Plant Nutr. 21, 475–488.

Shao, G.H., Song, J.Z., Liu, H.L., 1986. Preliminary studies on the evaluation of salt tolerance in soybean varieties. Acta Agron. Sinica 6, 30–35.

Shao, G.H., Wan, C.W., Chang, R.Z., Chen, Y.W., 1993. Preliminary study on the damage of plasma membrane caused by salt stress. Crops 1, 39–40.

Shao, G.H., Wan, C.W., Li, S.F., 1994. Preliminary study on the physiology of soybean tolerance to salt stress at germinating stage. Crops 6, 25–27.

Shen, B., Jensen, R.G., Bohnert, H.J., 1997a. Increased resistance to oxidative stress in transgenic plants by targeting mannitol biosynthesis to chloroplasts. Plant Physiol. 113, 1177–1183.

Shen, B., Jensen, R.G., Bohnert, H.J., 1997b. Mannitol protects against oxidation by hydroxyl radicals. Plant Physiol. 115, 527–532.

Sieberer, B.J., Timmers, A.C., Emons, A.M., 2005. Nod factors alter the microtubule cytoskeleton in *Medicago truncatula* root hairs to allow root hair reorientation. Mol. Plant Microbe Interact. 18, 1195–1204.

Silvente, S., Sobolev, A., Lara, M., 2012. Metabolite adjustments in drought tolerant and sensitive soybean genotypes in response to water stress. PLoS One 7, e38554.

Singleton, P., Bohlool, B., 1984. Effect of salinity on nodule formation by soybean. Plant Physiol. 74, 72–76.

Sobhanian, H., Razavizadeh, R., Nanjo, Y., Ehsanpour, A.A., Rastgar Jazii, F., Motamed, N., Komatsu, S., 2010. Proteome analysis of soybean leaves, hypocotyls and roots under salt stress. Proteome Sci. 8, 1–15.

Sobhanian, H., Aghaei, K., Komatsu, S., 2011. Changes in the plant proteome resulting from salt stress: toward the creation of salt-tolerant crops? J. Proteomics 74, 1323–1337.

Soussi, M., Lluch, C., Ocana, A., 1999. Comparative study of nitrogen fixation and carbon metabolism in two chick-pea *Cicer arietinum* L. cultivars under salt stress. J. Exp. Bot. 50, 1701–1708.

Soussi, M., Santamaria, M., Ocana, A., Lluch, C., 2001. Effects of salinity on protein and lipopolysaccharide pattern in a salt-tolerant strain of *Mesorhizobium ciceri*. J. Appl. Microbiol. 90, 476–481.

Toorchi, M., Yukawa, K., Nouri, M., Komatsu, S., 2009. Proteomics approach for identifying osmotic-stress-related proteins in soybean roots. Peptides 30, 2108–2117.

Tunnacliffe, A., Wise, M.J., 2007. The continuing conundrum of the LEA proteins. Naturwissenschaften 94, 791–812.

Valentine, A.J., Benedito, V.A., Kang, Y., 2011. Legume nitrogen fixation and soil abiotic stress: from physiology to genomics and beyond. Annu. Plant Rev. 42, 207–248.

Van Ha, C., Le, D., Nishiyama, R., Watanabe, Y., Sulieman, S., Tran, U., Mochida, K., Dong, N., Yamaguchi-Shinozaki, K., Shinozaki, K., Tran, L., 2013. The auxin response factor transcription factor family in soybean: genome-wide identification and expression analyses during development and water stress. DNA Res. 20, 511–524.

Wan, C., Shao, G., Chen, Y., Yan, S., 2002. Relationship between salt tolerance and chemical quality of soybean under salt stress. Chin. J. Oil Crop Sci. 24, 67–72.

Wan, J.R., Torres, M., Ganapathy, A., Thelen, J., DaGue, B.B., Mooney, B., Xu, D., Stacey, G., 2005. Proteomic analysis of soybean root hairs after infection by *Bradyrhizobium japonicum*. Mol. Plant Microbe Interact. 18, 458–467.

Wang, G., Zhang, C., Lu, J., Li, J., 1999. Structural comparison of *Glycine soja* in different ecological environments. Chin. J. Appl. Environ. Biol. 10, 696–698.

Wang, H., Zhou, L., Fu, Y., Cheung, M.Y., Wong, F.L., Phang, T.H., Sun, Z., Lam, H.M., 2012. Expression of an apoplast-localized BURP-domain protein from soybean (GmRD22) enhances tolerance towards abiotic stress. Plant Cell Environ. 35, 1932–1947.

Wang, X., Pan, Q., Chen, F., Yan, X., Liao, H., 2011. Effects of co-inoculation with arbuscular mycorrhizal fungi and rhizobia on soybean growth as related to root architecture and availability of N and P. Mycorrhiza 21, 173–181.

Wang, X., Zhao, X., Jiang, C., Li, C., Cong, S., Wu, D., Chen, Y., Yu, H., Wang, C., 2014. Effects of potassium deficiency on photosynthesis and photo-protection mechanisms in soybean (*Glycine max* (L.) Merr.). J. Integr. Agric. 14 (5), 856–863.

Wei, W., Li, Q., Chu, Y., Reiter, R., Yu, X., Zhu, D., Zhang, W., Ma, B., Lin, Q., Zhang, J., Chen, S., 2015. Melatonin enhances plant growth and abiotic stress tolerance in soybean plants. J. Exp. Bot. 66 (3), 695–707.

Xie, Q., Guo, H.S., Dallman, G., Fang, S., Weissman, A.M., Chua, N.H., 2002. SINAT5 promotes ubiquitin-related degradation of NaC1 to attenuate auxin signals. Nature 419, 167–170.

Xiong, L., Zhu, J.K., 2002. Molecular and genetic aspects of plant responses to osmotic stress. Plant Cell Environ. 25, 131–139.

Xu, X., Fan, R., Zheng, R., LI, C., Yu, D., 2011. Proteomic analysis of seed germination under salt stress in soybeans. J. Zhejiang Univ. Sci. B 12, 507–517.

Yamaguchi-Shinozaki, K., Shinozaki, K., 2006. Transcriptional regulatory networks in cellular responses and tolerance to dehydration and cold stresses. Annu. Rev. Plant Biol. 57, 781–803.

Yamamoto, E., Karakaya, H.C., Knap, H.T., 2000. Molecular characterization of two soybean homologs of *Arabidopsis thaliana* CLAVATA1 from the wild type and fasciation mutant. Biochim. Biophys. Acta 1491, 333–340.

Yamamoto, E., Knap, H.T., 2001. Soybean receptor-like protein kinase genes: paralogous divergence of a gene family. Mol. Biol. Evol. 18, 1522–1531.

Yoon, J., Hamayun, M., Lee, S., Lee, I.J., 2009. Methyl jasmonate alleviated salinity stress in soybean. Crop Sci. Biotechnol. 12, 63–68.

Yu, B.J., Lam, H.M., Shao, G.H., Liu, Y.L., 2005. Effects of salinity on activities of H^+–ATPase, H^+–PPase and membrane lipid composition in plasma membrane and tonoplast vesicles isolated from soybean (*Glycine max* L.) seedlings. J. Environ. Sci. 17, 259–262.

Zhang, G., Xu, S., Hu, Q., Mao, W., Gong, Y., 2014. Putrescine plays a positive role in Salt-tolerance mechanisms by reducing oxidative damage in roots of vegetable soybean. J. Integr. Agric. 13, 349–357.

Soybean production and drought stress

Tayyaba Shaheen*, Mahmood-ur- Rahman*, Muhammad Shahid Riaz*,
Yusuf Zafar†, Mehboob-ur- Rahman**
*Department of Bioinformatics and Biotechnology, Government College University,
Faisalabad, Pakistan; **Plant Genomics and Molecular Breeding Laboratory, Agricultural
Biotechnology Division, National Institute of Biotechnology and Genetic Engineering,
Faisalabad, Pakistan; †Minister Technical, Permanent Mission of Pakistan to IAEA,
Vienna, Austria

Introduction

Soybean (*Glycine max* (L.) Merr.), which is a rich source of protein and edible oil, has many local names, for example, large bean or yellow bean in China, "edamame" in Japan (Shurtleff and Aoyagi, 2009), and the miracle bean or golden bean in the USA. The genus *Glycine* comprises two subgenera, one being *Soja* and the other, *Glycine*. The subgenus *Glycine* consists of 25 perennial species, while the subgenus *Soja* contains almost all annual species and is the dominant wild ancestor of soybean species (Newell and Hymowitz, 1983; Singh, 2006). Soybean seeds contain 40% protein, 35% carbohydrate, 20% oils, and ~5% ash. Most proteins are heat stable. Due to their heat stability, soybean food products require high temperatures for cooking. Soybean is also a major source of secondary metabolite phytoestrogen, goitrogens, oligosaccharides, and isoflavones. The chief carbohydrates in soybean are disaccharides, trisaccharides, and tetrasaccharides (Ososki and Kennelly, 2003; Phillips and Smith, 1974; Zhang et al., 2010a). Also, soybean is a source of plastic, biodiesel, hydraulic fluids, and lubricants for industrial uses. Among oilseed crops, soybean is the most cultivated crop worldwide, and collectively occupies around 6% of the world's land under cultivation (Figure 8.1; Goldsmith 2008; Sulieman et al., 2015). The total area under soybean cultivation has been increasing year on year.

Drought stress is the most devastating of the environmental stresses, affecting the productivity of soybean by up to 40% (Le et al., 2012). The main causes of drought are low humidity in the atmosphere, high temperature, and water deficiency (Hirt and Shinozaki, 2004; Szilagyi, 2003). Drought affects the carbon assimilation and phenology of the plant (Hirt and Shinozaki, 2004). Various plant species show different levels of responses to drought stress (Ito et al., 2006) including physiological, biochemical, and molecular responses (Le et al., 2012; Mochida et al., 2010; Tran et al., 2007).

Drought stress affects the germination rate and seedling vigor of the plant (Kaya et al., 2006). Under water-limited conditions, the hypocotyl length, germination, and root and shoot dry and fresh weight are reduced, while the root length is increased (Zeid and Shedeed, 2006). Growth occurs by cell differentiation and cell

Abiotic and Biotic Stresses in Soybean Production. http://dx.doi.org/10.1016/B978-0-12-801536-0.00008-6

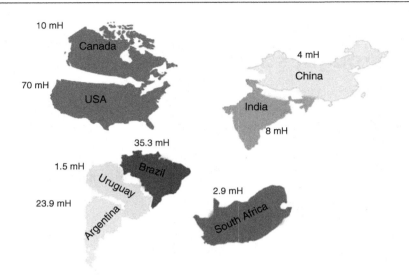

Figure 8.1 Total area under soybean crops in millions of hectares (mH).

division, which is negatively affected by the water deficiency. Under drought conditions, turgor pressure is reduced, which causes reduction in the cell elongation rate (Tripathy et al., 2000).

The important characteristics affecting plant water relations are the leaf temperature, transpiration rate, resistance of stomata, relative water content, and leaf water potential. The relative water content is high during the development of leaves and gradually decreases as leaves mature (Siddique et al., 2000). Plants exposed to water stress have a lower water potential than do plants without such stress. The ratio between the dry mass that is produced and water consumed is known as the water-use efficiency, which decreases under drought stress (Abbate et al., 2004).

For mitigating the negative impact of drought stress, the soybean plant adopts three mechanisms, that is, escape, tolerance, and avoidance (Turner et al., 2001). In the escape mechanism, the plant completes its life cycle before the onset of drought. Normally, the plants complete their life cycle very quickly and produce few seeds. For instance, early planting of soybean helps to avoid drought, and is largely practiced in the USA – planting in March to April affords escape from water stress (Boerma and Specht, 2004; Purcell and Specht, 2004). For drought avoidance, reduction in evapotranspiration rate or the efficient absorption of water from roots helps in overcoming any water stress. The tolerance mechanism is mainly dependent upon the synthesis of osmoprotectants, osmolytes, and compatible solutes (Kavar et al., 2008; Nguyen et al., 1997).

In soybean, the mathematical equation for yield under drought conditions is $Y = HI \times T \times WUE$. In this equation HI is harvest index, T is water transpired, and WUE is the water use efficiency (Turner et al., 2001). The optimum transpiration rate and increase in WUE are important factors contributing to increased yield. There are nine different traits including root density and depth, early vigor, osmotic adjustment, heat tolerance, leaf area, and phenology that are involved in increasing the T value under drought conditions. Other traits, such as leaf reflectance and the transpiration efficiency, are associated with the WUE (Boerma and Specht, 2004).

Development of drought-resistant cultivars is only possible if the material is screened under drought conditions, repeatedly for several years. A uniform drought imposed by holding irrigations is required for such experiments but is difficult to achieve (Pathan et al., 2007). To establish the relationship between the trait and water used in the soybean plant, modern facilities such as rainout shelter are used for breeding drought resistant types. It has been successfully demonstrated in China and the genetics of some complex drought resistance traits has also been described (Pennisi, 2008).

Improving drought tolerance in soybean

Morphological and physiological studies

Root traits

The soybean plant can mitigate drought by developing tap and fibrous roots as vital tools for scavenging more nutrients and water from the deeper layers of soil (Taylor et al., 1978). The depth of the rooting system in soybean is largely influenced by the elongation of the taproot. Therefore, studying the taproot elongation rate under well-watered conditions is helpful to determine the deeper rooting capability of the soybean plant. Also, under water-limited conditions, some lateral roots are produced and the number of these roots continually increases under extended periods of drought stress (Fehr et al., 1971).

The water-limited environment increases the root to shoot ratio by increasing the flow of biomass towards root. The nonirrigated plants have long root length as compared to irrigated plants (Huck et al., 1983). In earlier studies, a positive effect on taproot elongation rate was observed after the depletion of soil water even at 120 cm (Kaspar et al., 1984). In soybean, relatively high growth rate of roots was observed in plants grown under nonirrigated conditions compared with the plants grown under irrigated conditions (Hoogenboom et al., 1987). If soybean plants can develop a dense root system in early vegetative stages, they can withstand drought conditions efficiently (Hirasawa, 1994). The exotic plant introduction line PI416937 has been proved exceptionally drought resistant because it can limit transpiration during the periods of high evaporative demand (Fletcher et al., 2007).

A positive relationship exists between root traits and resistance to drought. These root traits have been used as indicators of drought resistance (Manavalan et al., 2009; Ying, 2005). Germplasm resources are available for undertaking breeding for improving root traits. However, the laborious conventional screening procedure is a major constraint on improving the root traits of soybean because of the difficulty in making selections of roots, which are not visible in the deeper layer of soils (Manavalan et al., 2009).

In this scenario, molecular and biotechnological approaches can facilitate the procedure by choosing a candidate gene and studying its association with the root traits (Manavalan et al., 2009). A well-known relationship exits between the circadian clock and response to different abiotic stresses in model plants, but little is known about the circadian clock in soybean. In soybean, under drought stress, rhythmic expression patterns of drought-responsive genes such as *RAB-18, GOLS, bZIP*, and *DREB* are changed. An *in silico* study of promoter regions of these genes has demonstrated that some *cis*-elements are associated with stress and circadian clock regulation (Marcolino-Gomes et al., 2014).

Nitrogen fixation

Drought stress causes reductions in nitrogen fixation activity of soybean plants along with the loss of CO_2 accumulation and reduced leaf area (Sinclair and Serraj, 1995), thus reducing protein synthesis and yield loss (Purcell and King, 1996). Reduction in the availability of oxygen, reduced carbon flux to nodules, deterioration in nodule sucrose synthase activity, and a rise in ureides and free amino acids cause the inhibition of N_2 fixation under drought (Arrese-Igor et al., 1999; Durand et al., 1987; González et al., 1995; King and Purcell, 2005).

In soybean, under drought conditions, nitrogenase activity has been found to be affected more than photosynthesis. Due to decreased oxygen diffusion to bacteroides, nodule activity is also affected (Naya et al., 2007). Water deficiency also causes ureide accumulation in soybean leaves, which is considered as an inhibitor of nodulation (Sinclair and Serraj, 1995). Under drought conditions, tissue turgidity, leaf and nodule function and petiole–ureide levels have been used for screening the germplasm against drought stress (Patterson and Hudak, 1996; Sall and Sinclair, 1991; Sinclair et al., 2000, 2003). On the basis of physiological screening, the N_2 fixation tolerant variety "Jackson" has been successfully utilized in breeding programs under drought conditions (Sinclair et al., 2007).

The regulation of N_2 fixation during drought has been found to be controlled at local level (Marino et al., 2007). In a study, soybean plants were grown symbiotically in a split-root system to grow half of the root system in irrigated environment and half in water-deficit conditions. A decrease in N_2 fixation was observed in the root system grown under water-limited conditions. It was found that under drought stress, the plant carbon metabolism, amino acid metabolism, protein synthesis, and cell growth are affected more in nodules. This study supports the theory of local regulation of N_2 fixation in soybean (Gil-Quintana et al., 2013). This physiological attribute can be utilized for selecting drought resistant genotypes.

Shoot traits

Stomatal conductance
Use of moisture by the whole plant under water-limited conditions depends upon stomatal conductance (Earl, 2003), which fluctuates significantly among various plant species. Effect of stomatal conductance on the exchange of water vapor and leaf gas has been studied extensively in soybean (Ray and Sinclair, 1998). In soybean, the in-

volvement of root-originated abscisic acid (ABA) in stomatal closure has been studied (Liu et al., 2003). Poor control of stomatal closure causes severe water deficiency in soybean plants (Liu et al., 2005). A feeble control of stomatal conductance was observed in soybean compared with other crops (Vignes et al., 1986). This physiological trait can potentially be used for selecting drought-tolerant soybean genotypes (Manavalan et al., 2009).

Epidermal conductance

Epidermal conductance under drought conditions plays a significant role in conferring drought tolerance, and thus can be used as one of the selection criteria in breeding soybean-conferring tolerance to drought stress (Van Gardingen and Grace, 1992). A low leaf epidermal conductance has been observed in varieties adapted to arid environments (Riederer and Schreiber, 2001). In soybean, significant diversity for epidermal conductance that was not associated with stomatal density was observed (Paje et al., 1988). Minor epidermal conductance has been found to be a trait of choice for developing drought-tolerant varieties (Hufstetler et al., 2007). Stomatal conductance and water potential were simultaneously found to be high in soybean genotypes adapted to temperate regions compared with perennial wild type genotypes (James et al., 2008a). It is a highly heritable trait (James et al., 2008b). This trait needs to be further explored for utilization in breeding programs.

Leaf pubescence density

Leaf pubescence density may improve drought tolerance in soybean by enhancing vegetative growth, reducing leaf temperature, reducing water loss by transpiration, and increasing photosynthesis (Garay and Wilhelm, 1983; Specht and Williams, 1985). However, the high pubescence trait, conferred by two alleles *Pd1* and *Pd2*, affects the important agronomic traits, for example, reduction in seed yield and increased plant height, making the crop vulnerable to lodging and delaying maturity (Pfeiffer et al., 2003; Specht et al., 1985). This negative correlation should also be considered before using this trait in breeding.

Water use efficiency

WUE is the total accumulated biomass per used unit of water. Soybean varieties show genetic variation in WUE (Mian et al., 1996). Enhancing WUE of a plant can result in better yields because a positive correlation exists between WUE and total biomass yield (Wright, 1996). In soybean, the WUE can be improved by simply selecting the genotypes based on mean yield (Hufstetler et al., 2007; Specht et al., 2001). An increase in WUE can be achieved by regulating stomatal partial closure at a specific level of water deficiency in the soil (Liu et al., 2005). Further research into this is needed.

Osmotic adjustment

Accumulation of solutes in response to enhanced water deficiency is called osmotic adjustment (OA). It is helpful in maintaining the stomatal conductance and photosynthesis under water stress, thus delaying leaf death, reducing the flower abortion and enhancing root growth (Turner et al., 2001). In multiple studies, genetic variations in

OA in soybean germplasms were found to be narrower than those of the other legumes (Cortes and Sinclair, 1986; James et al., 2008a; Turner et al., 2001).

It has been demonstrated that soybean varieties with high osmotic potential exhibit a slower decrease in relative water content, thus maintaining turgor for a longer period of time (James et al., 2008b). In other crops, inconsistent results and inefficient methods for measuring the OA are major obstacles in drawing comprehensive results (Serraj and Sinclair, 2002; Turner et al., 2007). The process for the identification of molecular markers linked with high OA requires time, as has been demonstrated in other model crops (Turner et al., 2001).

Molecular studies in soybean for improved drought tolerance

The genome sequencing era coupled with the availability of advanced microarray technology has revolutionized the research in understanding the dehydration-responsive genes in model plants as well as in other crops (Le et al., 2012). Under drought stress a number of genes are up- and downregulated to combat the stress (Kavar et al., 2008). Thus tolerance to drought is a very complex mechanism involving the action and interactions of multiple genes (Cattivelli et al., 2008). Different structural and functional genomic approaches have been used for exploring these mechanisms in soybean.

Structural genomic studies

Genetic mapping and marker-assisted selection (MAS) for drought tolerance

DNA markers can be divided into two categories: nonpolymerase chain reaction based markers and PCR-based markers. These markers are often linked with quantitative trait loci (QTLs) (Kumar, 2013). Accessibility of whole genome sequence (WGS) information has further improved the application of genomic information in breeding soybean (Schmutz et al., 2010). Thousands of simple sequence repeats (SSRs) and millions of SNPs have been developed on the basis of WGS (Sonah et al., 2013; Song et al., 2010).

Detailed genetic and physical maps of soybean were developed using SSRs derived from a bacterial artificial chromosome library comprising 92,160 clones with approximately 12 genome equivalents coverage (Gustafson et al., 2008). Moreover, a genetic map was integrated into this physical map that anchored 1000 SSRs and STS markers, which were associated with seed quality traits, disease resistance, and drought response (Wu et al., 2008).

Next generation sequencing assays have provided a cost effective means of sequencing-based genotyping. Restriction site associated DNA sequencing, reduced representation libraries, and genotyping-by-sequencing (GBS) are generally used, and GBS is considered the most simple and cost-effective method (Elshire et al., 2011; Sonah et al., 2012). For soybean GBS, the methodology has been upgraded (Sonah et al., 2013). SNP array based on thousands of available SNPs has been developed in soybean (Song et al., 2013).

These new techniques are very promising for the research aimed at improving the soybean genome against in relation to drought stress (Blum, 2005). As in many other crops, improvement of soybean genetics to in relation to drought is very difficult due to the complex nature of drought tolerance mechanisms, which has impaired the progress in breeding towards developing the high yielding lines under drought stress.

In this genomic era, progress towards the identification of drought-related QTLs has been accelerated (Pathan et al., 2007). In soybean, numerous QTLs have been identified for yield components and tolerance to biotic and abiotic stresses (www.soykb.org; www.soybase.org). The identification of QTLs related with drought resistance traits has also been done in soybean. Most of the identified QTLs associated with drought resistance were not applicable to unrelated populations because these were identified using small populations or a single population (Nicholas, 2006). Confirmation of QTLs and genes by screening across different environments and genetic backgrounds and development of dense genetic maps with sophisticated DNA markers are urgently needed to develop drought resistant soybean (Vuong et al., 2007).

Among the important traits associated with drought resistance, QTLs for WUE and leaf ash content were mapped in soybean using restriction fragment length polymorphism markers (Mian et al., 1996). The consistency of WUE QTLs was also validated in other populations (Mian et al., 1998). Specht et al. (2001) identified one QTL associated with yield and Bhatnagar et al. (2005) identified one associated with leaf wilting, while Monteros et al. (2006) identified three QTLs associated with yield and three associated with wilting.

Du et al. (2009) identified 19 QTLs associated with seed yield per plant under water stress (YP–WS) and seed yield per plant in well-watered (YP–WW) conditions, and 10 QTLs associated with drought susceptibility index (DSI) in soybean. Reports on mapping QTLs controlling biological-nitrogen fixation are very few in plants including soybean. Two QTLs for shoot dry weight (LGs namely E and L), three for nodule number (LGs B1, E, and I), and one for ratio nodule dry weight (LG I) were identified using the composite interval mapping for multiple traits method.

All QTLs contributed a small effect (R^2-values ranging from 1.7% to 10.0%) and described 15.4, 13.8, and 6.5% of total variation for shoot dry weight, nodule number, and ratio nodule dry weight, respectively (Santos et al., 2013) (Table 8.1). Inheritance of tolerance to abiotic stresses is complex in soybean and thus utilization of the information obtained from QTLs is difficult. "Meta-QTL analysis" – a statistical tool – has been developed, which can compile QTL data from different studies on the same linkage map (Deshmukh et al., 2012; Sosnowski et al., 2012).

Functional genomic studies for drought tolerance in soybean

Functional genomics is based on the study of various genes functions. The sequencing of soybean genome and the development of detailed genetic and physical maps have revolutionized the genomic research in soybean (Shinozaki, 2007). Initially, expressed sequence tag (EST)–SSRs based genetic maps were developed (Hisano et al., 2007; Xia et al., 2007).

Table 8.1 **Studies conducted in soybean for QTL mapping**

QTL numbers and traits	Mapping population	Linkage groups (LGs)	References
Five QTLs associated with water use efficiency	Young × PI416937, 120 F$_4$ population	LG-G, LG-H, LG-J, LG-J, and LG-C1	Mian et al. (1996)
Two QTLs associated with water use efficiency	S-100 × Tokyo, 116 F2 population	LG-L, unlinked	Mian et al. (1998)
One QTL associated with yield	Minsoy × Noir 1236 RILs	LG-C2	Specht et al. (2001)
One QTL associated with leaf wilting	Jackson × KS4895, 81 RILs	LG-K	Bhatnagar et al. (2005)
Three QTLs associated with yield and three associated with wilting	Hutcheson × PI471938, 140 F$_4$ population	LG-D2, LG-F1, and LG-F2	Monteros et al. (2006)
Nineteen QTLs associated with seed yield per plant under water stress (YP–WS) and seed yield per plant well-watered (YP–WW), and 10 QTLs associated with drought susceptibility index (DSI)	Kefeng1 × Nan-nong1138-2 F$_{2:7:11}$	LG-K, LG-C2, LG-H, and LG-A1	Du et al. (2009)
Two QTLs for shoot dry weight (LGs E and L), three for nodule number (LGs B1, E, and I), and one for ratio nodule dry weight (LG I)	Bossier × Embrapa 20, F 2:7 RILs	LGs E and L; LGs B1, E, and I; LG-I	Santos et al. (2013)

The availability of 66,153 protein-coding loci of soybean (http://www.phytozome.net/soybean #C) is a valuable source for studying the function of various genes. EST sequencing projects and availability of spotted cDNA microarrays for specific tissues and particular stresses are top-rate resources (Vodkin et al., 2004). The ESTs derived from root tips of soybean after exposure to drought stress are promising for exploring the genes conferring drought stress (Valliyodan and Nguyen, 2008). A total of 6570 new full-length gene cDNAs derived from abiotic stress treated tissues were captured to analyze gene function (Umezawa et al., 2008). A comparative analysis was conducted between the derived sequences from soybean and *Arabidopsis*, rice, and other legumes. This large set of full-length cDNA clones for soybean is a useful resource for the new functions found for a gene from soybean and will also provide assistance in a precise annotation of the soybean genome (Umezawa et al., 2008). Furthermore, metabolic profiling along with transcriptomic data and proteomic analyses can be valuable for understanding the physiology and biochemistry of cells (Morgenthal et al., 2007). It has been shown that the root is the most drought-responsive organ in the soybean plant under drought stress (Mohammadi et al., 2012), which can be exploited for selection of drought-tolerant genotypes.

The development of metabolite profiling technologies like gas chromatography or liquid chromatography coupled with mass spectrometry (GC-time-of-flight (TOF)/MS or UPLC-Q-TOF-MS) have made it possible to explore metabolomics profiles under stress conditions (Urano et al., 2009). Detailed temporal transcriptional and metabolic changes in developing soybean embryos were analyzed to classify the potential targets for metabolic engineering (Collakova et al., 2013).

Several regulators involved in various growth stages, especially in the grain filling stage, splice variants, and 3400 new genes, were identified (Collakova et al., 2013). These genes can be explored at length for validating their role in desiccation tolerance. Such kinds of novel genes can be introduced in other crops species (wheat, rice, and soybean) for enhancing the drought tolerance capabilities (de Ronde et al., 2004a, b; Seki et al., 2007; Umezawa et al., 2006).

Utilization of antisense and RNA interference (RNAi)-mediated transcriptional or posttranscriptional gene silencing in soybean has also been reported to study the function of genes (Buhr et al., 2002; Nunes et al., 2006; Subramanian et al., 2005). Ribozyme termination of RNA transcripts was used by Buhr et al. (2002) to downregulate the seed fatty acid genes in transgenic soybean. Subramanian et al. (2005) observed that RNAi of isoflavone synthase genes in soybean results in silencing of tissues distal to the transformation site and thus enhances the susceptibility to *Phytophthora sojae*. Using RNAi-mediated silencing of the myo-inositol-1-phosphate synthase gene (*GmMIPS1*) in transgenic soybean, Nunes et al. (2006) inhibited the development of seed and reduced the phytate content. These results are promising for the utilization of this technique to verify the involvement of specific genes in drought tolerance. Mutant plants and targeting-induced local lesions in genomes (TILLING) have been researched and used in four mutagenized soybean populations (Bhatia et al., 1999; Cooper et al., 2008; Men et al., 2002).

It was concluded that the substantial mutation density together with a supplementary method for screening closely related targets, specifies soybean as an appropriate organism for high-throughput mutation detection even with its extensively duplicated genome. Insertional mutagenesis like T-DNA mutant development is time consuming and labor-intensive in soybean for generating a satisfactory number of insertions (Parrott and Clemente, 2004). It can be overcome by coupling T-DNA regions and transposon-based elements (Brutnell, 2002). These strategies can be utilized to enhance drought tolerance in soybean.

Identification of loss of mutant functions in paleopolyploid like soybean is difficult due to duplications of various genes (Mathieu et al., 2009). This can be avoided by a gain-of-function approach to develop mutants by introducing multiple enhancer sequences into plants using transformation (Jeong et al., 2006). Mathieu et al. (2009) generated a transposon-based mutant collection of soybean. The Ds transposon system was used to create activation tagging, gene, and enhancer trap elements. This is an example demonstrating that although the nature of soybean is tetraploid, it is possible to disrupt soybean gene function by insertional mutagenesis.

Expression levels of multiple genes including *gmdreb1*, *gmGOLS*, *gmpip1b*, *Gmereb*, and *Gmdefensin* were studied in soybean genotypes (tolerant as well as sensitive) after exposing them to drought stress. The experimental data showed that the expression of *gmdreb1* and *GmGOLS* increased in the roots and leaves of both type of genotypes when exposed to drought stress. These drought-responsive genes can also be used to develop drought-tolerant soybean plants (Stolf-Moreira et al., 2010).

At present, 66 K Affymetrix soybean array gene chip is available to study the genome expression profiling of the soybean plant. A total of almost 1818 genes were upregulated while 1688 were downregulated in leaf tissues of soybean plant after exposure to water-limited conditions (Le et al., 2012). Using the *in silico* analysis, four drought-responsive genes involved in ABA-independent and ABA-dependent pathways were studied. The expression pattern of these genes was studied using quantitative PCR (Neves-Borges et al., 2012), illustrating that gene expression was changed under drought conditions.

One of the drought-responsive genes is *LEA-D11*, which is involved in the synthesis of dehydrin protein, stabilizing the membrane structure. This gene is a potential target for improving the drought tolerance in other crop species (Savitri et al., 2013). Transcriptome analysis coupled with microarray technology is very effective in dissecting the abiotic stress-responsive pathway. Use of an Affymetrix GeneChip with 61 K probe sets is now a routine for conducting the transcriptome profiling under abiotic stress (Haerizadeh et al., 2011; Le et al., 2012). In another study, it was indicated that the expression levels of GmNAC genes (*GmNAC011, 085, 092, 095, 101,* and *109*) can be exploited for improving drought tolerance in soybean (Thao et al., 2013).

Utilization of genetic engineering technology in soybean to study drought tolerance

To date, considerable research has been conducted on the utilization of genetic engineering methods for engineering various traits including resistance to drought in many crop species (Hu et al., 2006, 2008; Nakashima et al., 2007; Tran et al., 2007a, b; Valliyodan and Nguyen, 2006; Yamaguchi-Shinozaki and Shinozaki, 2006). The first drought-tolerant soybean plant was engineered by introducing a *P5CR* gene, isolated from *Arabidopsis*, showing enhanced tolerance to drought (de Ronde et al., 2004a, b; Kocsy et al., 2005).

More emphasis has been given on isolation of drought-related genes and transcription factors followed by studying their expression in model plant species (Table 8.2). A gene, *GmDREB2*, (homolog of *AtDREB*) was isolated from the soybean, and this gene enhanced the survival rate of transgenic plants under drought and salinity stresses (Chen et al., 2007). Expression of basic-leucine zipper genes encoding bZIP transcription factors isolated from soybean has shown enhanced freezing and drought tolerance in *Arabidopsis* transgenic plants (Liao et al., 2008). When expressed in tobacco transgenic plants, the *GmERF3* gene (isolated from soybean) showed an improved tolerance to drought (Zhang et al., 2009). A chilling inducible *GmCHI* gene, which

Table 8.2 Transcription factors identified in soybean found to be upregulated under drought stress

S. no.	Transcription factors	Plant systems used for expression	References
1	GmDREB	Wheat	Gao et al. (2010)
2	GmDREB2	*A. thaliana*	Chen et al. (2007)
3	GmERF3 b	Tobacco	Zhang et al. (2009)
4	GmERF4	Tobacco	Zhang et al. (2010b)
5	GmERF089	Tobacco	Liao et al. (2008)
6	GmbZIP1	*A. thaliana*, wheat	Gao et al. (2011)
7	GmGT-2A	*A. thaliana*	Xie et al. (2010)
8	GmGT-2B b	*A. thaliana*	Xie et al. (2010)
9	GsZFP1	*A. thaliana*	Luo et al. (2012)
10	GmWRKY54	*A. thaliana*	Zhou et al. (2008)
11	GsWRKY20	*A. thaliana*	Luo et al. (2013)
12	GmMYBJ1	*A. thaliana*	Su et al. (2014)

belongs to the GmERF transcription factor family, enhanced drought tolerance when overexpressed in *Arabidopsis* (Cheng et al., 2009).

Tran et al. (2009) used 31 *GmNAC* genes, including six genes isolated by Meng et al. (2007), in soybean for studying their response to drought stress. They demonstrated that the expression of nine genes out of 31 was induced by dehydration (Tran et al. 2009). The NAC gene family has been found to be a major group of transcription factors that plays a role in the development of roots and stress-resistance in plants, thus these genes are promising for improving drought stress tolerance (Hu et al., 2006, 2008; Nakashima et al., 2007; Tran et al., 2004).

Transcription factors such as MYB play vital roles in conferring tolerance to drought. The *GmMYBJ1* gene was isolated from soybean plant and its expression in *Arabidopsis thaliana* was investigated. This gene is 1296 base pairs long, encoding 271 amino acids. The expression of the *GmMYBJ1* gene is induced under drought stress (Su et al., 2014). The *NTR1* gene isolated from *Brassica campestris* (encoding jasmonic acid carboxyl methyltransferase) was tested in soybean, resulting in the enhanced concentration of methyl jasmonate and improved tolerance toward desiccation during seed germination (Xue et al., 2007).

In a study, the *DREB1D* transcription factor in *A. thaliana* along with the constitutive and ABA-inducible promoters was inserted into soybean using *Agrobacterium tumefaciens*-mediated gene transfer methodology. Transgenic plants with an ABA-inducible promoter indicated a 1.5- to twofold increase of transgene expression under severe stress conditions. Transgenic plants exhibited high drought tolerance compared with nontransgenic plants. This study enhances the possibility of engineering soybean for enhanced drought tolerance by expressing stress-responsive genes (Guttikonda et al., 2014).

In another study, *GsWRKY20* was identified as a stress response gene. This gene was overexpressed in *Arabidopsis*, conferring enhanced drought tolerance and up-regulation of ABA signaling compared with wild species. Microarray and quantitative real-time PCR assays showed that *GsWRKY20* facilitated ABA signaling by stimu-lating the expression of negative regulators of ABA signaling, such as AtWRKY40, ABI1, and ABI2, while suppressing the expression of the positive regulators of ABA, for example ABI5, ABI4, and ABF4. It was also found that *GsWRKY20* upregulated the expression of a group of wax biosynthesis genes (Luo et al., 2013). These studies have shown the proficiency of soybean drought-responsive genes and transcription factors. There is a need to emphasize the importance of the development of transgenic soybean plants with improved characters against drought stress.

Conclusions

The challenges of higher production with less availability of vital resources of land, water, and other expensive inputs of fertilizers and agrochemicals need to be addressed. Among these, drought is a major limiting factor, which substantially hampers produc-tion annually worldwide. In the present genomic era, exploring the resilient genetic re-sources followed by their utilization in developing drought-tolerant cultivars has been one of the alternative approaches adopted for combating drought. Though modest ef-forts have been made for exploring the genetics of drought tolerance, there is a need to explore various genetic pathways using genomic tools and genetic resources that would pave the way for developing resilient cultivars, which was not possible previously. Ex-ploring the transcriptomes, proteomes, and metabolomes of the developing seed under drought would help in the identification of genes and their regulators, and this infor-mation could supplement the breeding programs in soybean. Moreover, TILLING ap-proaches can be used to identify SNPs, which could help in the identification of genes – a potential source for genetic engineering. Lastly, a blend of conventional breeding, marker-assisted breeding, and genetic engineering approaches will be required for im-proving the genetics of soybean for mitigating drought stress in the future.

References

Abbate, P.E., Dardanelli, J.L., Cantarero, M.G., Maturano, M., Melchiori, R.J.M., Suero, E.E., 2004. Climatic and water availability effects on water-use efficiency in wheat. Crop Sci. 44, 474–483.

Arrese-Igor, C., González, E., Gordon, A., Minchin, F., Gálvez, L., Royuela, M., Cabrerizo, P., Aparicio-Tejo, P, 1999. Sucrose synthase and nodule nitrogen fixation under drought and other environmental stresses. Symbiosis 27, 189–212.

Bhatia, C., Nichterlein, K., Maluszynski, M, 1999. Oilseed cultivars developed from induced mutations and mutations altering fatty acid composition. Mutat. Breed. Rev. 11, 1–36.

Bhatnagar, S., King, C.A., Purcell, L., Ray, J.D., 2005. Identification and mapping of quantita-tive trait loci associated with crop responses to water-deficit stress in soybean [Glycine max (L.) Merr.]. The ASACSSA-SSSA International annual meeting poster abstract, November 6–10, 2005, Salt Lake City, UT, USA.

Blum, A., 2005. Drought resistance, water-use efficiency, and yield potential – are they compatible, dissonant, or mutually exclusive? Aust. J. Agric. Res. 56, 1159–1168.

Boerma, H.R., Specht, J.E., 2004 Soybeans: Improvement, Production and Uses. American Society of Agronomy and Crop Science, Society of America Soil Science, Madison, WI.

Brutnell, T.P., 2002. Transposon tagging in maize. Funct. Integr. Genomics 2, 4–12.

Buhr, T., Sato, S., Ebrahim, F., Xing, A., Zhou, Y., Mathiesen, M., Schweiger, B., Kinney, A., Staswick, P., 2002. Ribozyme termination of RNA transcripts down-regulate seed fatty acid genes in transgenic soybean. Plant J. 30, 155–163.

Cattivelli, L., Rizza, F., Badeck, F.-W., Mazzucotelli, E., Mastrangelo, A.M., Francia, E., Mare, C., Tondelli, A., Stanca, A.M., 2008. Drought tolerance improvement in crop plants: an integrated view from breeding to genomics. Field Crops Res. 105, 1–14.

Chen, M., Wang, Q.-Y., Cheng, X.-G., Xu, Z.-S., Li, L.-C., Ye, X.-G., Xia, L.-Q., Ma, Y.-Z., 2007. GmDREB2, a soybean DRE-binding transcription factor, conferred drought and high-salt tolerance in transgenic plants. Biochem. Biophys. Res. Commun. 353, 299–305.

Cheng, L., Huan, S., Sheng, Y., Hua, X., Shu, Q., Song, S., Jing, X., 2009. GMCHI, cloned from soybean [Glycine max (L.) Merr.], enhances survival in transgenic Arabidopsis under abiotic stress. Plant Cell Rep. 28, 145–153.

Collakova, E., Aghamirzaie, D., Fang, Y., Klumas, C., Tabataba, F., Kakumanu, A., Myers, E., Heath, L., Grene, R., 2013. Metabolic and transcriptional reprogramming in developing soybean (Glycine max) embryos. Metabolites 3, 347–372.

Cooper, J.L., Till, B.J., Laport, R.G., Darlow, M.C., Kleffner, J.M., Jamai, A., El-Mellouki, T., Liu, S., Ritchie, R., Nielsen, N., 2008. TILLING to detect induced mutations in soybean. BMC Plant Biol. 8, 9.

Cortes, P., Sinclair, T., 1986. Water relations of field-grown soybean under drought. Crop Sci. 26, 993–998.

De Ronde, J., Cress, W., Krüger, G., Strasser, R., Van Staden, J., 2004a. Photosynthetic response of transgenic soybean plants, containing an Arabidopsis P5CR gene, during heat and drought stress. J. Plant Physiol. 161, 1211–1224.

De Ronde, J., Laurie, R., Caetano, T., Greyling, M., Kerepesi, I., 2004b. Comparative study between transgenic and non-transgenic soybean lines proved transgenic lines to be more drought tolerant. Euphytica 138, 123–132.

Deshmukh, R.K., Sonah, H., Kondawar, V., Tomar, R.S.S., Deshmukh, N.K., 2012. Identification of meta quantitative trait loci for agronomical traits in rice (Oryza sativa). Indian J. Genet. Plant Breed. 72, 264–270.

Du, W., Wang, M., Fu, S., Yu, D., 2009. Mapping QTLs for seed yield and drought susceptibility index in soybean (Glycine max L.) across different environments. J. Genet. Genomics 36, 721–731.

Durand, J., Sheehy, J., Minchin, F., 1987. Nitrogenase activity, photosynthesis and nodule water potential in soyabean plants experiencing water deprivation. J. Exp. Bot. 38, 311–321.

Earl, H.J., 2003. A precise gravimetric method for simulating drought stress in pot experiments. Crop Sci. 43, 1868–1873.

Elshire, R.J., Glaubitz, J.C., Sun, Q., Poland, J.A., Kawamoto, K., Buckler, E.S., Mitchell, S.E., 2011. A robust, simple genotyping-by-sequencing (GBS) approach for high diversity species. PLoS One 6, e19379.

Fehr, W., Caviness, C., Burmood, D., Pennington, J., 1971. Stage of development descriptions for soybeans, Glycine max (L.) Merrill. Crop Sci. 11, 929–931.

Fletcher, A.L., Sinclair, T.R., Allen JR, L.H., 2007. Transpiration responses to vapor pressure deficit in well watered "slow-wilting" and commercial soybean. Environ. Exp. Bot. 61, 145–151.

Gao, F., Xiong, A., Peng, R., Jin, X., Xu, J., Zhu, B., Chen, J., Yao, Q., 2010. OsNAC52, a rice NAC transcription factor, potentially responds to ABA and confers drought tolerance in transgenic plants. PCTOC 100, 255–262.

Gao, S.-Q., Chen, M., Xu, Z.-S., Zhao, C.-P., Li, L., Xu, H.-J., Tang, Y.-M., Zhao, X., Ma, Y.-Z., 2011. The soybean GmbZIP1 transcription factor enhances multiple abiotic stress tolerances in transgenic plants. Plant Mol. Biol. 75, 537–553.

Garay, A., Wilhelm, W., 1983. Root system characteristics of two soybean isolines undergoing water stress conditions. Agron. J. 75, 973–977.

Gil-Quintana, E., Larrainzar, E., Seminario, A., Díaz-Leal, J.L., Alamillo, J.M., Pineda, M., Arrese-Igor, C., Wienkoop, S., González, E.M., 2013. Local inhibition of nitrogen fixation and nodule metabolism in drought-stressed soybean. J. Exp. Bot. 64, 2171–2182.

Goldsmith, P.D., 2008. Economics of soybean production, marketing, and utilization. In: Johnson, L.P., White, P.A., Galloway, R. (Eds.), Soybeans: Chemistry, Production, Processing, and Utilization. AOCS Press, Urbana, IL, pp. 117–150.

González, E., Gordon, A., James, C., Arrese-Lgor, C., 1995. The role of sucrose synthase in the response of soybean nodules to drought. J. Exp. Bot. 46, 1515–1523.

Gustafson, P., Shoemaker, R.C., Grant, D., Olson, T., Warren, W.C., Wing, R., Yu, Y., Kim, H., Cregan, P., Joseph, B., 2008. Microsatellite discovery from BAC end sequences and genetic mapping to anchor the soybean physical and genetic maps. Genome 51, 294–302.

Guttikonda, S.K., Valliyodan, B., Neelakandan, A.K., Tran, L.-S.P., Kumar, R., Quach, T.N., Voothuluru, P., Gutierrez-Gonzalez, J.J., Aldrich, D.L., Pallardy, S.G., 2014. Overexpression of AtDREB1D transcription factor improves drought tolerance in soybean. Mol. Biol. Rep. 14 (12), 7995–8008.

Haerizadeh, F., Singh, M.B., Bhalla, P.L., 2011. Transcriptome profiling of soybean root tips. Funct. Plant Biol. 38, 451–461.

Hirasawa, T., 1994. Effects of pre-flowering soil moisture deficits on dry matter production and ecophysiological characteristics in soybean plants under drought conditions during grain filling. Jpn. J. Crop Sci. 63, 721–730.

Hirt, H., Shinozaki, K., 2004. Plant Responses to Abiotic Stress. Springer, Berlin.

Hisano, H., Sato, S., Isobe, S., Sasamoto, S., Wada, T., Matsuno, A., Fujishiro, T., Yamada, M., Nakayama, S., Nakamura, Y., 2007. Characterization of the soybean genome using EST-derived microsatellite markers. DNA Res. 14, 271–281.

Hoogenboom, G., Huck, M., Peterson, C.M., 1987. Root growth rate of soybean as affected by drought stress. Agron. J. 79, 607–614.

Hu, H., Dai, M., Yao, J., Xiao, B., Li, X., Zhang, Q., Xiong, L., 2006. Overexpressing a NAM, ATAF, and CUC (NAC) transcription factor enhances drought resistance and salt tolerance in rice. Proc. Natl Acad. Sci. USA 103, 12987–12992.

Hu, H., You, J., Fang, Y., Zhu, X., Qi, Z., Xiong, L., 2008. Characterization of transcription factor gene SNAC2 conferring cold and salt tolerance in rice. Plant Mol. Biol. 67, 169–181.

Huck, M.G., Ishihara, K., Peterson, C.M., Ushijima, T., 1983. Soybean adaptation to water stress at selected stages of growth. Plant Physiol. 73, 422–427.

Hufstetler, E.V., Boerma, H.R., Carter, T.E., Earl, H.J., 2007. Genotypic variation for three physiological traits affecting drought tolerance in soybean. Crop Sci. 47, 25–35.

Ito, Y., Katsura, K., Maruyama, K., Taji, T., Kobayashi, M., Seki, M., Shinozaki, K., Yamaguchi-Shinozaki, K., 2006. Functional analysis of rice DREB1/CBF-type transcription factors involved in cold-responsive gene expression in transgenic rice. Plant Cell Physiol. 47, 141–153.

James, A., Lawn, R., Cooper, M., 2008. Genotypic variation for drought stress response traits in soybean. I. Variation in soybean and wild *Glycine* spp. for epidermal conductance, osmotic potential, and relative water content. Aust. J. Agric. Res. 59, 656–669.

Jeong, D.H., An, S., Park, S., Kang, H.G., Park, G.G., Kim, S.R., Sim, J., Kim, Y.O., Kim, M.K., Kim, S.R., 2006. Generation of a flanking sequence-tag database for activation-tagging lines in japonica rice. Plant J. 45, 123–132.

Kaspar, T., Taylor, H., Shibles, R., 1984. Taproot-elongation rates of soybean cultivars in the glasshouse and their relation to field rooting depth. Crop Sci. 24, 916–920.

Kavar, T., Maras, M., Kidrič, M., Šuštar-Vozlič, J., Meglič, V., 2008. Identification of genes involved in the response of leaves of *Phaseolus vulgaris* to drought stress. Mol. Breed. 21, 159–172.

Kaya, M.D., Okçu, G., Atak, M., Çikili, Y., Kolsarici, Ö., 2006. Seed treatments to overcome salt and drought stress during germination in sunflower (*Helianthus annuus* L.). Eur. J. Agron. 24, 291–295.

King, C.A., Purcell, L.C., 2005. Inhibition of N_2 fixation in soybean is associated with elevated ureides and amino acids. Plant Physiol. 137, 1389–1396.

Kocsy, G., Laurie, R., Szalai, G., Szilágyi, V., Simon-Sarkadi, L., Galiba, G., De Ronde, J.A., 2005. Genetic manipulation of proline levels affects antioxidants in soybean subjected to simultaneous drought and heat stresses. Physiol. Plant. 124, 227–235.

Kumar, S., 2013. Potential of molecular markers in plant biotechnology. Int. J. Plant Sci. 8, 426–444.

Le, D.T., Nishiyama, R., Watanabe, Y., Tanaka, M., Seki, M., Yamaguchi-Shinozaki, K., Shinozaki, K., Tran, L.-S.P., 2012. Differential gene expression in soybean leaf tissues at late developmental stages under drought stress revealed by genome-wide transcriptome analysis. PLoS One 7, e49522.

Liao, Y., Zhang, J.S., Chen, S.Y., Zhang, W.K., 2008. Role of soybean GmbZIP132 under abscisic acid and salt stresses. J. Integr. Plant Biol. 50, 221–230.

Liu, F., Andersen, M.N., Jensen, C.R., 2003. Loss of pod set caused by drought stress is associated with water status and ABA content of reproductive structures in soybean. Funct. Plant Biol. 30, 271–280.

Liu, Y., Gai, J.-Y., Lu, H., Wang, Y.-J., Chen, S.-Y., 2005. Identification of drought tolerant germplasm and inheritance and QTL mapping of related root traits in soybean [*Glycine max* (L.) Merr.]. Yi Chuan Xue Bao. 32, 855–863.

Luo, X., Bai, X., Sun, X., Zhu, D., Liu, B., Ji, W., Cai, H., Cao, L., Wu, J., Hu, M., 2013. Expression of wild soybean *WRKY20* in *Arabidopsis* enhances drought tolerance and regulates ABA signalling. J. Exp. Bot. 64, 2155–2169.

Luo, X., Bai, X., Zhu, D., Li, Y., Ji, W., Cai, H., Wu, J., Liu, B., Zhu, Y., 2012. GsZFP1, a new Cys2/His2-type zinc-finger protein, is a positive regulator of plant tolerance to cold and drought stress. Planta 235, 1141–1155.

Manavalan, L.P., Guttikonda, S.K., Tran, L.-S.P., Nguyen, H.T., 2009. Physiological and molecular approaches to improve drought resistance in soybean. Plant Cell Physiol. 50, 1260–1276.

Marcolino-Gomes, J., Rodrigues, F.A., Fuganti-Pagliarini, R., Bendix, C., Nakayama, T.J., Celaya, B., Molinari, H.B.C., De Oliveira, M.C.N., Harmon, F.G., Nepomuceno, A., 2014. Diurnal oscillations of soybean circadian clock and drought responsive genes. PLoS One 9, e86402.

Marino, D., Frendo, P., Ladrera, R., Zabalza, A., Puppo, A., Arrese-Igor, C., González, E.M., 2007. Nitrogen fixation control under drought stress. Localized or systemic? Plant Physiol. 143, 1968–1974.

Mathieu, M., Winters, E.K., Kong, F., Wan, J., Wang, S., Eckert, H., Luth, D., Paz, M., Donovan, C., Zhang, Z., 2009. Establishment of a soybean (*Glycine max* Merr. L) transposon-based mutagenesis repository. Planta 229, 279–289.

Men, A.E., Laniya, T.S., Searle, I.R., Iturbe-Ormaetxe, I., Gresshoff, I., Jiang, Q., Carroll, B.J., Gresshoff, P.M., 2002. Fast neutron mutagenesis of soybean (Glycine soja L.) produces a supernodulating mutant containing a large deletion in linkage group H. Genome Lett. 1, 147–155.

Meng, Q., Zhang, C., Gai, J., Yu, D., 2007. Molecular cloning, sequence characterization and tissue-specific expression of six NAC-like genes in soybean (Glycine max (L.) Merr.). J. Plant Physiol. 164, 1002–1012.

Mian, M., Ashley, D., Boerma, H., 1998. An additional QTL for water use efficiency in soybean. Crop Sci. 38, 390–393.

Mian, M., Bailey, M., Ashley, D., Wells, R., Carter, T., Parrott, W., Boerma, H., 1996. Molecular markers associated with water use efficiency and leaf ash in soybean. Crop Sci. 36, 1252–1257.

Mochida, K., Yoshida, T., Sakurai, T., Yamaguchi-Shinozaki, K., Shinozaki, K., Tran, L.-S.P., 2010. Genome-wide analysis of two-component systems and prediction of stress-responsive two-component system members in soybean. DNA Res. 17, 303–324.

Mohammadi, P.P., Moieni, A., Hiraga, S., Komatsu, S., 2012. Organ-specific proteomic analysis of drought-stressed soybean seedlings. J. Proteomics 75, 1906–1923.

Monteros, M., Lee, G., Missaoui, A., Carter, T., Boerma, H., 2006. Identification and confirmation of QTL conditioning drought tolerance in Nepalese soybean PI471938. The 11th Biennial Conference on the Molecular and Cellular Biology of the Soybean, August, 2006, pp. 5–8.

Morgenthal, K., Wienkoop, S., Wolschin, F., Weckwerth, W., 2007. Integrative profiling of metabolites and proteins: Improving pattern recognition and biomarker selection for systems level approaches. Methods Mol. Biol. 358, 57–75.

Nakashima, K., Tran, L.S.P., Van Nguyen, D., Fujita, M., Maruyama, K., Todaka, D., Ito, Y., Hayashi, N., Shinozaki, K., Yamaguchi-Shinozaki, K., 2007. Functional analysis of a NAC-type transcription factor OsNAC6 involved in abiotic and biotic stress-responsive gene expression in rice. Plant J. 51, 617–630.

Naya, L., Ladrera, R., Ramos, J., González, E.M., Arrese-Igor, C., Minchin, F.R., Becana, M., 2007. The response of carbon metabolism and antioxidant defenses of alfalfa nodules to drought stress and to the subsequent recovery of plants. Plant Physiol. 144, 1104–1114.

Neves-Borges, A.C., Guimarães-Dias, F., Cruz, F., Mesquita, R.O., Nepomuceno, A.L., Romano, E., Loureiro, M.E., Grossi-De-Sá, M.D.F., Alves-Ferreira, M., 2012. Expression pattern of drought stress marker genes in soybean roots under two water deficit systems. Genet. Mol. Biol. 35, 212–221.

Newell, C.A., Hymowitz, T., 1983. Hybridization in the genus Glycine subgenus Glycine Willd. (Leguminosae, Papilionoideae). Am. J. Bot. 70, 334–348.

Nguyen, H.T., Babu, R.C., Blum, A., 1997. Breeding for drought resistance in rice: physiology and molecular genetics considerations. Crop Sci. 37, 1426–1434.

Nicholas, F., 2006. Discovery, validation and delivery of DNA markers. Aust. J. Exp. Agric. 46, 155–158.

Nunes, A.C., Vianna, G.R., Cuneo, F., Amaya-Farfán, J., DE Capdeville, G., Rech, E.L., Aragão, F.J., 2006. RNAi-mediated silencing of the myo-inositol-1-phosphate synthase gene (Gm-MIPS1) in transgenic soybean inhibited seed development and reduced phytate content. Planta 224, 125–132.

Ososki, A.L., Kennelly, E.J., 2003. Phytoestrogens: a review of the present state of research. Phytother. Res. 17, 845–869.

Paje, M., Ludlow, M., Lawn, R., 1988. Variation among soybean (Glycine max (L.) Merr.) accessions in epidermal conductance of leaves. Aust. J. Agric. Res. 39, 363–373.

Parrott, W.A., Clemente, T.E., 2004. Transgenic soybean. In: Boerma, H.R., Specht, J.E. (Eds.), Soybeans: Improvement, Production, and Uses. American Society of Agronomy, Crop Science Society of America, Soil Science Society of America Publishers, Madison, WI, pp. 265–302.

Pathan, M.S., Lee, J.-D., Shannon, J.G., Nguyen, H.T., 2007. Recent advances in breeding for drought and salt stress tolerance in soybean. In: Jenks, M.A., Hasegawa, P.M., Jain, S.M. (Eds.), Advances in Molecular Breeding Toward Drought and Salt Tolerant Crops. Springer, Netherlands.

Patterson, R.P., Hudak, C.M., 1996. Drought-avoidant soybean germplasm maintains nitrogen-fixation capacity under water stress. Plant Soil 186, 39–43.

Pennisi, E., 2008. Plant genetics. The blue revolution, drop by drop, gene by gene. Science, 320, 171–173.

Pfeiffer, T.W., Peyyala, R., Ren, Q., Ghabrial, S.A., 2003. Increased soybean pubescence density. Crop Sci. 43, 2071–2076.

Phillips, D., Smith, A., 1974. Soluble carbohydrates in soybean. Can. J. Bot. 52, 2447–2452.

Purcell, L.C., King, C.A., 1996. Drought and nitrogen source effects on nitrogen nutrition, seed growth, and yield in soybean. J. Plant Nutr. 19, 969–993.

Purcell, L.C., Specht, J.E., 2004. Physiological traits for ameliorating drought stress. In: Boerma, H.R., Specht, J.E. (Eds.), Soybeans: Improvements, Production and Uses. Agronomy monograph no. 16. 3rd ed. American Society of Agronomy and Crop Science, Society of America Soil Science, Madison, WI, pp. 569–620.

Ray, J.D., Sinclair, T.R., 1998. The effect of pot size on growth and transpiration of maize and soybean during water deficit stress. J. Exp. Bot. 49, 1381–1386.

Riederer, M., Schreiber, L., 2001. Protecting against water loss: analysis of the barrier properties of plant cuticles. J. Exp. Bot. 52, 2023–2032.

Sall, K., Sinclair, T., 1991. Soybean genotypic differences in sensitivity of symbiotic nitrogen fixation to soil dehydration. Plant Soil 133, 31–37.

Santos, M.A., Geraldi, I.O., Garcia, A.A.F., Bortolatto, N., Schiavon, A., Hungria, M., 2013. Mapping of QTLs associated with biological nitrogen fixation traits in soybean. Hereditas 150, 17–25.

Savitri, E.S., Basuki, N., Aini, N., Arumingtyas, E.L., 2013. Identification and characterization drought tolerance of gene LEA-D11 soybean (Glycine max L. Merr.) based on PCR-sequencing. Am. J. Mol. Biol. 3, 32.

Schmutz, J., Cannon, S.B., Schlueter, J., MA, J., Mitros, T., Nelson, W., Hyten, D.L., Song, Q., Thelen, J.J., Cheng, J., 2010. Genome sequence of the palaeopolyploid soybean. Nature 463, 178–183.

Seki, M., Umezawa, T., Urano, K., Shinozaki, K., 2007. Regulatory metabolic networks in drought stress responses. Curr. Opin. Plant Biol. 10, 296–302.

Serraj, R., Sinclair, T., 2002. Osmolyte accumulation: can it really help increase crop yield under drought conditions? Plant Cell Environ. 25, 333–341.

Shinozaki, K., 2007. Acceleration of soybean genomics using large collections of DNA markers for gene discovery. DNA Res. 14, 235–1235.

Shurtleff, W., Aoyagi, A., 2009. History of Soybeans and Soyfoods in Africa (1857–2009). Bibliography and Source Book. Soyinfo Center, Lafayette, CA.

Siddique, M., Hamid, A., Islam, M., 2000. Drought stress effects on water relations of wheat. Bot. Bull. Acad. Sinica 41, 35–39.

Sinclair, T., Purcell, L., Vadez, V., Serraj, R., King, C.A., Nelson, R., 2000. Identification of soybean genotypes with N fixation tolerance to water deficits. Crop Sci. 40, 1803–1809.

Sinclair, T., Serraj, R., 1995. Legume nitrogen-fixation and drought. Nature 378, 344.

Sinclair, T., Vadez, V., Chenu, K., 2003. Ureide accumulation in response to Mn nutrition by eight soybean genotypes with N fixation tolerance to soil drying. Crop Sci. 43, 592–597.

Sinclair, T.R., Purcell, L.C., King, C.A., Sneller, C.H., Chen, P., Vadez, V., 2007. Drought tolerance and yield increase of soybean resulting from improved symbiotic Nsub 2/sub fixation. Field Crops Res. 101, 68–71.

Singh, R.J., 2006. Genetic Resources, Chromosome Engineering, and Crop Improvement: Vegetable Crops. CRC Press, London, UK.

Sonah, H., Bastien, M., Iquira, E., Tardivel, A., Légaré, G., Boyle, B., Normandeau, É., Laroche, J., Larose, S., Jean, M., 2013. An improved genotyping by sequencing (GBS) approach offering increased versatility and efficiency of SNP discovery and genotyping. PLoS One 8, e54603.

Sonah, H., Deshmukh, R., Chand, S., Srinivasprasad, M., Rao, G.J., Upreti, H., Singh, A., Singh, N., Sharma, T., 2012. Molecular mapping of quantitative trait loci for flag leaf length and other agronomic traits in rice (*Oryza sativa*). Cereal Res. Commun. 40, 362–372.

Song, Q., Hyten, D.L., Jia, G., Quigley, C.V., Fickus, E.W., Nelson, R.L., Cregan, P.B., 2013. Development and evaluation of SoySNP50K, a high-density genotyping array for soybean. PLoS One 8, e54985.

Song, Q.J., Jia, G., Zhu, Y., Grant, D., Nelson, R.T., Hwang, E.Y., Hyten, D.L., Cregan, P., 2010. Abundance of SSR motifs and development of candidate polymorphic SSR markers (BARCSOYSSR_1.0) in soybean. Crop Sci. 50, 1950–1960.

Sosnowski, O., Charcosset, A., Joets, J., 2012. BioMercator v3: an upgrade of genetic map compilation and quantitative trait loci meta-analysis algorithms. Bioinformatics 28, 2082–2083.

Specht, J., Chase, K., Macrander, M., Graef, G., Chung, J., Markwell, J., Germann, M., Orf, J., Lark, K., 2001. Soybean response to water. Crop Sci. 41, 493–509.

Specht, J., Williams, J. 1985. Breeding for drought and heat resistance: some prerequisites and examples. In: Shibles, R.M. (Ed.), Proceedings of the III World Soybean Conference held in Ames, Iowa, USA, 12–17 August 1983. Westview Press, Boulder, CO, USA. pp. 468–475.

Specht, J., Williams, J., Pearson, D., 1985. Near-isogenic analyses of soybean pubescence genes. Crop Sci. 25, 92–96.

Stolf-Moreira, R., Medri, M., Neumaier, N., Lemos, N., Brogin, R., Marcelino, F., De Oliveira, M., Farias, J., Abdelnoor, R., Nepomuceno, A., 2010. Cloning and quantitative expression analysis of drought-induced genes in soybean. Genet. Mol. Res. 9, 858–867.

Su, L.-T., Li, J.-W., Liu, D.-Q., Zhai, Y., Zhang, H.-J., Li, X.-W., Zhang, Q.-L., Wang, Y., Wang, Q.-Y., 2014. A novel MYB transcription factor, GmMYBJ1, from soybean confers drought and cold tolerance in *Arabidopsis thaliana*. Gene 538, 46–55.

Subramanian, S., Graham, M.Y., Yu, O., Graham, T.L., 2005. RNA interference of soybean isoflavone synthase genes leads to silencing in tissues distal to the transformation site and to enhanced susceptibility to *Phytophthora sojae*. Plant Physiol. 137, 1345–1353.

Sulieman, S., Ha, C.V., Nasr Esfahani, M., Watanabe, Y., Nishiyama, R., Pham, C.T.B., Nguyen, D.V., Tran, L.-S.P., 2015. DT2008: A promising new genetic resource for improved drought tolerance in soybean when solely dependent on symbiotic N_2 fixation. BioMed Res. Int. 2015:687213, 1–7.

Szilagyi, L., 2003. Influence of drought on seed yield components in common bean. Bulg. J. Plant Physiol., Special Issue, 320–330.

Taylor, H., Burnett, E., Booth, G., 1978. Taproot elongation rates of soybeans. Z. Acker Pflanzenbau 146, 33–39.

Thao, N.P., Thu, N.B.A., Hoang, X.L.T., Ha, C.V., Tran, L.-S.P., 2013. Differential expression analysis of a subset of drought-responsive GmNAC genes in two soybean cultivars differing in drought tolerance. Int. J. Mol. Sci. 14, 23828–23841.

Tran, L.-S.P., Nakashima, K., Sakuma, Y., Simpson, S.D., Fujita, Y., Maruyama, K., Fujita, M., Seki, M., Shinozaki, K., Yamaguchi-Shinozaki, K., 2004. Isolation and functional analysis of *Arabidopsis* stress-inducible NAC transcription factors that bind to a drought-responsive *cis*-element in the early responsive to dehydration stress 1 promoter. Plant Cell 16, 2481–2498.

Tran, L.-S.P., Quach, T.N., Guttikonda, S.K., Aldrich, D.L., Kumar, R., Neelakandan, A., Valliyodan, B., Nguyen, H.T., 2009. Molecular characterization of stress-inducible GmNAC genes in soybean. Mol. Genet. Genomics 281, 647–664.

Tran, L.-S.P., Urao, T., Qin, F., Maruyama, K., Kakimoto, T., Shinozaki, K., Yamaguchi-Shinozaki, K., 2007. Functional analysis of AHK1/ATHK1 and cytokinin receptor histidine kinases in response to abscisic acid, drought, and salt stress in *Arabidopsis*. Proc. Natl Acad. Sci. USA 104, 20623–20628.

Tripathy, J., Zhang, J., Robin, S., Nguyen, T.T., Nguyen, H., 2000. QTLs for cell-membrane stability mapped in rice (*Oryza sativa* L.) under drought stress. Theor. Appl. Genet. 100, 1197–1202.

Turner, N.C., Abbo, S., Berger, J.D., Chaturvedi, S., French, R.J., Ludwig, C., Mannur, D., Singh, S., Yadava, H., 2007. Osmotic adjustment in chickpea (*Cicer arietinum* L.) results in no yield benefit under terminal drought. J. Exp. Bot. 58, 187–194.

Turner, N.C., Wright, G.C., Siddique, K., 2001. Adaptation of grain legumes (pulses) to water-limited environments. Adv. Agron. 71, 194–233.

Umezawa, T., Fujita, M., Fujita, Y., Yamaguchi-Shinozaki, K., Shinozaki, K., 2006. Engineering drought tolerance in plants: discovering and tailoring genes to unlock the future. Curr. Opin. Biotechnol. 17, 113–122.

Umezawa, T., Sakurai, T., Totoki, Y., Toyoda, A., Seki, M., Ishiwata, A., Akiyama, K., Kurotani, A., Yoshida, T., Mochida, K., 2008. Sequencing and analysis of approximately 40,000 soybean cDNA clones from a full-length-enriched cDNA library. DNA Res. 15, 333–346.

Urano, K., Maruyama, K., Ogata, Y., Morishita, Y., Takeda, M., Sakurai, N., Suzuki, H., Saito, K., Shibata, D., Kobayashi, M., 2009. Characterization of the ABA-regulated global responses to dehydration in *Arabidopsis* by metabolomics. Plant J. 57, 1065–1078.

Valliyodan, B., Nguyen, H.T., 2006. Understanding regulatory networks and engineering for enhanced drought tolerance in plants. Curr. Opin. Plant Biol. 9, 189–195.

Valliyodan, B., Nguyen, H.T., 2008. Genomics of abiotic stress in soybean. In: Stacey, G. (Ed.), Soybean Genomics. Springer, USA. pp. 342–373.

Van Gardingen, P.R., Grace, J., 1992. Vapour pressure deficit response of cuticular conductance in intact leaves of *Fagus sylvatica* L. J. Exp. Bot. 43, 1293–1299.

Vignes, D., Djekoun, A., Planchon, C., 1986. Reponses de differents genotypes *de soja au* deficit hydrique. Can. J. Plant Sci. 66, 247–255.

Vodkin, L.O., Khanna, A., Shealy, R., Clough, S.J., Gonzalez, D.O., Philip, R., Zabala, G., Thibaud-Nissen, F., Sidarous, M., Strömvik, M.V., 2004. Microarrays for global expression constructed with a low redundancy set of 27,500 sequenced cDNAs representing an array of developmental stages and physiological conditions of the soybean plant. BMC Genom. 5, 73.

Vuong, T.D., Wu, X., Pathan, M.S., Valliyodan, B., Nguyen, H.T., 2007. Genomics approaches to soybean improvement. In: Varshney, P.K., Tuberosa, R. (Eds), Genomics-assisted crop improvement. Springer, USA.

Wright, G., 1996. Review of ACIAR selection for water use efficiency in legumes project recommends further research. ACIAR Food Legume Newslett. 1996, 2–3.

Wu, X., Zhong, G., Findley, S.D., Cregan, P., Stacey, G., Nguyen, H.T., 2008. Genetic marker anchoring by six-dimensional pools for development of a soybean physical map. BMC Genom. 9, 28.

Xia, Z., Tsubokura, Y., Hoshi, M., Hanawa, M., Yano, C., Okamura, K., Ahmed, T.A., Anai, T., Watanabe, S., Hayashi, M., 2007. An integrated high-density linkage map of soybean with RFLP, SSR, STS, and AFLP markers using a single F2 population. DNA Res. 14, 257–269.

Xie, Z.-M., Zou, H.-F., Lei, G., Wei, W., Zhou, Q.-Y., Niu, C.-F., Liao, Y., Tian, A.-G., Ma, B., Zhang, W.-K., 2010. Soybean trihelix transcription factors GmGT-2A and GmGT-2B improve plant tolerance to abiotic stresses in transgenic *Arabidopsis*. PLoS One 4, e6898.

Xue, R.-G., Zhang, B., Xie, H.-F., 2007. Overexpression of a NTR1 in transgenic soybean confers tolerance to water stress. Plant Cell Tiss. Organ Cult. 89, 177–183.

Yamaguchi-Shinozaki, K., Shinozaki, K., 2006. Transcriptional regulatory networks in cellular responses and tolerance to dehydration and cold stresses. Annu. Rev. Plant Biol. 57, 781–803.

Ying, L., Gai, J.-Y., Lü, H.-N., Wang, Y.-J., Chen, S.-Y., 2005. Identification of drought tolerant germplasm and inheritance and QTL mapping of related root traits in soybean (*Glycine max* (L.) Merr.). Acta Genet. Sin. 32 (8), 855–863.

Zeid, I., Shedeed, Z., 2006. Response of alfalfa to putrescine treatment under drought stress. Biol. Plant. 50, 635–640.

Zhang, B., Chen, P., Florez-Palacios, S.L., Shi, A., Hou, A., Ishibashi, T., 2010a. Seed quality attributes of food-grade soybeans from the US and Asia. Euphytica 173, 387–396.

Zhang, G., Chen, M., Chen, X., Xu, Z., Li, L., Guo, J., Ma, Y., 2010b. Isolation and characterization of a novel EAR-motif-containing gene GmERF4 from soybean (*Glycine max* L.). Mol. Biol. Rep. 37, 809–818.

Zhang, G., Chen, M., Li, L., Xu, Z., Chen, X., Guo, J., Ma, Y., 2009. Overexpression of the soybean *GmERF3* gene, an AP2/ERF type transcription factor for increased tolerances to salt, drought, and diseases in transgenic tobacco. J. Exp. Bot., 60 (13), 3781–3796.

Zhou, Q.Y., Tian, A.G., Zou, H.F., Xie, Z.M., Lei, G., Huang, J., Wang, C.M., Wang, H.W., Zhang, J.S., Chen, S.Y., 2008. Soybean WRKY-type transcription factor genes, *GmWRKY13*, *GmWRKY21*, and *GmWRKY54*, confer differential tolerance to abiotic stresses in transgenic *Arabidopsis* plants. Plant Biotechnol. J. 6, 486–503.

Soybean production and heavy metal stress

Mohammad Miransari
Department of Book & Article, AbtinBerkeh Ltd. Company, Isfahan, Iran

Introduction

Metals with the density higher than 5 g m^{-3} are defined as heavy metals. Among the 90 metals, 53 of them are heavy metals. However, only a few of them are biologically important. According to the solubility of heavy metals, 17 are of importance for the activity of living cells, and the environment. Five of them including iron (Fe), zinc (Zn), manganese (Mn), copper (Cu), and molybdenum (Mo) are essential micronutrients for plant growth and activities. Nickel (Ni), vanadium (V), cobalt (Co), cadmium (Cd), and chromium (Cr) are toxic and have adverse effects on the growth and activities of plants and microbes (Nies, 1999; Salt et al., 1995; Zenk, 1996).

The two important sources of heavy metals derive from the soil and from the atmosphere. Natural anthropogenic activities result in the increased concentration of heavy metals in the soil, which has mobile and immobilized fractions of heavy metal attached to the organic and inorganic parts thereof. Industrial activities are also an important source of atmospheric contamination adversely affecting the climate, including the ozone layer (Wang and Chen, 2009). Accordingly, the following are the most important sources of heavy metal contamination in the soil: (1) industrial and mining activities, (2) polluted water, and (3) sewage sludge (Lebeau et al., 2008).

Metal concentration is determined by the amount of metal in the bedrock and deposition from the atmosphere. Contamination of soil by heavy metals reduces the fertility of soil, and is not favorable to the environment and human health. Because wide areas of fields are contaminated or subjected to the stress of heavy metal contamination, using a suitable remediation strategy to alleviate the stress of heavy metals is of great importance (Kachenko and Singh, 2006).

Plants are subjected to different kinds of stresses including the stress of heavy metals such as Cu, Cd, Co, and Mn, which are the result of anthropogenic activities. Although such metals may be essential to plant growth and activity, at higher amounts they cause stress and decrease plant growth and yield. The most important reasons for the adverse effects of heavy metals on the growth and yield of plants are (1) the interference of such metals with the activity of nutrients, which are essential for plant growth and activities and are chemically similar; (2) adverse effects on plant morphology and physiology; (3) the negative effects of heavy metals on the growth of soil microbes, including the symbiotic microbes such as rhizobia; and (4) adverse effects on the properties of soils (Jing et al., 2007; Mittler, 2006; Xiong et al., 2010).

Abiotic and Biotic Stresses in Soybean Production. http://dx.doi.org/10.1016/B978-0-12-801536-0.00009-8

Different methods and techniques have so far been examined and used for the alleviation of heavy metal stress, among which the use of biological methods may be the most effective. In the biological methods, tolerant plants and bacterial species, as well as plant hyperaccumulators, which are able to absorb high amounts of heavy metals, while their growth remains unaffected, are used. With respect to the properties of the related site, a favorable method of remediating the contaminated field must be selected and used. Accordingly, the remediation of contaminated fields is site specific (Smith et al., 1995). The following biochemical processes have been tested and used for bioremediation: (1) bioleaching using fungi and bacteria; (2) bioreduction of heavy metals using bacteria; (3) biosorption of metals from water using bacteria or algae; and (4) biomethylation of metals (Ali et al., 2013; Kavamura and Esposito, 2010; Khan, 2005).

Some *in situ* methods have also been tested and used such as the extraction of heavy metals by organic, inorganic, as well as complexation products. However, such methods are expensive, difficult, and not practicable in a large area. As a result, the biological methods including the use of plants (phytoremediation) and microbes are being tested and used (Kavamura and Esposito, 2010; Ngah and Hanafiah, 2008; Zhuang et al., 2007a). The advantages of bioremediation over the use of physical and chemical methods are (1) preserving the natural properties of the environment; (2) use of the sunlight; (3) inexpensive; (4) significant increase in the population of soil microbes, especially in the rhizosphere; and (5) can become rapid (Huang et al., 2004).

Soybean and its symbiotic rhizobia are not tolerant to the stress of heavy metals and hence under such a stress the growth and activity of plant and bacteria decrease. Table 9.1 illustrates the tolerance limits of soybean and some other plants to heavy metals (Cheng, 2003). As it is evident from the table, soybean is able to tolerate higher contents of zinc and copper relative to the other heavy metals. Accordingly, soybean may have a higher tolerance to some heavy metals than other crops, vegetables, and fruits.

Huang et al. (1974) investigated soybean growth and activity under heavy metal stress. They found that heavy metals decreased the fresh weight of soybean pods by

Table 9.1 **Tolerance limit (mg kg^{-1} DW) of fruits, vegetables, and some crop plants including soybean to heavy metal stress in China (Cheng, 2003)**

Elements	Fruits	Vegetables	Soybeans	Crops
As	0.5	0.5	n.d.	0.7
Al	n.d.	n.d.	n.d.	100
Cd	0.03	0.05	0.05	0.05
Cr	0.5	0.5	1.0	1.0
Cu	10	10	20	10
Hg	0.01	0.01	0.01	0.02
Pb	0.2	0.2	0.8	0.4
Zn	5	20	100	50

n.d., Not determined.

35%. At day 52 (the time of maximum enzyme activity), the activity of N-fixing enzyme was diminished by 30%, increasing to a 71% diminution at day 59 under the stress. At the peak of photosynthetic activity the photosynthetic rate decreased by 60%, and as a result carbohydrate was not accumulated in the nodules. The other plant tissues and factors such as root, leaf, nodule weight (dry) as well as nodule ammonia, and the content of soybean carbohydrate and protein, also decreased under the stress.

As previously mentioned, because heavy metal contamination is widespread in different parts of the world, it is essential to use methods and techniques which may result in the alleviation of heavy metal stress. The use of plants for the bioremediation of contaminated fields is applicable to those places where the concentration of heavy metals is not high. This method is a suitable alternative to the previously mentioned methodologies. Such plants are able to accumulate heavy metals in their tissues or affect the behavior in the rhizosphere (Das et al., 2008; Wenzel et al., 1999). Soybean is not a hyperaccumulator and is not able to accumulate a high rate of heavy metals in its tissues. However, with respect to the importance of soybean as a source of food and oil, and its plantation worldwide, some of the most important details related to the response of soybean under the stress of heavy metals have been presented here.

Using plants for bioremediation

Among the most useful methods for the bioremediation of heavy metal stress is the use of plants that are able to absorb heavy metals while their growth remains unaffected. Such plants are called hyperaccumulators. The following mechanisms have been defined for the use of bioremediating plants (Khan, 2005).

- Phytoextraction: Uptake and accumulation of heavy metals in the harvestable parts of plants.
- Phytostabilization: Decreased mobility and bioavailability as well as immobilization of heavy metals by plant roots and the related microbes.
- Phytodegradation: The degradation of contaminants by plant roots and the related microbes.
- Phytovolatilization: Contaminants are volatilized from the soil by plants.
- Rhizofiltration: Remediation of contaminated soils by the root absorption of heavy metals.

The process of phytoremediation is usually long because plant growth is not fast and plant volume is low. A hyperaccumulator must have an extensive root network and the ability to grow rapidly. It can also make the process of phytoremediation more efficient if hyperaccumulation genes are used in the plant. Some plants such as vetiver and hemp, which are fast growing and produce high biomass, have the ability of tolerating high concentrations of heavy metals but are not able to accumulate them. Accordingly, because such plants are also able to grow under different climatic conditions with high biomass, they are ideal as hyperaccumulators if their related traits are modified (Khan, 2003; Linger et al., 2002).

It has been indicated that it is possible to enhance the phytoremediation ability of nonhyperaccumulating plants by enhancing their growth and biomass using plant hormones including auxin and cytokinins (Fuentes et al., 2000; Pe et al., 2000). Plant growth promoting rhizobacteria (PGPR) such as *Pseudomonas* and *Acinetobacter* may also increase plant biomass and hence their absorbing potential by enhancing the availability of zinc, iron, magnesium, potassium, phosphorus, and calcium and their subsequent uptake by the plant (Lippmann et al., 1995).

One of the techniques to enhance the phytoremediating potential of plants is to increase their growth by using soil microbes such as arbuscular mycorrhizal (AM) fungi and plant growth promoting rhizobacteria (PGPR) (Miransari, 2011a). Using soil microbes as the only source of nutrients for plant growth under nutrient deficient conditions may not efficiently provide plants with their essential nutrients. As a result, their combined use with chemical fertilizers may be an efficient method to enhance their growth and hence their phytoremediating potential (Shetty et al., 1995; Tyagi et al., 2011). However, more research is essential to enhance the phytoremediating potential of hyperaccumulators by combining the right combination of plants, microbes, and fertilizer.

The tolerant plant species must be able to absorb heavy metals in their tissues, including roots and the aerial parts, without affecting their growth. Under such conditions plants cells have the ability to allocate the heavy metals to the cell vacuoles and hence alleviate the stress of heavy metals. This can also be the case for the tolerant bacteria, and sequestration of heavy metals to their cellular vacuoles may be an effective method of alleviating the stress (Miransari, 2011b; Zhang and Shu, 2006).

The following mechanisms are used by plants to alleviate the stress of heavy metals. (1) Cell wall binding; (2) decreased uptake of heavy metals; (3) pumping of heavy metal outside the cell by the plasma membrane; (4) phytochelatins, which bind heavy metals; (5) improving the structure of damaged proteins; (6) mycorrhization; and (7) metal compartmentation in the vacuole related to the activity of tonoplast carriers (Hall, 2002; Yang et al., 2005; Yadav, 2010).

A few species of hyperaccumulators are able to develop a symbiotic association with the soil microbes such as AM fungi (Miransari, 2011b). However, for the alleviation of heavy metal stress, different mechanisms are used by the hyperaccumulators such as: (1) high absorbance of heavy metals by the plant; (2) use of root products, which may result in the alteration of heavy metal behavior in the soil such as solubility, stability, and availability; and (3) use of plant products that change the properties of soil such as pH and the activity of soil microbes (Behra, 2014; Saikia, 2008; Visioli and Marmiroli, 2012).

Soil microbes such as AM fungi are able to alleviate the stress of heavy metals on the growth and activities of plant by (1) absorbing a high rate of heavy metals; (2) dedicating the absorbed heavy meals to their hyphae and cellular vacuole; (3) increasing the tolerance of the host plant to heavy metals by increasing the uptake of water and nutrients; (4) interacting with the other soil microbes such as *Rhizobium* and enhancing their tolerance to the stress; and (5) alteration of soil environment (Chatterjee et al., 2013; Miransari, 2015).

The symbiotic association between AM fungi and the host plant results in the production of an extensive hyphal network with the high capability of absorbing water and nutrients. The fungal hyphal network is able to grow into the finest pores of the soil and absorb water and nutrients, however, even, root hairs are not able to grow in such pores and absorb water and nutrients. Such ability is important for the growth and activity of the fungi and the host plant, especially under stress (Miransari, 2010).

Tolerant plant species are able to tolerate the unfavorable effects of heavy metal stress by altering their morphological and physiological properties. The production of different organic products by such plants may help them to sequester heavy metals to cellular organelles such as vacuoles, which are more tolerant to stress and can absorb high amounts of heavy metals without affecting their growth and activities (Yadav, 2010).

Hyperaccumulators

High concentrations of heavy metals may adversely affect plant functioning by influencing the structure of enzymes and protein as well as by substituting the elements essential for plant growth. Accordingly, the sensitivity of cellular membrane proteins such as H^+–ATPases to heavy metals can negatively affect the functionality of plant cells. Plants utilize different mechanisms to keep ion homeostasis and hence alleviate the adverse effects of heavy metals including: (1) chelation of heavy metals by root products; (2) binding of heavy metals by cellular walls; and (3) the production of compounds such as metallothioneins and phytochelatins, which have a high affinity for heavy metals and hence enable adjustment of the cytoplasmic concentrations of heavy metals by transporting them across the tonoplast resulting in their subsequent sequestration in the vacuole (Hall, 2002).

The use of hyperaccumulators is among the most efficient methods for the remediation of heavy metal stress. Such plants are able to accumulate a high concentration of heavy metals in their cells and tissues without affecting their growth. This technology is called phytoremediation. Just 0.2% of angiosperms (450 species) are able to hyperaccumulate heavy metals, most of which are Ni hyperaccumulators. The most important activity of hyperaccumulators in remediating the stress of heavy metal is ion homeostasis (Verbruggen et al., 2009).

The Thlaspi family consists of hyperaccumulating plants, among which 23, 10, 3, and 1 plant species are able to hyperaccumulate Ni, Zn, Cd, and Pb, respectively (Vogel-Mikuš et al., 2005, 2008). Some important mechanisms are used by hyperaccumulators to alleviate the stress of heavy metals including (1) the transport of heavy metals; (2) the formation of complexes with organometallic products including glutathione, hystidine, cysteine, nicotinamine, and the other thiols, which are not of high molecular weight; (3) the storage of heavy metals in the vacuoles; and (4) compartmentation potential (Callahan et al., 2006; Kupper et al., 1999; Miransari, 2011b). For example *Thlaspi caerulescens* is able to accumulate Zn and Cd in the vacuoles of its epidermal and mesophyll cells, while the leaf trichomes of *Brassica juncea* and *Arabidopsis thaliana* are able to accumulate Zn and Cd (Wójcik et al., 2005).

Soybean and heavy metal stress

Soybean is not a hyperaccumulator; however, it may be possible to enhance its hyperaccumulating capability under the stress of heavy metals by inserting the related gene providing that the quality of its seeds is not adversely affected. Accordingly, if the related properties of soybean including (1) high accumulation of heavy metals in the vacuole;

(2) production of products, which may enhance the transfer of heavy metals in the rhizosphere and in plant; and (3) higher soybean tolerance under heavy metal stress are improved, soybean will be able to grow and produce higher yields under such stress (Miransari, 2011b).

Another important point which may affect the alleviating process of heavy metal stress is the interaction between plants and soil microbes (Glick, 2010). In soybean, its interaction with *Bradyrhizobium japonicum* is an important aspect affecting soybean behavior under heavy metal stress. If the behavior of both plants and symbiotic microbes is improved under the stress, it is possible to increase their tolerance under stress. However, as previously mentioned, under the stress of heavy metals the quality of seed is an important factor, which is affected by plant uptake of heavy metals.

The interaction of soybean with AM fungi and PGPR can also affect soybean growth and yield under heavy metal stress. Such microbes are able to enhance plant behavior under stress and if a consortium of microbes including rhizobia, PGPR, and AM fungi are used it may be possible to increase soybean tolerance under heavy metal stress. The right combination of microbes can efficiently increase soybean tolerance under stress (Miransari, 2011b).

The following affect the ability of soil microbes, under heavy metal stress, to interact with the host plant: (1) type of association between soil microbes and the plant; (2) conditions of plant growth; (3) metal species and concentrations; (4) the combination of heavy metals; (5) properties of the soil; and (6) density and volume of the plant roots (Bi et al., 2003).

In plants under heavy metal stress, a related signaling pathway can enable the plant to survive the stress. Under stress, reactive oxygen species (ROS) are produced, which are damaging to plant cellular structure at high concentrations. ROS result in lipid oxidation. As a result oxylipins are produced, which are endogenous signals or plant response to the stress (Mithofer et al., 2004). As a result, plants may alleviate the stress by activating antioxidant enzymes and products, which are able to scavenge the ROS produced and hence alleviate the stress (Sajedi et al., 2010, 2011).

The following render the transport of heavy metals in plants, including soybean, possible: (1) the macrophage protein; (2) the ATPases of heavy metals (CPx-type); and (3) the cation diffusion facilitator (Williams et al., 2000). Phenolic compounds are among the most important plant compounds, which enable the plant to tolerate the stress of heavy metal. Such compounds are able to act as heavy metal chelators and also scavenge the ROS. Under the stress of heavy metals, phenolics such as flavonoid and phenylopropanoids are oxidized by peroxidase and can scavenge H_2O_2. The antioxidant activity of phenolic compounds is mainly related to their biochemical structure (Michalak, 2006). The production of plant hormonal signals including those related to ethylene and jasmonate are also among plant responses to the stress of heavy metal (Maksymiec, 2007; Miransari, 2014).

Chen et al. (2003) investigated the effects of cadmium stress on the growth of soybean, biological N fixation, nodulation, nodule structure, and cadmium distribution in plants. Cadmium significantly decreased nodulation, especially at levels of 10 and 20 mg kg^{-1} in soil. With increasing cadmium concentrations plant growth, especially root growth, decreased more significantly.

Cadmium resulted in a lower root/leaf weight ratio, which may be the reason for less nodulation weight. Biological N fixation increased at the low level of cadmium; however, a higher level of cadmium significantly decreased the process of biological N fixation. High levels of cadmium reduced the N-fixing area in the nodules as well as the number of N-fixing cells. The concentration of cadmium in different plant parts was in the following order: root > stem > seed (Chen et al., 2003).

In an interesting study, soybean plants were subjected to concentrations of 50 and 200 μM of $CdCl_2$ and their N metabolism in nodules and roots was evaluated. The concentration of 200 μM Cd^{2+} significantly decreased the activity of N-fixation enzyme. Under a concentration of 50 μM Cd^{2+} the amount of NH_4^+ was similar to that in control treatments, but it significantly increased in the roots. However, the concentration of 200 μM Cd^{2+} resulted in a higher NH_4^+ concentration in both tissues. The concentration of 50 μM did not alter protein content in nodules but the higher cadmium content decreased the level of proteins in nodules (Balestrasse et al., 2003).

At both concentrations of cadmium, polyamine content increased. The content of polyamine and protein in roots decreased significantly under both concentrations of cadmium. The concentration of N assimilating enzymes moderately increased in nodules and roots at the concentration of 50 μM and was equal to that of the control treatment at the higher concentration. The lower concentration of cadmium caused no change in the protease activity in either tissue, while the higher concentration decreased it in nodules and increased it in roots (Balestrasse et al., 2003).

Alleviation of heavy metal stress

Different techniques and strategies have been used for bioremediation and alleviation of heavy metal stress in contaminated sites and on the growth of plants, respectively. Among biological methods, using tolerant plant species as well as using free-living and symbiotic microbes are the most suitable and effective. The most relevant details are presented here.

Using tolerant soybean genotypes

Tolerant plant species use different morphological and physiological responses under stress. Under heavy metal stress, the following are resulted. (1) Reactive oxygen species are produced, especially when a plant is subjected to iron and copper contamination, adversely affecting plant cellular structure by lipid peroxidation, H_2O_2 production, and oxidative burst. (2) The activity of biomolecules in the plant decreases, particularly under cadmium stress. (3) The structure of biomolecules in the plant is negatively affected as essential nutrients for plant growth are displaced from biomolecules (Ali et al., 2013; Cobbett, 2000; Hall, 2002; Zenk, 1996).

ROS may cause the oxidation of proteins and lipids, and adversely affect DNA. As a result, the affected tissue may have increased levels of carbonylated proteins. Although the increased level of ROS adversely affects plant morphology and physiology, ROS are

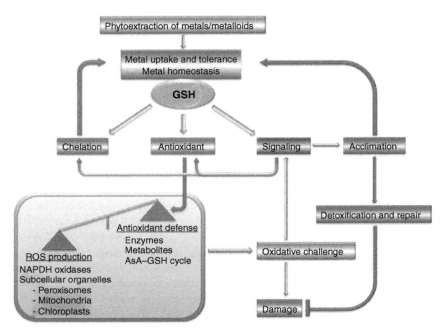

Figure 9.1 The general model indicating the role of glutathione in the phytoextraction of metals as affected by metal uptake and tolerance and hence by metal homeostasis.
Adopted from Seth et al. (2012), with permission from Wiley, license number 3603661305893.

essential for (1) controlling the activities of pathogenic microorganisms; (2) production of certain plant structures involving lignification and cell wall production; and (3) acting as signaling molecules to regulate the activity and expression of genes in plant. Accordingly, ROS must be kept at a suitable level by the plant (Figure 9.1, Lin and Aarts, 2012; Seth et al., 2012).

In tolerant plants, mechanisms such as production of antioxidant products including the related enzymes can alleviate the negative effects of ROS produced, by catabolizing such products under stress. However, if under stress hydroxyl ions are produced, the antioxidant products may not be able to alleviate the stress, as they are unable to decrease the levels of these ions (Sajedi et al., 2010, 2011). Production of tolerant soybean genotypes may be a suitable method for planting soybean under heavy metal stress. However, more research is essential to produce soybean genotypes that are able to grow under heavy metal stress and give a healthy yield, including that of seed and oil.

Using soil microbes

Among the most important biological methods used for the bioremediation of contaminated fields is the use of soil microbes, such as AM fungi, PGPR, and endophytic bacteria.

Soil microbes are found in the environment around plant roots termed the rhizosphere. In such an environment, the density and activity of soil microbes more greatly affect plant activities and its bioremediating capabilities. As a result, the microbial community in the rhizosphere, including the free-living and the symbiotic, are of great significance for the bioremediation of fields contaminated with heavy metals (Lasat, 2002).

PGPR

Although the use of plants has been promising for the alleviation of heavy metal stress, their use alone may not be as effective as combining their use with soil microbes including PGPR. Soil microbes are able to change the root environment by affecting the mobility and uptake of heavy metals by plant roots (Lasat, 2002; Zhuang et al., 2007b). The three main functions of PGPR are: (1) production of specific compounds affecting plant activities; (2) enhancing the uptake of nutrients by plant; and (3) controlling the activities of pathogens (Artursson et al., 2006; Berg and Zachow, 2011; Glick, 1995).

PGPR alleviate the stress of heavy metal by altering the mobility and availability of heavy metals via producing different compounds (e.g., chelating products and siderophores) and changing soil pH (Abou-Shanab et al., 2003a; Kuiper et al., 2004). For example, EDGA and EDTA can chelate heavy metals and enhance their uptake by plants; however, they can also increase the leaching of heavy metals and decrease the activity of soil microbes. The availability of heavy metals in the soil is an important parameter affecting the process of bioremediation. In most cases the metals are bound to the inorganic and organic phases of the soil, and hence are unavailable for plant uptake and microbial activities (Ernst, 1996).

The two other important products (as peptides) are metallothioneins and phytochelatins, which have a high affinity for the absorption of a wide range of heavy metals. However, due to their attractive structure, phytochelatins bind more heavy metals (Cobbett, 2000). Accordingly, if the genetic component of soil microbes is modified, it is possible to enable them to synthesize such products and hence increase their bioremediation potential.

Another important role of a PGPR is to alleviate the adverse effects of stress by the production of ACC-deaminase. Although ethylene is an important hormone for plant growth and activities, under stress the enhanced production of this enzyme can adversely affect plant growth. However, PGPR such as *Pseudomonas* and *Bacillus* are able to alleviate the adverse effects of ethylene on plant growth and activity by the production of enzyme ACC deaminase, which is able to catabolize ACC (1-aminocyclopropane-1-carboxylate), the immediate precursor of ethylene synthesis in plants (Glick et al., 1999; Jalili et al., 2009).

PGPR, including the free-living and symbiotic PGPR, are able to enhance plant growth and activities by (1) N symbiotic fixation; (2) increasing the availability of phosphorus for plant use; (3) production of plant hormones; (4) production of enzymes; (5) controlling unfavorable microbes; (6) production of antibiotics; (7) production of siderophores and chelating products; and (8) controlling the level of stress hormone, ethylene (Abou-Shanab et al., 2003a; Miransari, 2014). Soil microbes are able to grow fast by sensing the presence of other microbes and plant roots through the process of quorum sensing. Such microbes can create an active community in the rhizosphere, affecting different plant activities and the availability of heavy metals (Miller and Bassler, 2001).

PGPR are now widely used in the process of phytoremediation and for the detoxification of soil. The following indicates how the properties of hyperaccumulator plants, which are used for the process of phytoremediation, are improved by PGPR. (1) Increased growth and biomass; (2) adjusted uptake and degradation of contaminants including heavy metals; and (3) improved plant nutrition and health. However, the important point related to the use of PGPR for the bioremediation of fields contaminated with heavy metals is the survival of PGPR under such a stress. The combined use of PGPR and AM fungi is a suitable method for increasing the phytoremediating ability of soil microbes (Lucy et al., 2004).

Contamination of heavy metals in the soil may negatively affect microbial growth and activities by: (1) decreasing the total microbial biomass; (2) reducing some more sensitive microbial communities; and (3) altering the structure of microbial communities and hence diversity (Cobbett, 2000; Zhuang et al., 2007b). Using simulating models, Pishchik et al. (2005) simulated the conditions relating to cadmium stress, which is initiated with the production of plant hormones such as auxin and ethylene followed by the high uptake of heavy metals. Due to the properties of the rhizosphere and production of products by plant roots, the strains and species of microbes in the rhizosphere are able to tolerate higher concentrations of heavy metals (Dell'Amico et al., 2005).

Abou-Shanab et al. (2008) investigated the effects of different bacterial isolates including *Pseudomonas* and *Bacillus* on the uptake of heavy metals by corn (*Zea mays* L.). *Bacillus* increased the rate of chromium and copper extraction from the polluted soil. The bacteria resulted in the highest uptake of Zn (4 g kg^{-1}) and Cu (2 g kg^{-1}) by corn. Accordingly, the researchers indicated that such PGPR are able to significantly increase the availability of heavy metals and their subsequent uptake by corn.

Nogueira et al. (2004) evaluated the effects of AM fungi on the alleviation of manganese toxicity in soybean plants at two different growth stages, 45 and 90 days. They also used a control treatment by using extra P to be able to investigate the dilution effects resulting from the higher uptake of P in mycorrhizal plants. At 45 days, soybean plants indicated the toxicity symptoms of Mn; however, the fungi were able to alleviate the stress at 90 days even relative to the control treatment with extra P, which had a similar biomass. The conditions that contributed to alleviated Mn toxicity were lower Mn uptake and higher P uptake by the roots and aerial parts of mycorrhizal soybean. The researchers speculated that the higher activity of Mn-oxidizing bacteria and the lower activity of Mn-reducing bacteria in the rhizosphere of soybean plants, might have contributed to the reduced uptake of Mn by soybean plants.

The interactions between the host plant and the rhizosphere microbes can affect the activity of soil microbes in relation to the alleviation of contaminants including heavy metal. This process is called rhizoremediation and is affected by the rhizosphere environment including the root products. Such exudates can increase the activity and survival of rhizosphere microbes by positively affecting their alleviating activities in the contaminated rhizosphere. Plant roots can also enable the microbes to become distributed in the soil in places which are not accessible by the microbes themselves. Another effective method for the alleviation of heavy metal stress is the inoculation of seeds with the related microbes (Kuiper et al., 2004).

Although there has been considerable work on the use of PGPR for the alleviation of heavy metal stress, most research has been conducted in the lab or under greenhouse conditions. More has yet to be investigated on the role of PGPR in the alleviation of heavy metals by performing *in situ* experiments under field conditions, as the soil is a complicated ecosystem, and the actions and interactions between the soil microbes and plant roots can significantly affect the process of bioremediation (Zhuang et al., 2007b).

Mycorrhizal fungi

AM fungi are able to develop a symbiotic association with most terrestrial plants and increase the uptake of water and nutrients by the host plant. The host plants provide the fungi with its essential carbon. Although the symbiotic association is non-specific, combination of the host plant with some specific fungal species may result in a more efficient symbiosis. Although the fungal spores are able to germinate in the absence of the host plant, for the symbiotic association to proceed and produce an extensive hyphal network, the presence of the host plant is essential. During the initiation of the symbiotic process between AM fungi and the host plant, signals are produced, resulting in the germination of fungal spores (Miransari, 2010).

AM fungi are among the most influential microbes affecting the process of bioremediation of heavy metals. The fungi are able to produce an extensive network of hyphae, significantly affecting the uptake of water and nutrients by the host plant. Such capability is also effective under stress, alleviating the adverse effects of stress on the growth of the host plant. The fungi are also able to alleviate the stress of heavy metals on the growth of the host plant and can also be used for the bioremediation of fields contaminated with heavy metals (Kuiper et al., 2004).

However, these fungi are able to develop a symbiotic association with only a few species of hyperaccumulators, which is due to the production of some root exudates by the hyperaccumulators. The families including Cruciferaceae, Chenopodiaceae, Plumbaginaceae, Juncaceae, Juncaginaceae, Amaranthaceae, and a few species from Fabaceae, are not able to develop a symbiotic association with AM fungi (Miransari, 2011a; Vierheilig et al., 2003).

Most hyperaccumulators are from the Brassicaceae family and do not establish a symbiotic association with their host plant. In some cases, although the fungal hyphae are produced, the arbuscule, which is the main site of nutrient exchange and heavy metal absorption, is not produced and hence such mycorrhizal symbiosis is not functional (Vierheilig et al., 2003).

The fungi may alleviate the adverse effects of heavy metals on plant growth by affecting the interactions between the host plant and heavy metals (Marschner, 1995). The extensive network of fungal hyphae is able to grow in a wider area of soil, compared with even the finest root hairs, and exploit the resources of water and nutrients. This may positively affect the growth of plants in contaminated fields.

Using a consortium of soil microbes from the rhizosphere of plants including AM fungi, PGPR, rhizobia, and mycorrhizal-friendly bacteria may positively affect the alleviation of heavy metal stress on the growth of plants (Khan, 2004a, b). It is desirable to isolate the strains and species of soil microbes from the stress conditions created by the presence of heavy metals in order to alleviate the stress more efficiently.

Mycorrhizal plants are able to absorb higher rates of heavy metals without affecting their growth as compared to plants without mycrorrhizal symbiosis. The density and abundance of fungal hyphae is an important factor affecting the ability of fungi to mitigate heavy metal stress. Cell wall components of mycorrhizal fungi such as cellulose and chitin are able to bind heavy metals. Metallothionein-like peptides are also among the binding factors of heavy metals (Galli et al., 1994).

Endophytic bacteria

It is now more than 50 years since bacteria were found to reside in plant tissues without showing any unfavorable effects. Endophytic microbes are defined as microbes which colonize the internal parts of different plant tissues without causing any negative symptoms. Accordingly, it was speculated that there might be some kind of mutual or symbiotic association between the plant and the endophytic microbes. Bacteria and fungi are the two types of most commonly occurring endophytic microbes (Sessitsch et al., 2002).

The interactions of endophytic diastrophic bacteria with the host plant have been widely investigated under sterile conditions. The bacteria enter the plant via the roots and colonize the sites at the connection of root hairs and root epidermis. As a result the internal parts of the root and the aerial parts are colonized by the bacteria (James et al., 1994; Zakria et al., 2007).

Endophytic microbes include bacteria, fungi, and actinomycetes. Such microbes can establish the symbiosis with almost all plant species. Some of the most well-known species belong to the genera *Colletotrichum* sp., *Enterobacter* sp., *Phomopsis* sp., *Cladosporium* sp., *Phyllosticta* sp., etc. The population of endophytic bacteria is affected by the environment and climatic conditions. The biochemical compounds produced by the endophytic microbes are useful for plants and are of economic significance for humans. Such products can be used, for example, as antibiotics for the production of medicine and food. The role of such products is also of significance for the cycling of nutrients and for the bioremediation of the environment from contaminants (Nair and Padmavathy, 2014).

With the exception of seed endophytic microbes, which enter the plant via seed growth, the other microbes must enter the plant by colonizing the roots. Such microbes may also enter the plant via the stomata, and the areas of lateral root emergence. After entering the plant, these microbes may stay in the tissue or may colonize the other plant tissues. Plants may be colonized simultaneously by different varieties of endophytic bacteria including both Gram-positive and Gram-negative (Sessitsch et al., 2004).

The endophytic bacteria may affect plant growth and health by producing plant hormones and antifungal biochemicals, as well as by inducing systemic resistance and biological N fixation. The other effects of endophytic bacteria on the growth are (1) increased uptake of N and P by the host plant; (2) production of siderophores; (3) production of plants with their essential vitamins; (4) adjustment of osmotic potential; (5) effects in stomatal activities; (6) modifications in the morphology of roots; (7) effects in nitrogen metabolism; and (8) higher uptake of nutrients (Newman and Reynolds, 2005; Rosenblueth and Martínez-Romero, 2006; Zhuang et al., 2007b).

The actions and interactions of the host plant and the endophytic microbes are beneficial for both parties. Depending on the strains and species of entophytic microbes and the host plant genotype, the populations of such microbes differ in plant tissues. The endophytic bacteria are also able to enhance plant resistance under different stresses including drought, salinity, acidity, suboptimal root heat, microbial infections, etc. (Kuiper et al., 2004; Zhang et al., 2006).

Endophytic bacteria reside in and activate the tissues of their host plants, favorably affecting plant growth and health. Endophytic bacteria are also able to reside in and activate the tissues of hyperaccumulators. Such bacteria may be more efficient than the soil bacteria in the bioremediation of heavy metals. However, the related mechanisms are yet to be investigated (Kuiper et al., 2004; Pishchik et al., 2005).

Although endophytic microbes can colonize different plant tissues, roots are the most suitable sites for their growth and activities. This is because roots are in intimate contact with the soil, and especially the rhizosphere, which continuously provide water and nutrients for the plant and hence the endophytic microbes. During the symbiotic process between endophytic microbes and host plant, tissues such as nodules are not produced; there are few details relating to the symbiotic process existing between the endophytic microbes and the host plant (Sessitsch et al., 2002, 2004).

Another important fact is that endophytic microbes are selected by the plant niche, and hence there are certain kinds of preferences made by the host plant for the establishment of a symbiotic association with the endophytic microbes. Such selections by plants are in accordance with the microbial activities, which favor plant growth and health, and also are able to tolerate plant systemic resistance. During the process of endophytic symbiosis, enzymes, which are able to degrade the plant cell wall, are produced. Such enzymes include polygalacturonase, cellulose, pectinase, lyase, and pectate (Hallmann et al., 1997).

Old and Nicolson (1978) submitted a model in which they have stated that roots are an important part of the soil–root, providing the apoplastic pathways to the plant's aerial parts, hence facilitating the entry of microbes into these aerial parts including the xylem. Such an intimate association establishes a pathway from the soil to the plant roots and aerial parts, for the activities and growth of endophytic microbes. Accordingly, the highest number of endophytic microbes is found in the roots in contrast to the aerial parts. The number of endophytic microbes is controlled by the plant and the environment, ranging from 10^3 CFU g^{-1} to 10^6 CFU g^{-1} fresh weight (Hallmann et al., 1997).

One of the most important activities of endophytic bacteria is N fixation; however, this only provides part of the plant's N requirement. This activity is in fact determined by plant genotype and growth stage, the bacterial strains, the environment, and the method of inoculation. It has also been indicated that some of the endophytic bacterial strains are not able to fix N. Some of the endophytic strains are wide ranging and can colonize different hosts suggesting nonspecificity, especially for the Gramineae plants, of the symbiotic association between endophytic bacteria and host plants (Sessitsch et al., 2002, 2004).

Favorable effects of endophytic diazotrophic strains such as *Azospirillum* sp. and *Entrobacter* sp. on the growth and health of the host plants have been indicated including: (1) biologically fixing N; (2) increasing crop yield; (3) production of plant hormone; (4) enhancing plant health, etc. (Bashan, 1986; Dalton et al., 2004; Sessitsch et al., 2004). However, it has been suggested that bacterial growth and its enzymatic activities are more sensitive to heavy meal stress than bacterial diversity (Kandeler et al., 2000).

A higher number of endophytic microbes were found in the rhizosphere of hyperaccumulator plants (Mengoni et al., 2001, 2004). The authors analyzed the serpentine sites in Italy and found that the microbes dominant there are *Pseudomonas* sp. and *Streptomyces* sp., in the rhizosphere of the metal hyperaccumulator *Alyssum bertolonii*. They also indicated that plant species rather than the place determined the diversity of microbial strains in the rhizosphere. The *Methylobacterium* strains were isolated from the rhizosphere and the aerial parts of *Thlaspi goesingense and* Zn-hyperaccumulator *T. caerulescens*, respectively. Such strains are able to tolerate high concentrations of heavy metals (James, 2000; Rosenblueth and Martínez-Romero, 2006).

The positive correlation between the activities of endophytic bacteria and the availability of heavy metals suggest the role of such bacteria in the behavior of heavy metals. The endophytic bacteria are able to increase plant tolerance in the presence of heavy metals (Wenzel et al., 2003). Enhanced plant activity and root growth are among the most important reasons for the increased tolerance of the host plant under heavy metal stress. Abou-Shanab et al. (2003b) indicated that the presence of endophytic bacteria in the soil was the most important cause for the increased availability and uptake of Ni by *Alyssum murale*.

The direct effects of heavy metal stress on the growth of plants and the subsequent hyperaccumulation by rhizospheric and endophytic microbes are the increased availability of heavy metals and the higher iron uptake by plants. The indirect mechanisms are the enhanced plant growth and the alleviation of stress. Accordingly, if efficient strains of rhizospheric and endophytic bacteria with suitable colonizing potential are selected, they can be used to bioremediate the stress of heavy metals (Clemens, 2001; Sessitsch et al., 2002).

The use of genetically modified strains of endophytic bacteria may also be a good practical method for the bioremediation of contaminated soil, provided that the related details have been investigated in the field. However, their competition with the native soil bacteria can adversely affect their activities and hence their bioremediation potential. Hence, the important aspects related to the use of endophytic bacteria are biosafety, persistence, and competition. Such aspects with respect to the properties of bacterial inoculants including stress alleviation, plant growth promotion, and increased uptake of heavy metals are of significance for the technique of bioremediation (Glick, 1995; Gutierrez-Zamora and Martinez-Romero, 2001).

Conclusions and future perspectives

Soybean, as an important legume crop, is able to establish a symbiotic association with its symbiotic N-fixing bacterium, *B. japonicum*, and fixes significant amounts of atmospheric N. Accordingly, it is one of the most important sources of food and oil for human beings. However under stress, soybean and rhizobia are not tolerant, and as a result the amount of fixed N and yield production significantly decreases. Due to anthropogenic activities the concentrations of heavy metals in the soil increase, and hence plants face the stress of heavy metals. Although some of the heavy metals such as Fe, Zn, Mn, and Cu are essential for plant growth and yield production, at high levels such elements, along with other elements including Cd, Co, Cr, V, and Ni, can adversely affect plant and microbial activities. Under such stresses the production of reactive oxygen species can have negative effects on the plant cellular structure. The plant response, including that of soybean under heavy metal stress, is the activation of antioxidant enzymes and products and the related signaling pathway, which scavenge the produced ROS. Different methods have been so far tested to alleviate the stress of heavy metals, among which the biological ones have been the most effective. Among the biological methods, plants and microbes are used for the alleviation of stress. Some plants called hyperaccumulators are able to absorb high amounts of heavy metals, while their growth remains unaffected. Soil microbes including AM fungi, PGPR, and endophytic bacteria are also able to enhance plants ability to grow and produce yields under stress via different mechanisms such as increased nutrient uptake. If it is essential to plant soybean in fields contaminated with heavy metals, more tolerant genotypes must be produced by using relevant genes. Future research may focus on the production of soybean and rhizobial species which are more tolerant under the stress of heavy metals. The produced species must be tested under field conditions and the quality of their seed must also be tested after being subjected to the stress of heavy metal. Using the right combination of plant and microbes may be the most effective method of alleviating such stress. The capability of hyperaccumulators, which are not strong symbionts with AM fungi, must also be improved in order to enhance the efficiency of alleviating heavy metal stress.

References

Abou-Shanab, R.A., Delorme, T.A., Angle, J.S., Chaney, R.L., Ghanem, K., Moawad, H., 2003a. Phenotypic characterization of microbes in the rhizosphere of *Alyssum murale*. Int. J. Phytoremediat. 5, 367–379.

Abou-Shanab, R.A., Angle, J.S., Delorme, T.A., Chaney, R.L., van Berkum, P., Moawad, H., Ghanem, K., Ghozlan, H.A., 2003b. Rhizobacterial effects on nickel extraction from soil and uptake by *Alyssum murale*. New Phytol. 158, 219–224.

Abou-Shanab, R.A., Ghanem, K., Ghanem, N., Al-Kolaibe, A., 2008. The role of bacteria on heavy-metal extraction and uptake by plants growing on multi-metal-contaminated soils. World J. Microbiol. Biotechnol. 24, 253–262.

Ali, H., Khan, E., Sajad, M., 2013. Phytoremediation of heavy metals – concepts and applications. Chemosphere 91, 869–881.

Artursson, V., Finlay, R., Jansson, J., 2006. Interactions between arbuscular mycorrhizal fungi and bacteria and their potential for stimulating plant growth. Environ. Microbiol. 8, 1–10.

Balestrasse, K., Benavides, M., Gallego, S., Tomaro, M., 2003. Effect of cadmium stress on nitrogen metabolism in nodules and roots of soybean plants. Funct. Plant Biol. 30, 57–64.

Bashan, Y., 1986. Inoculation of rhizosphere bacteria *Azospirillum brasilense* and *Pseudomonas fluorescens* towards wheat roots in the soil. J. Gen. Microbiol. 132, 3407–3414.

Behra, K., 2014. Phytoremediation, transgenic plants and microbes. Sustain. Agric. Rev. 13, 65–85.

Berg, G., Zachow, C., 2011. PGPR interplay with rhizosphere communities and effect on plant growth and health. In: Maheshwari, D.K. (Ed.), Bacteria in Agrobiology: Crop Ecosystems. Springer, New York, pp. 97–109.

Bi, Y.L., Li, X.L., Christie, P., 2003. Influence of early stages of arbuscular mycorrhiza on uptake of zinc and phosphorus by red clover from a low phosphorus soil amended with zinc and phosphorus. Chemosphere 50, 831–837.

Callahan, D.L., Baker, A.J.M., Kolev, S.D., Wedd, A.G., 2006. Metal ion ligands in hyperaccumulating plants. J. Biol. Inorg. Chem. 11, 2–12.

Chatterjee, S., Mitra, A., Datta, S., Veer, V., 2013. Phytoremediation protocols: an overview. In: Chatterjee, S., Mitra, A., Datta, S., Veer, V. (Eds.), Plant-Based Remediation Processes, vol. 35. Springer, Berlin, Heidelberg, pp. 1–18.

Chen, Y.X., He, Y.F., Yang, Y., Yu, Y.L., Zheng, S.J., Tian, G.M., Luo, Y.M., Wong, M.H., 2003. Effect of cadmium on nodulation and N_2-fixation of soybean in contaminated soils. Chemosphere 50, 781–787.

Cheng, S., 2003. Heavy metal pollution in China: origin, pattern and control. Environ. Sci. Pollut. Res. 10, 192–198.

Clemens, S., 2001. Molecular mechanisms of plant metal tolerance and homeostasis. Planta 212, 475–486.

Cobbett, C.S., 2000. Phytochelatins and their roles in heavy metal detoxification. Plant Physiol. 123, 825–832.

Dalton, D.A., Kramer, S., Azios, N., Fusaro, S., Cahill, E., Kennedy, C., 2004. Endophytic nitrogen fixation in dune grasses (*Ammophila arenaria* and *Elymus mollis*) from Oregon. FEMS Microbiol. Ecol. 49, 469–479.

Das, N., Vimala, N., Karthika, P., 2008. Biosorption of heavy metals – an overview. Indian J. Biotechnol. 7, 159–169.

Dell'Amico, E., Cavalca, L., Andreoni, V., 2005. Analysis of rhizobacterial communities in perennial Graminaceae from polluted water meadow soil, and screening of metal-resistant, potentially plant growth-promoting bacteria. FEMS Microbiol. Ecol. 52, 153–162.

Ernst, W., 1996. Bioavailability of heavy metals and decontamination of soil by plants. Appl. Geochem. 11, 163–167.

Fuentes, H.D., Khoo, C., Pe, T., Muir, S., Khan, A.G., 2000. Phytoremediation of a contaminated mine site using plant growth regulators to increase heavy metal uptake. In: Sanches, M.A., Vergara, F., Castro, S.H. (Eds.), Waste Treatment and Environmental Impact in the Mining Industry. University of Concepcion Press, Victor Lamas, Concepcion, Chile, pp. 427–435.

Galli, U., Schuepp, H., Brunold, C., 1994. Heavy metal binding by mycorrhizal fungi. Physiol. Plant. 92, 364–368.

Glick, B.R., 1995. The enhancement of plant growth by free-living bacteria. Can. J. Microbiol. 41, 109–117.

Glick, B.R., 2010. Using soil bacteria to facilitate phytoremediation. Biotechnol. Adv. 28, 367–374.

Glick, B.R., Patten, C.L., Holguin, G., Penrose, D.M., 1999. Biochemical and Genetic Mechanisms Used by Plant Growth-Promoting Bacteria. Imperial College Press, London, UK.

Gutierrez-Zamora, M.L., Martinez-Romero, E., 2001. Natural endophytic association between *Rhizobium etli* and maize (*Zea mays* L.). J. Biotechnol. 91, 117–126.

Hall, J.L., 2002. Cellular mechanisms for heavy metal detoxification and tolerance. J. Exp. Bot. 53, 1–11.

Hallmann, J., Quadt-Hallmann, A., Mahaffee, W.F., Kloepper, J.W., 1997. Bacterial endophytes in agricultural crops. Can. J. Microbiol. 43, 895–914.

Huang, C., Bazzaz, F., Vanderhoef, L., 1974. The inhibition of soybean metabolism by cadmium and lead. Plant Physiol. 54, 122–124.

Huang, X.D., El-Alawi, Y., Penrose, D.M., Glick, B.R., Greenberg, B.M., 2004. A multiprocess phytoremediation system for removal of polycyclic aromatic hydrocarbons from contaminated soils. Environ. Pollut. 130, 465–476.

Jalili, F., Khavazi, K., Pazira, E., Nejati, A., Asadi Rahmani, H., Rasuli Sadaghiani, H., Miransari, M., 2009. Isolation and characterization of ACC deaminase producing fluorescent pseudomonads, to alleviate salinity stress on canola (*Brassica napus* L.) growth. J. Plant Physiol. 166, 667–674.

James, E.K., 2000. Nitrogen fixation in endophytic and associative symbiosis. Field Crops Res. 65, 197–209.

James, E., Reis, V., Olivares, F., Baldani, J., Dobereiner, J., 1994. Infection and colonization of sugarcane by the nitrogen fixing bacterium *Acetobacter diazotrophicus*. J. Exp. Bot. 45, 757–766.

Jing, Y., He, Z., Yang, X., 2007. Role of soil rhizobacteria in phytoremediation of heavy metal contaminated soils. J. Zhejiang Univ. Sci. B 8, 192–207.

Kachenko, A., Singh, B., 2006. Heavy metals contamination in vegetables grown in urban and metal smelter contaminated sites in Australia. Water Air Soil Pollut. 169, 101–123.

Kandeler, E., Tscherko, D., Bruce, K.D., Stemmer, M., Hobbs, P.J., Bardgett, R.D., Amelung, W., 2000. Structure and function of the soil microbial community in microhabitats of a heavy metal polluted soil. Biol. Fertil. Soils 32, 390–400.

Kavamura, V., Esposito, E., 2010. Biotechnological strategies applied to the decontamination of soils polluted with heavy metals. Biotechnol. Adv. 28, 61–69.

Khan, A.G., 2003. Vetiver grass as an ideal phytosymbiont for Glomalian fungi for ecological restoration of derelict land. In: Truong, P., Hanping, X. (Eds.), Proceedings of the Third International Conference on Vetiver and Exhibition: Vetiver and Water, China Agricultural Press. Guangzou, Beijing. pp. 466–474.

Khan, A.G., 2004a. Co-inoculum of vesicular-arbuscular mycorrhizal fungi (AMF), mycorrhiza-helping bacteria (MBF), and plant-growth-promoting rhizobacteria (PGPR) for phytoremediation of heavy metal contaminated soils. Proceedings of the Fifth International Conference on Environmental Geochemistry in the Tropics, Chinese Academy of Science, March 21–26. Nanjing, China. p. 68.

Khan, A.G., 2004b. Co-inoculum of vesicular arbuscular mycorrhizal fungi (AMF), mycorrhiza-helping bacteria (MHB) and plant growth promoting rhizobacteria (PGPR) for phytoremediation of heavy metal contaminated soils. In: Proceedings of the Fifth International Conference on Environmental Geochemistry in the Tropics, Chinese Academy of Science, March 21–26. Nanjing, China.

Khan, A.G., 2005. Role of soil microbes in the rhizospheres of plants growing on trace metal contaminated soils in phytoremediation. J. Trace Elem. Med. Biol. 18, 355–364.

Kuiper, I., Lagendijk, E., Bloemberg, G., Lugtenberg, B., 2004. Rhizoremediation: a beneficial plant-microbe interaction. Mol. Plant Microbe Interact. 17, 6–15.

Kupper, H., Zhao, F., McGrath, S., 1999. Cellular compartmentation of zinc in leaves of the hyperaccumulator *Thlaspi caerulescens*. Plant Physiol. 119, 305–311.

Lasat, M., 2002. Phytoextraction of toxic metals: a review of biological mechanisms. J. Environ. Qual. 31, 109–120.

Lebeau, T., Braud, A., Jezequel, K., 2008. Performance of bioaugmentation-assisted phytoextraction applied to metal contaminated soils: a review. Environ. Pollut. 153, 497–522.

Lin, Y., Aarts, M., 2012. The molecular mechanism of zinc and cadmium stress response in plants. Cell. Mol. Life Sci. 69, 3187–3206.

Linger, P., Mussing, J., Fischer, H., Kobert, J., 2002. Industrial hemp (*Cannabis sativa* L.) growing on heavy metal contaminated soil: fiber quality and phytoremediation potential. Ind. Crops Prod. 16, 33–42.

Lippmann, B., Leinhos, V., Bergmann, H., 1995. Influence of auxin producing rhizobacteria on root morphology and nutrient accumulation of crops. 1. Changes in root morphology and nutrient accumulation in maize (*Zea mays* L.) caused by inoculation with indol-3 acetic acid (IAA) producing *Pseudomonas* and *Acinetobacter* strains of IAA applied exogenously. Angew. Bot. 69, 31–36.

Lucy, M., Reed, E., Glick, B.R., 2004. Application of free living plant growth promoting rhizobacteria. Antonie von Leeuwenhoek 86, 1–25.

Maksymiec, W., 2007. Signaling responses in plants to heavy metal stress. Acta Physiol. Plant. 29, 177–187.

Marschner, H., 1995. Mineral Nutrition of Higher Plants. Academic Press, London.

Mengoni, A., Barzanti, A., Gonnelli, C., Gabrielli, R., Bazzicalupo, M., 2001. Characterization of nickel-resistant bacteria isolated from serpentine soil. Environ. Microbiol. 3, 691–698.

Mengoni, A., Grassi, E., Brazanti, A., Biondi, E.G., Gonnelli, C., Kim, C.K., Bazzicalupo, M., 2004. Genetic diversity of microbial communities of serpentine soil and of rhizosphere of the Ni hyperaccumulator plant *Alyssum bertolonii*. Microb. Ecol. 48, 209–217.

Michalak, A., 2006. Phenolic compounds and their antioxidant activity in plants growing under heavy metal stress. Polish J. Environ. Stud. 15, 523–530.

Miller, M.B., Bassler, B.L., 2001. Quorum sensing in bacteria. Annu. Rev. Microbiol. 55, 165–199.

Miransari, M., 2010. Contribution of arbuscular mycorrhizal symbiosis to plant growth under different types of soil stresses. (Review article.) Plant Biol. 12, 563–569.

Miransari, M., 2011a. Soil microbes and plant fertilization. Review article. Appl. Microbiol. Biotechnol. 92, 875–885.

Miransari, M., 2011b. Hyperaccumulators, arbuscular mycorrhizal fungi and stress of heavy metals. Biotechnol. Adv. 29, 645–653.

Miransari, M., 2014. Plant Hormonal Signaling Under Stress. AbtinBerkeh Ltd. Company and Creative Space, Isfahan, Iran and USA, 98 pp.

Miransari, M., 2015. Phytoremediation using microbial communities. In: Ansari, A., Gill, S., Gill, R., Lanza, G., Newman, L. (Eds.), Phytoremediation, Management of Environmental Contaminates, vol. 2. Springer International Publishing, Switzerland, pp. 177–182.

Mithofer, A., Schulze, B., Boland, W., 2004. Biotic and heavy metal stress response in plants: evidence for common signals. FEBS Lett. 566, 1–5.

Mittler, R., 2006. Abiotic stress, the field environment and stress combination. Trends Plant Sci. 11, 15–19.

Nair, D.N., Padmavathy, S., 2014. Impact of endophytic microorganisms on plants, environment and humans. Scien. World J. 14, 1–11.

Newman, L., Reynolds, C., 2005. Bacteria and phytoremediation: new uses for endophytic bacteria in plants. Trends Biotechnol. 23, 6–8.

Ngah, W., Hanafiah, M., 2008. Removal of heavy metal ions from wastewater by chemically modified plant wastes as adsorbents: a review. Bioresour. Technol. 99, 3935–3948.

Nies, D.H., 1999. Microbial heavy-metal resistance. Appl. Microbiol. Biotechnol. 51, 730–750.

Nogueira, M., Magalhaes, G., Cardoso, E., 2004. Manganese toxicity in mycorrhizal and phosphorus-fertilized soybean plants. J. Plant Nutr. 27, 141–156.

Old, K.M., Nicolson, T.H., 1978. The root cortex as part of a microbial continuum. In: Loutit, M.V., Miles, J. (Eds.), Microbial Ecology. Springer, Berlin, pp. 291–294.

Pe, T., Fuentes, H.D., Khoo, C., Muir, S., Khan, A.G., 2000. Preliminary experimental results in phytoremediation of a contaminated mine site using plant growth regulators to increase heavy metal uptake. Handbook and abstracts 15th Australian statistical conference, July 3–7. Adelaide Hilton International, South Australia. pp. 143–144.

Pishchik, V.N., Vorob'ev, N.I., Provorov, N.A., 2005. Experimental and mathematical simulation of population dynamics of rhizospheric bacteria under conditions of cadmium stress. Microbiology 74, 735–740.

Rosenblueth, M., Martínez-Romero, E., 2006. Bacterial endophytes and their interactions with hosts. Mol. Plant Microbe Interact. 19, 827–837.

Saikia, R. (Ed.), 2008. Microbial Biotechnology. New India Publishing, India, p. 422.

Sajedi, N.A., Ardakani, M.R., Rejali, F., Mohabbati, F., Miransari, M., 2010. Yield and yield components of hybrid corn (Zea mays L.) as affected by mycorrhizal symbiosis and zinc sulfate under drought stress. Physiol. Mol. Biol. Plants 16, 343–351.

Sajedi, N.A., Ardakani, M.R., Madani, H., Naderi, A., Miransari, M., 2011. The effects of selenium and other micronutrients on the antioxidant activities and yield of corn (Zea mays L.) under drought stress. Physiol. Mol. Biol. Plants 17, 215–222.

Salt, D.E., Blaylock, M., Kumar, N.P.B.A., Dushenkov, V., Ensley, B.D., Chet, I., Raskin, I., 1995. Phytoremediation: a novel strategy for the removal of toxic metals from the environment using plants. Biotechnology 13, 468–474.

Sessitsch, A., Reiter, B., Pfeifer, U., Wilhelm, E., 2002. Cultivation-independent population analysis of bacterial endophytes in three potato varieties based on eubacterial and Actinomycetes specific PCR of 16S rRNA genes. FEMS Microbiol. Ecol. 39, 23–32.

Sessitsch, A., Reiter, B., Berg, G., 2004. Endophytic bacterial communities of field-grown potato plants and their plant growth-promoting and antagonistic abilities. Can. J. Microbiol. 50, 239–249.

Seth, C.S., Remans, T., Keunen, E., Jozefczak, M., Gielen, H., Opdenakker, K., Weyens, N., Vangronsveld, J., Cuypers, A., 2012. Phytoextraction of toxic metals: a central role for glutathione. Plant Cell Environ. 35, 334–346.

Shetty, K.G., Hetrick, B., Schwab, A.P., 1995. Effects of mycorrhizae and fertilizer amendments on zinc tolerance of plants. Environ. Pollut. 88, 307–314.

Smith, L.A., Means, J.L., Chen, A., Alleman, B., Chapma, C.C., Tixier, Jr., J.S., Brauning, S.E., Gavaskar, A.R., Royer, M.D., 1995. Remedial Options for Metal Contaminated Sites. Lewis, Boca Raton, FL.

Tyagi, M., da Fondeca, M., de Carvalho, C., 2011. Bioaugmentation and biostimulation strategies to improve the effectiveness of bioremediation processes. Biodegradation 22, 231–241.

Verbruggen, N., Hermans, C., Schat, H., 2009. Molecular mechanisms of metal hyperaccumulation in plants. New Phytol. 181, 759–776.

Vierheilig, H., Lerat, S., Piche, Y., 2003. Systemic inhibition of arbuscular mycorrhiza development by root exudates of cucumber plants colonized by Glomus mosseae. Mycorrhiza 13, 167–170.

Visioli, G., Marmiroli, N., 2012. Proteomics of plant hyperaccumulators. In: Gupta, D., Sandalio, L. (Eds.), Metal Toxicity in Plants: Perception, Signaling and Remediation. Springer, Berlin, Heidelberg, pp. 165–186.

Vogel-Mikuš, K., Drobne, D., Regvar, M., 2005. Zn, Cd and Pb accumulation and arbuscular mycorrhizal colonisation of pennycress *Thlaspi praecox* Wulf. (Brassicaceae) from the vicinity of a lead mine and smelter in Slovenia. Environ. Pollut. 133, 233–242.

Vogel-Mikuš, K., Regvar, M., Mesjasz-Przybylowicz, J., Przybylowicz, W., Simcic, J., Pelicon, P., Budnar, M., 2008. Spatial distribution of cadmium in leaves of metal hyperaccumulating *Thlaspi praecox* using micro-PIXE. New Phytol. 179, 712–721.

Wang, J., Chen, C., 2009. Biosorbents for heavy metals removal and their future. Biotechnol. Adv. 27, 195–226.

Wenzel, W.W., Adriano, D.C., Salt, D., Smith, R., 1999. Phytoremediation: a plant-microbe-based remediation system. In: Adriano, D.C. (Ed.), Bioremediation of Contaminated Soils. Agronomy Monographs, vol. 37, ASA, CSSA and SSSA, Madison, WI, pp. 457–508.

Wenzel, W.W., Bunkowski, M., Puschenreiter, M., Horak, O., 2003. Rhizosphere characteristics of indigenously growing nickel hyperaccumulator and tolerant plants on serpentine soil. Environ. Pollut. 123, 131–138.

Williams, L., Pittman, J., Hall, J.L., 2000. Emerging mechanisms for heavy metal transport in plants. Biochim. Biophys. Acta 1465, 104–126.

Wójcik, M., Vangronsveld, J., D'Haen, J., Tukiendorf, A., 2005. Cadmium tolerance in *Thlaspi caerulescens* II. Localization of cadmium in *Thlaspi caerulescens*. Environ. Exp. Bot. 53, 163–171.

Xiong, J., Fu, G., Tao, L., Zhu, C., 2010. Roles of nitric oxide in alleviating heavy metal toxicity in plants. Arch. Biochem. Biophys. 497, 13–20.

Yadav, S., 2010. Heavy metals toxicity in plants: An overview on the role of glutathione and phytochelatins in heavy metal stress tolerance of plants. S. Afr. J. Bot. 76, 167–179.

Yang, X., Jin, X., Feng, Y., Islam, E., 2005. Molecular mechanisms and genetic basis of heavy metal tolerance/hyperaccumulation in plants. J. Integr. Plant Biol. 47, 1025–1035.

Zakria, M., Njoloma, J., Saeki, Y., Akao, S., 2007. Colonization and nitrogen-fixing ability of *Herbaspirillum* sp. strain B501 gfp1 and assessment of its growth-promoting ability in cultivated rice. Microbes Environ. 22, 197–206.

Zenk, M.H., 1996. Heavy metal detoxification in higher plants – a review. Gene 179, 21–30.

Zhang, J., Shu, W.S., 2006. Mechanisms of heavy metal cadmium tolerance in plants. J. Plant Physiol. Mol. Biol. 32, 1–8.

Zhang, X.X., George, A., Bailey, M.J., Rainey, P.B., 2006. The histidine utilization (*hut*) genes of *Pseudomonas fluorescens* SBW25 are active on plant surfaces, but are not required for competitive colonization of sugar beet seedlings. Microbiology 152, 1867–1875.

Zhuang, P., Yang, Q., Wang, H., Shu, W., 2007a. Phytoextraction of heavy metals by eight plant species in the field. Water Air Soil Pollut. 184, 235–242.

Zhuang, X., Chen, J., Shim, H., Bai, Z., 2007b. New advances in plant growth-promoting rhizobacteria for bioremediation. Environ. Int. 33, 406–413.

Soybean production and suboptimal root zone temperatures

Narjes H. Dashti*, Vineetha Mariam Cherian*, Donald L. Smith**
*Department of Biological Sciences, Faculty of Science, Kuwait University, Safat, Kuwait;
**James McGill Professor Department of Plant Science, McGill University, Montréal, Canada

Soybean: structure, benefits, and cultivation

Soybean (*Glycine max* (L.) Merr.) belongs to the family Fabaceae, subfamily Faboideae, and the genus *Glycine*. The genus *Glycine* is divided into two subgenera: *Glycine* and *Soja* based on the nature of their growth, that is, perennial or annual, respectively. The cultivated soybean belongs to the subgenus *Soja* and is predominantly cultivated during summer. The height of the plants varies from 0.2 m to 2 m. Soybean has trifoliated leaves, with each node of the raceme interleaved individually with flowers, a five-toothed calyx, a glabrous corolla with long clawed petals, a keel, and seed (Hymowitz and Newell, 1981).

The pigments of the seed hull vary from plant to plant and commonly observed colors include black, brown, blue, yellow, green, and mottled. Two distinct stem growth and floral initiation pattern are observed in soybean. In the first pattern, the stem is indeterminate, in which the terminal bud continues vegetative activity during most of the growing season. In the other pattern the stem is determinate, that is, the vegetative activity of the terminal bud ceases after the flowering (Dashti et al., 1997). Like all leguminous crops, the soybean is also capable of fixing nitrogen by symbiotically associating with beneficial soil microorganisms and forming nodules. The most commonly found microbe associated with the soybean rhizosphere is *Bradyrhizobium japonicum*. Together, they fix about 100–200 kg ha^{-1} year^{-1} of atmospheric nitrogen (Smith and Hume, 1987).

Soybean has numerous beneficial properties, which has made it one of the most extensively cultivated legume crops worldwide. Soybean comprises about 40% proteins making it the cheapest, richest, and most easily accessible vegetable protein source (Dugje et al., 2009). Additionally it is also an excellent source of oil, 85% of which is unsaturated and cholesterol-free, carbohydrates, and dietary fiber (Bellaloui et al., 2011). Soymilk is used as an alternative to cow's milk. Soybeans are a concentrated source for isoflavones, a class of polyphenolic organic compounds that have diverse biological functions (Messina, 1999).

The two chief isoflavones of soybeans, that is, genistein and daidzein, have attained significant global importance due to their potential role in preventing and treating cancer

Abiotic and Biotic Stresses in Soybean Production. http://dx.doi.org/10.1016/B978-0-12-801536-0.00010-4

(Messina, 1999). Daidzein was found to inhibit the proliferation of HL-60 tumor cells implanted in the subrenal capsules of mice, while genistein was found to inhibit a wide array of hormone-dependent and hormone-independent cancer cells with an IC_{50} between ~5 μM and 40 μM, including breast, colon, prostate, and skin cells (Messina, 1999).

Soybean isoflavones are also classified as phytoestrogens due to their ability to exhibit estrogen-mimicking properties and their ability to bind to the estrogen receptor in the body. The estrogenic properties of soybean isoflavones have raised speculation that this might prevent osteoporosis and promote bone health in women (Messina, 1999). Soybean isoflavones are structurally similar to 7-isopropoxyisoflavone, a synthetic isoflavone that has been shown to increase bone mass in postmenopausal women.

Furthermore, soy proteins have low sulfur amino acid content (SAA), which improves calcium retention. Generally, metabolism of SAA causes demineralization and excretion of calcium ions from the bones in the urine due to hydrogen ions produced by the buffering action of the skeletal system (Anderson et al., 1987). Soy protein is associated with a noticeable decline of calcium excretion in the urine relative to other vegetable and animal proteins. Aside from the vitalizing health benefits of consuming soybean, it has been observed that the isoflavone genistein is also integral in nodulation and signal transduction (Dashti et al., 1997; Messina, 1999).

Soybean is a subtropical legume, native to east Asia. It was first domesticated by Chinese farmers around 1100 BC. By the first century AD, soybean cultivation had pervaded the surrounding regions of China as well. Soybean was introduced to the temperate parts of the world such as Europe and North America as recently as the twentieth century. Although initially used in these regions as fodder for cattle, it quickly became established as an important protein substitute after World War II. At present the chief producers of soybean are the USA, Argentina, Brazil, China, and India. Together, they represent about 90% of the total soybean production. Soybean cultivation is increasingly becoming popular in cooler regions such as Canada. As of 1993, about 676,000 ha of cultivated land in Ontario was dedicated solely for soybean cultivation (Dashti, 1996).

Soybeans require mean temperatures ranging from 20°C to 30°C with optimal growth, nodulation, and nitrogen fixing (Dashti et al., 1997). They are capable of growing on a wide range of soils, with an optimum growth in moist alluvial soils replete with good organic content. Although soybean can autonomously fix nitrogen by forming symbiotic alliance with beneficial soil microbes, effective plant–microbe association requires optimum temperature conditions at the zone of interaction near the plant roots (Dashti et al., 2000). In many of the cooler soybean growing regions of Canada, two of the major limiting factors to successful soybean cultivation are the suboptimal root zone temperature (RZTs) and short season conditions (Miransari and Smith, 2008).

Effect of suboptimal root zone temperatures on the growth and development of soybean

When temperatures drop below the optimum RZT, a stress is created because of which many biological functions are adversely affected including plant growth, yield, nodulation, and nitrogen fixation (Miransari, 2012; Zhang and Smith, 2002). At low RZT, the ability of the plants to acquire nutrients is greatly hindered due to several factors such as poor chemical nutrient availability, retarded root and shoot growth, and the failure of the root systems to uptake sufficient nutrients (Engels and Marschner, 1996; Liu et al., 2004).

Nitrogen fixation, nodulation, and nodule function are severely affected at a low RZT (Younesi et al., 2013). At optimum temperatures, during the initial stages of infection with *B. japonicum*, the host plant produces the signal molecule genistein, which trigger the bacterial genes to initiate the process of nodulation, by which the bacteria enter the plant roots and alter the plant cellular behavior to form nodules (Miransari et al., 2013). Bacterial infection and node initiation are the most sensitive low RZT.

The hindrance to effective nitrogen fixation has been partly attributed to factors such as (1) changes in the nodule O_2 permeability; (2) low solubility and the reduced translocation of ureide, the processed form of nitrogen utilized by the plant as the temperature drops (Dashti et al., 2013); (3) progressive decrease in the bacteroid tissue and decrease in its formation rate; and (4) disruption of signal exchange between the *Bradyrhizobium* and the host plant (Dashti, 1996; Panwar and Laxmi, 2005).

Various experiments were conducted to check the variations occurring in nitrogen-fixing ability of soybean plants over a range of RZT under controlled environmental conditions. The parameters tested include nodulation, nitrogen fixation, and total nitrogen yield (Zhang et al., 1996). To determine the changes in nitrogen fixation rate and nodulation of soybean, two simultaneous experiments were conducted on two adjacent sites using a randomized split plot design with three replications.

In the first experiment the plants were grown in nitrogen-free soil, and in the second experiment nitrogen fertilizer was prefed to the plants in addition to autonomous nitrogen fixation. The main plot units consisted of three RZTs: 25 (optimal temperature for soybean nodulation and nitrogen fixation), 17.5, and 15°C. *B. japonicum* was added by pipette to the rooting medium to the base of the plant.

From the aforementioned studies, the following observations were made: (1) low RZT affects nitrogen fixing soybean plants more than nitrogen-fed soybean plants (Figure 10.1); (2) the time between soybean inoculation with *B. japonicum* and the beginning of nitrogen fixation increases by 2–2.5 days for every degree between 25°C and 17°C. However, a drop in temperature below 17°C delayed the nitrogen fixation by a week per degree from the time of inoculation (Figure 10.1). This delay is believed to be caused by the inability of the plants to excrete the plant-to-bacterium isoflavone signal molecule genistein at the beginning of symbiosis establishment; (3) slow nodule development in cool soils prolongs the period of nitrogen deficiency that occurs between the depletion of cotyledonal nitrogen reserves and the nitrogen fixation (Figure 10.1); and (4) sluggish growth of the plants was observed throughout the growth phase.

Recent studies have revealed in detail the molecular mechanism of legume–rhizobia interaction, which leads to the establishment of a symbiotic relationship between the bacteria and the plant (Duzan et al., 2004; Zhang and Smith, 2002). During the early stages of infection and nodule organogenesis, bacteria attach themselves to the roots in response to root exudates, mainly flavonoids such as genistein, which trigger the bacterial NodD protein causing transcription of bacterial *nod* genes.

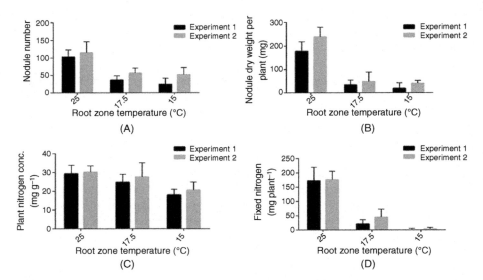

Figure 10.1 The effect of varying root zone temperatures. Effect on (A) nodule number, (B) nodule dry weight per plant, (C) plant nitrogen concentration, and (D) fixed nitrogen in soybean plants. Vertical lines on top of each bar represent the standard error unit ($n = 24$).

Expression of *nod* genes results in the synthesis of Nod factors as bacteria-to-plant signal molecules composed of tri- to pentachitin backbones with an *N*-acyl group at the nonreducing end and a selection of substitutions along the chitin backbone. The application of an appropriate Nod factor to soybean roots results in root hair deformation and can cause initiation of the nodule primordia or the formation of complete nodule structures. Inhibition of the *nod* genes at low suboptimal RZT causes decline of the nodulation and nitrogen fixation (Duzan et al., 2004).

The soybean growth and physiology were also found to be delayed at low RZTs (Zhang et al., 1997). An experimental study arranged in completely randomized split-plot designs with three replications with the main plot units maintained at three temperatures, namely 25, 17.5, and 15°C, drew the conclusions that when the plants were harvested 50 days after inoculation with *B. japonicum*: (1) soybean growth diminished with the decrease in temperature. The highest growth was observed at 25°C and the lowest at 15°C (Figure 10.2). Growth parameters such as plant leaf number, plant leaf area, plant dry weight, and pod number were evaluated. All the growth parameters progressively decreased with decrease in temperature (Zhang et al., 1997). (2) Plant physiology was also considerably altered at low RZT (Figure 10.3). Parameters evaluated were stomatal conductance, transpiration, and photosynthetic rate. All of these values dropped considerably as temperatures declined. The changes in photosynthetic rate over time at varying RZTs are indicated in Figure 10.4.

Results showed that at 25°C, the plant photosynthetic rate increased steadily from day 17 to day 37 and declined sharply from day 37 to day 47. At 17.5°C, the photosynthetic

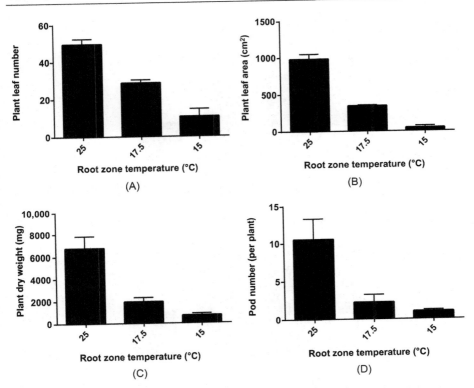

Figure 10.2 The effect of varying root zone temperatures on soybean growth parameters. (A) Plant leaf number, (B) plant leaf area, (C) plant dry weight, and (D) pod number. Vertical lines on top of each bar represent the standard error unit ($n = 24$).

rate sharply declined from day 17 to day 27, and then increased steeply up to day 37 before nominally increasing up to day 47. At 15°C, there was a sharp decline in the photosynthetic rate of the plants from week 17 to 47 (Figure 10.4). These changes occurring in the photosynthetic rate were concordant with the various stages of plant growth, with values obtained at 25°C being the most optimal.

At 17.5°C, although at the initial growth phases there is a decline in the photosynthetic activity, it picks up once the plants mature sufficiently to overcome the unfavorable stress present in the root system. The values however are lower compared with 25°C. At 15°C, the plants were unable to recover their photosynthetic activity even when mature. Photosynthesis and transpiration rates are controlled by leaf water potential, which is drastically lowered at low RZT (Duke et al., 1979).

All leaf enzyme assays related to photosynthesis showed higher activities at higher temperatures, which indicates that soybean photosynthetic enzymes are severely affected by suboptimal RZT during the middle of the growing season (Duke et al., 1979). Suboptimal RZTs affect the plants more drastically *in vitro* than in the greenhouse (Liu et al., 2004).

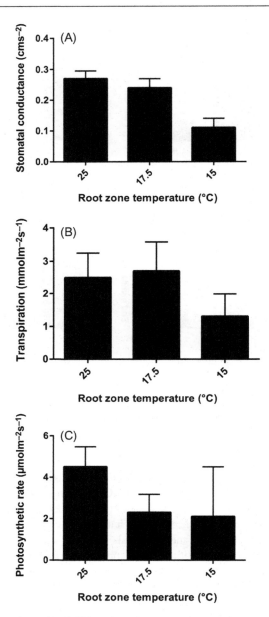

Figure 10.3 Effects of varying RZTs. Effect on (A) stomatal conductance, (B) transpiration, and (C) photosynthetic rate of plants. Vertical lines on top of each bar indicate one standard error unit ($n = 24$).

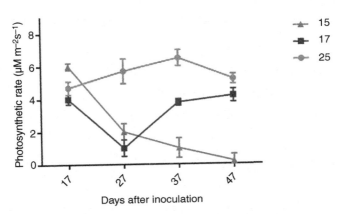

Figure 10.4 Changes occurring in the photosynthetic rate over time at varying root zone temperatures. Each point represents the mean value of six observations.

The origin of the host plant species is independent of the impact of the suboptimal RZTs on plant root colonization by nodule forming bacteria (Liu et al., 2004). Previous studies have indicated a strong repression of colonization on soybean (Dashti et al., 2013; Duke et al., 1979; Miransari and Smith, 2008; Zhang et al., 1996) and sorghum (Liu et al., 2004), both of which are tropical crops, and in barley (Baon et al., 1994) and tomato (Van Der Ploeg and Heuvelink, 2005), as temperate climate crops.

Smith and Bowen (1979) reported similar repression in *Medicago truncatula* and *Trifolium subterraneum* temperate climate plants at 16°C. In contrast, Volkmar and Woodbury (1989) found no effect on colonization by microbes at low RZTs. In fact, they found higher colonization rates at 12°C compared to 16 and 20°C, when the inoculum was dispersed through the soil at 5 cm subterranean depth.

They also concluded that even reduced root elongation at suboptimal RZT could bring about effective colonization as long as the microbial percentages and infection rates remain steady (Volkmar and Woodbury, 1989). Contrasting results occurring in the same plant species might be suggestive of the fact that colonization is at least partly reliant on metabolic activity of the microbial isolates at suboptimal RZTs and their ability to uphold the symbiotic alliance between the plant and the microbe (Liu et al., 2004).

Spatial discrepancies in the inoculum placement may change the host plant's response to the impact of suboptimal RZT (Liu et al., 2004). Volkmar and Woodbury (1989) found no effect of root zone temperature when the inoculum was dispersed throughout the soil but noted changes when the inoculum was placed 5 cm below the soil surface. Zhang et al. (1996) observed a drop in nodulation when *B. japonicum* was interspersed with the rooting medium before mechanically applying them to the base of the plants (Figure 10.1).

The varied plant response occurring due to the deviations made in the locus of the inoculum placement might be because of the varied concentration of the inoculum quantity. Effective colonization can occur only if a sufficient quantity of the inoculum is present, to saturate the roots (Liu et al., 2004). The effects of suboptimal RZT on the growth and development of soybean can be minimized by coinoculation of certain plant growth promoting rhizobacteria (PGPR) with *B. japonicum* and addition of the isoflavone molecule genistein.

Plant growth promoting rhizobacteria: importance and mechanism of action

PGPR is a beneficial group of soil bacteria that is capable of colonizing plant roots and promote plant growth and development (Saharan and Nehra, 2011). Most of the identified strains of rhizobacteria occur within the Gram-negative genera of which *Pseudomonas* are the most characterized (Dashti et al., 1997). *Pseudomonas* is a diverse group of Gram-negative, aerobic heterotrophic microorganisms commonly found in the soil. Some species have been isolated from aquatic environments and even from animals.

Individual *Pseudomonas* strains may have biocontrol activity, plant growth promoting capability, and the ability to induce systemic resistance in plants (Adesemoye and Ugoji, 2009). Many fluorescent *Pseudomonas* strains have also been found to confer protection on the plants by the secretion of *in situ* antibiotic compounds that inhibit the growth of pathogenic microbes. Along with *Pseudomonas*, some of the other commonly recognized PGPR genera include *Azospirillum, Alcaligenes, Arthrobacter, Acinetobacter, Bacillus, Burkholderia, Enterobacter, Erwinia, Flavobacterium, Rhizobium, Serratia, Azotobacter, Acetobacter diazotrophicus, Azoarcus, Allorhizobium, Azorhizobium, Bradyrhizobium, Mesorhizobium, Methylobacterium,* and *Sinorhizobium* (Saharan and Nehra, 2011).

The ability of the PGPR to promote plant growth and protection is dependent on: (1) genetic traits such as their motility; (2) chemotaxis to seed and root exudates; (3) pili and fimbriae production; (4) production of components related to the cell surface; (5) ability to use certain components of cell surface, root exudates, and protein secretion; and (6) quorum sensing (Nelson, 2004). Studies have shown that a single PGPR is capable of displaying multiple modes of action (Ahemad and Kibret, 2014). Recent studies have also shown that PGPR associations vary in the intimacy and degree of bacterial proximity to the roots (Ahemad and Kibret, 2014). In this regard, they can be broadly classified as (a) extracellular PGPR: those that live in the rhizosphere, rhizoplane, or spaces between the cells of the root cortex, and (b) intracellular PGPR: those that exist within the root cells, normally in a node, to form an effective association with the roots (Ahemad and Kibret, 2014).

The major applications of PGPR include biological nitrogen fixation, symbiotic nitrogen fixation, cyanide production, plant growth promotion, siderophore production, phosphate solubilization, biocontrol activity, antifungal activity, and alleviation of stressed conditions in the soil. The exact mechanism by which PGPR exert their beneficial effects on plants is not fully understood.

However, most PGPR are believed to protect plants and promote growth by one or more mechanisms such as (1) suppression of plant disease and by induction of systemic resistance or antibiotic production (bioprotectants); (2) improved nutrition acquisition (biofertilizers); (3) production of phytohormones (biostimulants); and (4) degrading organic pollutants (rhizomediators) (Ahemad and Kibret, 2014; Nelson, 2004; Saharan and Nehra, 2011).

PGPR influence plants both directly and indirectly. The direct influence of the PGPR is promotion of plant growth (Ahemad and Kibret, 2014; Vacheron et al., 2013). Promotion of plant growth is achieved by the production of plant growth regulators, such as indolacetic acid (IAA), ethylene, and gibberellic acid, which in turn trigger specific plant growth promoting phytohormones such as auxins and cytokinins (Joseph et al., 2007).

In addition to this, the presence of PGPR raises the nitrogen fixation rate and phosphate uptake, two integral nutrients in plant growth (Saharan and Nehra, 2011; Vacheron et al., 2013). Indirectly, they influence the plants by reinforcing the indigenous systemic resistance therein (Nelson, 2004). Generally, the induced systemic resistance (ISR) response against pathogens, elicited by a local infection, involves a cascade of salicylic acid (SA) signaling pathways (Saharan and Nehra, 2011).

High levels of SA found in the phloem of plants following an infection, is highly suggestive of the fact that the *de novo* production of SA in noninfected plant plants may lead to a systemic expression of ISR. In contrast to the pathogen-elicited response, the PGPR activate a different transduction pathway depending on the precipitation of ethylene and jasmonic acid, instead of SA and the pathogenesis-related (PR) genes (Saharan and Nehra, 2011).

The ISR response usually occurs in phytosystems, wherein the PGPR and the pathogen are spatially separated (e.g., PGPR in the roots and pathogens on the leaf) or in split root systems, with PGPR disease-suppression channeled through the plant intermediary as no direct interface between the microbial populations occurs (Bakker et al., 2007). Alternatively, PGPR are also able to promote plant protection against pathogens by the secretion of certain inhibitory secondary metabolites (Dashti et al., 2012, 2014).

Siderophores are an important secondary metabolite compound secreted by the PGPR, which are capable of inhibiting pathogenic bacteria by chelating iron (Dashti et al., 2012). Iron is a soil element whose accessibility for direct assimilation by the microorganism limits growth. The concentration of iron is much too limited to support microbial growth. Some of the PGPR secrete siderophores to bind to the available Fe^{3+} and transport them back to the cell. By binding all the available iron molecules in the rhizosphere, the PGPR limit the iron source in the soil thereby ensuring that the pathogenic microbes do not proliferate (O'Sullivan and O'Gara, 1992).

Antibiotics such as bacteriocins and phenyacetic acid (PAA), an auxin-like molecule with antimicrobial activities, are secreted by numerous PGPR species such as *Bacillus*, *Pseudomonas*, and *Azospirillum*, to diminish deleterious microbes in the soil, especially root associated pathogenic fungi (Adesemoye and Ugoji, 2009; Dashti, 1996; Saharan and Nehra, 2011). Certain fungi such as the arbuscular mycorrhizal (AM) fungi suppress diseases by improving the mineral uptake in plants (Liu et al., 2004).

PGPR action under stressed conditions of suboptimal root zone temperatures

PGPR alleviate the stress conditions caused by the suboptimal RZT by enhancing various biological functions of the plants (Younesi et al., 2013). The type of PGPR that interact with plant roots is chosen and enhanced by the plant itself and is dependent on the nature and the concentration of organic constituents exuded by the plants and the analogous ability of the bacteria to utilize these compounds (Saharan and Nehra, 2011).

As mentioned previously, one of the integral biological functions disrupted by suboptimal RZTs is the independent nitrogen fixing ability of leguminous plants (Dashti et al., 2013). Studies have shown that mixtures of PGPR and coinoculation of PGPR with AM fungi or *B. japonicum* can enhance the nitrogen fixation rate in both leguminous and nonleguminous plants (Dashti, 1996; Dashti et al., 2013; Liu et al., 2004; Zhang et al., 1996).

PGPR generally enhance nitrogen fixation by increasing the root nodule number, nodule mass, and the nitrogenase activity (Younesi et al., 2013). PGPR nitrogen fixers can be broadly classified into symbiotic nitrogen fixers and nonsymbiotic nitrogen fixers. Symbiotic nitrogen fixers have to form a symbiotic relationship with their host plants by forming root nodules before fixing nitrogen. Symbiotic nitrogen fixers consist of groups such as rhizobia in leguminous plants and frankiae in nonleguminous plants (Saharan and Nehra, 2011).

Rhizobia are comprised of seven genera of bacteria: *Allorhizobium*, *Bradyrhizobium*, *Methylobacterium*, *Azorhizobium*, *Mesorhizobium*, *Rhizobium*, and *Sinorhizobium*. *Frankia* is able to produce root nodules on more than 280 species of woody plants belonging to 8 different families (Saharan and Nehra, 2011). Nonsymbiotic nitrogen fixers are free-living, associative, and endophytic bacteria that can fix nitrogen without forming nodules in the plant roots. Species belonging to this group include cyanobacteria, *Azospirillum*, *Azotobacter*, *Acetobacter*, *Diazotrophicus*, *Azoarcus*, etc. (Saharan and Nehra, 2011).

The coinoculation of different nonsymbiotic nitrogen fixing PGPR such as *Bacillus* or *Pseudomonas* in synergy with a "helper" symbiotic nitrogen fixing bacteria enhances the overall performance of the selected microbes (Yadegari et al., 2010). A significant increase in the root and biomass was observed in chick pea plants when coinoculated with *Mesorhizobium* and *Pseudomonas* (Sindhu et al., 2002). Younesi et al. (2013) demonstrated that coinoculation of PGPR with *Sinorhizobium meliloti* produced a wide range of effects on alfalfa plants depending on the PGPR inoculum applied. PGPR were found to increase plant weight, nodule number, and nodule mass. Similar effects were also noted in fodder galega (Egamberdieva et al., 2010).

Zhang et al. (1996, 1997) demonstrated the effects of coinoculating PGPR with *B. japonicum* on soybean crops under suboptimal conditions in relation to the nitrogen

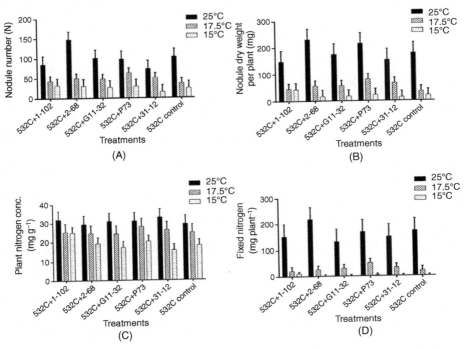

Figure 10.5 The effect of plant growth promoting rhizobacteria. Effect on (A) nodule number, (B) nodule dry weight per plant, (C) plant nitrogen concentration, and (D) fixed nitrogen at varying RZTs on soybean plants. Vertical lines on top of each bar represent the standard error unit ($n = 24$).

fixation, physiology, and growth of the plants. The results of the study are presented in Figures 10.5–10.7. Five different PGPR strains were used along with *B. japonicum* (532C), as shown in Table 10.1. All PGPR strains were cultured in *Pseudomonas* medium in 250 ml flasks shaken at 250 rpm at room temperature. *B. japonicum* was cultured in yeast extraction mannitol broth (Zhang et al., 1996).

Once the stationary phase was attained (7 days for *B. japonicum* and 1.5 days for PGPR), the cell density was adjusted to a known value, followed by admixing and cooling down to the corresponding RZT and addition by pipette to the rooting medium. The experiments were set up using completely randomized split-plot designs under three RZTs: 25, 17.5, and 15°C. The plants were harvested at 50 days after inoculation and the following data were collected: nodule number, nodule dry weight, plant nitrogen concentration, and the total fixed nitrogen.

The results indicated that the effects of the five PGPR strains on the nodule number varied with RZT (Figure 10.5A). At 15°C, PGPR did not increase nodule number compared with the non-PGPR controls (Zhang et al., 1996). Treatment containing *Pseudomonas fluorescens* (31-12) decreased the nodule number. This was due to the increased nodule size. Nodule dry weight per plant was not affected by coinoculation with PGPR.

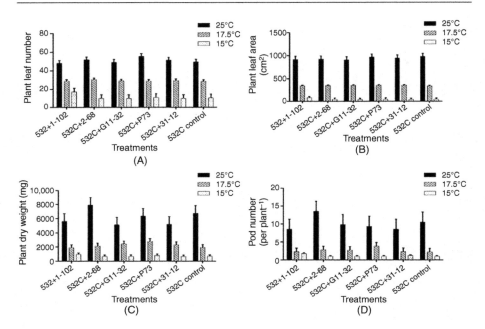

Figure 10.6 The effect of PGPR at varying RZTs on soybean growth parameters.
(A) Plant leaf number, (B) plant leaf area, (C) plant dry weight, and (D) pod number. Vertical lines on top of each bar represent the standard error unit ($n = 24$).

At 17.5°C RZT, none of the PGPR decreased nodule number. *Aeromonas hydrophilia* (P73) increased nodule number, nodule size, and nodule dry weight per plant, while *Serratia proteomaculans* increased nodule dry weight per plant (Figure 10.5).

At optimal RZT, only coinoculation of *B. japonicum* and *Serratia liquefaciens* (2-68) yielded a positive increase in the nodule number compared to the non-PGPR controls. PGPRs also increased the nodule mass per plant at both optimal and suboptimal RZT. *A. hydrophila* (P73) increased the total nodule weight per plant at 17.5°C RZT, with these increases being partially due to increased nodule number (Figure 10.5).

S. liquefaciens (2-68) increased nodule number and total nodule weight per plant at the optimal RZT (25°C). However, the nodule size remained unchanged at this temperature. The positive influences of *S. liquefaciens* (2-68) at 25°C and *Serratia proteamaculans* (1-102) at both 17.5 and 15°C were consistent across experiments. At 15°C, the plant nitrogen fixation was increased by PGPR strain *S. proteamaculans* (1-102). The increase observed was partly due to an increase in the plant tissue nitrogen concentration (Figure 10.5) and partly due to the increase in total dry matter, especially for plants treated with 1-102 (45.9%).

At 17.5°C RZT, the nitrogen concentration in plant tissue was not affected even in the presence of the PGPR. However, there was a substantial increase in the dry matter, as a result of which there was a marked increase in the total nitrogen fixed (Zhang et al., 1996). The onset of nitrogen fixation was much earlier for plants treated with the

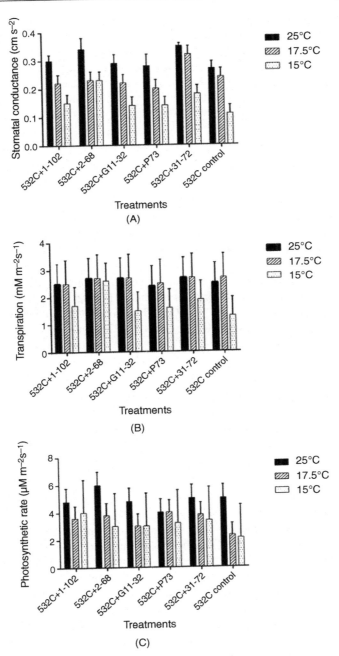

Figure 10.7 Effects of PGPR. Effect on (A) stomatal conductance, (B) transpiration, and (C) photosynthetic rate of plants at varying RZTs. Vertical lines on top of each bar indicate one standard error unit ($n = 24$).

Table 10.1 **PGPR tested for their effect on nitrogen fixation, growth, and physiology in soybean crops**

Strain numbers	Strain identification	Sources*
P73	*A. hydrophila*	Unknown
31-12	*P. fluorescens*	Mould Bay
2-68	*S. liquefaciens*	James Bay, NWT
G11-32	*P. putida*	Grise Fiord, NWT
1-102	*S. proteamaculans*	Yellowknife, NWT

* All five strains were provided by Cominco Fertilizers Ltd., Ag Biologicals (402-15 Innovation Boulevard, Saskatoon, Canada).

PGPR strains 1-102 and 2-68 compared to non-PGPR inoculated controls. The effects of the PGPRs on plant dry weight, leaf number, and leaf area was dependent on the PGPRs strains used (Figure 10.6).

At optimal RZT, *S. liquefaciens* (2-68) increased the plant total dry weight. The leaf number and the leaf area remained unchanged. *Pseudomonas putida* (G11-32), *Pseudomonas flourescens* (31-12), and *S. proteamaculans* (1-102) decreased the plant total dry weight at the optimal RZT. At 17.5°C RZT, *A. hydrophilia* (P73) increased total plant dry weight and pod number (Zhang et al., 1997; Figure 10.6). At 15°C, 1-102 increased plant leaf area and total plant dry weight compared with the non-PGPR controls. Plant photosynthetic rates, averaged across four measurements (17–47 days after inoculation), were affected by PGPR strains and varied with the different RZTs (Zhang et al., 1997; Figure 10.7).

At optimal RZT, plants treated with *S. liquefaciens* elevated the photosynthetic rates 24.2% more than the noninoculated healthy controls. All the strains that increased photosynthesis at optimal RZT also increased transpiration and stomatal conductance of the plants (Zhang et al., 1997). At 17.5°C RZT, *S. liquefaciens*, *P. fluorescens*, and *A. hydrophila* all increased the photosynthetic rates. At 15°C, *S. proteamaculans* increased leaf photosynthetic rates by 37%. *S. liquefaciens* was found to increase the plant transpiration and stomatal conductance. However, the photosynthetic rates remained unchanged in these plants (Zhang et al., 1997).

Based on the aforementioned results, the following conclusions were drawn from the study: (1) the beneficial effects of the PGPR vary with type and combination of the strains used. Some PGPR stimulated plant growth, development, nitrogen fixation, and some physiological processes, while the others inhibited these processes (Figures 10.5–10.7). (2) The physiological effects of the PGPR are altered with different RZTs. That is, some PGPR strains stimulated plant growth at one RZT but inhibited it at an other. For example, *S. proteamaculans* (1-102) increased all plant growth and nitrogen fixation variable at 15°C but reduced plant shoot and dry matter at the 25°C RZT. Due to varied activities at different temperatures, the PGPR intended for use as nodulation stimulators in different areas should be screened for and chosen according to the topographical conditions prevailing in a particular location, especially the RZT. (3) The enhancement of nodule number/size is independent of the nitrogen fixation. Some PGPR such as 1-102 were able to increase nodule size and nitrogen fixation without increasing the nodule number at 15°C RZT. (4) Nodule dry weight per plant dry weight was not influenced by PGPR at any RZT. (5) There is a linear relationship between RZT and time from inoculation with *B. japonicum* to the onset of nitrogen fixation between 25°C and 17°C RZT and in this range the time elapsed between inoculation and nitrogen fixation increased by 2 days for each degree drop in the RZT. (6) Photosynthetic rate was more sensitive to the use of PGPR at both suboptimal and optimal RZTs compared with transpiration and stomatal conductance (Figure 10.7). (7) The change in photosynthetic rate over time showed that plant photosynthesis was increased by PGPR applications over a wide range of plant growth stages. (8) Since photosynthesis, transpiration, and stomatal conductance (data not shown) were increased by stimulatory strains before the onset of nitrogen fixation at 17.5 and 15°C, the improvement of plant growth, development, and physiological activities must be due to the direct effects of PGPR on overall physiology rather than the specific effects on nitrogen fixation.

The survival and growth of rhizosphere populations of seven PGPR strains inoculated on soybean (c. Maple Glen) in sterile rooting medium under low RZTs were studied (Dashti, 1996). Three RZTs were tested: 25, 17.5, and 15°C. The results are presented in Tables 10.2–10.4. Population densities were found to vary with temperature. At each temperature, populations of some PGPR strains increased either on the root or in the rooting medium (rhizosphere).

RZT affected the distribution of PGPR populations between the root surface and the rooting medium. The strains with higher population densities on the root, which reflects their ability to colonize the root more rapidly, were: *S. proteamaculans* (1-102 at 15°C), *P. putida* (G11-32 at 17.5°C), and *S. liquefaciens* (2-68 at 25°C). The population densities of these strains in the rooting medium were lower at these temperatures.

Table 10.2 Root-associated PGPR colony-forming units (cfu) for different PGPR strains at different temperatures

PGPR strains	Log cfu g dry root^{-1}		
	15°C	17.5°C	25°C
31-34	8.73	8.86	8.49
G11-32	8.81	8.92	8.71
36-43	8.55	7.93	8.29
63-49	8.11	8.55	8.59
2-68	8.21	8.80	8.61
1-102	8.90	8.29	7.61
1-104	8.42	8.38	7.85
LSD$_{0.05a}$	0.45		
LSD$_{0.05b}$	0.48		

PGPR strains used were: 31-34, *P. putida*; G11-32, *P. putida*; 36-43, *P. fluorescens*; 63-49, *P. fluorescens*; 2-68, *S. liquefaciens*; 1-102, *S. proteamaculans*; 1-104, *P. putida*.
Means with the same column were analyzed by an ANOVA protected LSD test. LSD$_{0.05a}$ is for comparisons of means within the same main-plot and LSD$_{0.05b}$ is for comparisons of means across levels of the main-plot factor.

Table 10.3 Rhizosphere PGPR colony-forming units (log cfu g dry rooting media^{-1}) for different PGPR strains at different temperatures

PGPR strains	Log cfu g dry rooting media^{-1}		
	15°C	17.5°C	25°C
31-34	5.08	4.48	5.45
G11-32	5.30	5.40	5.56
36-43	4.99	4.99	5.35
63-49	4.92	4.86	4.67
2-68	5.10	5.29	5.38
1-102	4.82	3.96	4.58
1-104	5.05	5.10	5.21
$LSD_{0.05a}$	0.17		
$LSD_{0.05b}$	0.16		

PGPR strains used were: 31-34, *P. putida*; G11-32, *P. putida*; 36-43, *P. fluorescens*; 63-49, *P. fluorescens*; 2-68, *S. liquefaciens*; 1-102, *S. proteamaculans*; 1-104, *P. putida*.
Means with the same column were analyzed by an ANOVA protected LSD test. $LSD_{0.05a}$ is for comparisons of means within the same main-plot and $LSD_{0.05b}$ is for comparisons of means across levels of the main-plot factor.

Table 10.4 Root-associated and rhizosphere PGPR colony-forming units (log cfu g dry root^{-1} and rooting media^{-1}) for different PGPR strains at different temperatures

PGPR strains	Log cfu g dry root^{-1} and rooting media^{-1}		
	15°C	17.5°C	25°C
31-34	8.7	8.8	8.5
G11-32	8.8	8.9	8.7
36-43	8.6	7.9	8.3
63-49	8.1	8.6	8.6
2-68	8.5	8.7	8.6
1-102	8.9	8.4	7.6
1-104	8.4	8.4	7.9
$LSD_{0.05a}$	0.4		
$LSD_{0.05b}$	0.5		

PGPR strains used were: 31-34, *P. putida*; G11-32, *P. putida*; 36-43, *P. fluorescens*; 63-49, *P. fluorescens*; 2-68, *S. liquefaciens*; 1-102, *S. proteamaculans*; 1-104, *P. putida*.
Means with the same column were analyzed by an ANOVA protected LSD test. $LSD_{0.05a}$ is for comparisons of means within the same main-plot and $LSD_{0.05b}$ is for comparisons of means across levels of the main-plot factor.

These observations suggest that a PGPR inoculated onto the soybean root at its optimum temperature will tend to colonize the roots extensively but will proliferate relatively less in the rooting medium, while outside the optimum temperature range the reverse may be true. Other PGPR strains were not able to effectively colonize the roots of the soybean plants and their population densities remained very high in the rooting medium.

The strains that were able to colonize the roots at particular RZTs were also able to promote soybean growth at these temperatures (Dashti, 1996). Several factors are believed to influence PGPR colonization. One is the ability of the microbes to attach to the root (Nelson, 2004). The presence of polysaccharides on the cell surface seems to play an integral role in plant–microbe associations such as the formation of crown–gall by *Agrobacterium tumefaciens* (Thomashow et al., 1987) and nodulation of the legumes by *Rhizobium* species (Smit et al., 1987).

The ability of the PGPR to survive and grow in the rhizosphere may be dependent on the plant species and even plant cultivar (Beauchamp et al., 1993; Weller, 1986). One method of increasing the rhizosphere colonization by certain PGPR strains may be through maximizing the bacterial inoculum load on the seed or the rhizosphere (Dashti et al., 2012). This, however, is not always true as some PGPR such as *Pseudomonas* on maize are independent of the initial inoculum level (Dashti, 1996).

Beneficial bacteria which are introduced into the rhizosphere undergo a myriad of interactions with the host plants. They obtain nourishment from the root exudate and as a result are dependent on the host for survival. In exchange, they alter host plant metabolism by inducing certain physical changes (Saharan and Nehra, 2011). It has been observed that not all PGPR that colonize the roots successfully are able to promote growth and nitrogen fixing in their hosts. Rather, it seems likely that this may be dependent on plant type and environmental conditions.

Dashti (1996) demonstrated the effects of PGPR application on early symbiosis at low RZTs. Two controlled experiments were conducted. In the first experiment, the effects of PGPR on the early stages of symbiosis establishment between soybean and *B. japonicum* at low RZT were tested. The test was conducted for three RZTs at: 25, 17.5, and 15°C. The inoculum comprised *B. japonicum* alone (control), and *B. japonicum* with either *S. liquefaciens* (2-68) or *Serratia proteamonas* (1-102). In the second experiment, PGPR cells were centrifuged from log phase cultures and the supernatant obtained was first filter sterilized and tested for PGPR associated stimulation of the early stages of symbiosis development at 15 and 25°C.

At 15°C, soybean plants were inoculated with *B. japonicum* strain 532C alone or with *B. japonicum* added to the PGPR growth media with PGPR *S. proteamaculans* applied only once every day. At 25°C, all the conditions remained the same except for the PGPR strain. PGPR 1-102 was replaced by PGPR 2-68. The reason for this exchange is that 1-102 was previously found to be the most effective at 15°C while PGPR 2-68 was the most effective at 25°C (Dashti, 1996; Zhang et al., 1997). Early symbiotic establishment between soybean and *B. japonicum* was examined microscopically.

The results showed that (1) at 25°C, PGPR 2-68 reduced the time required for hair curling, infection thread initiation and spread of the infection thread to the base of the root hair; (2) at 15°C, PGPR 1-102 shortened the duration of all the measured steps of the infection process; (3) at both 15 and 25°C, daily watering of the soybean plants with PGPR supernatant decreased the time required for root hair curling, infection initiation, and spread of the infection threads to root base, much more effectively than direct inoculation with PGPR; and (4) the frequency of occurrence of every measured infection stage was the highest for those plants watered with the PGPR supernatants.

In brief, from the aforementioned results, it is possible to establish that PGPR accelerate the early stages of *B. japonicum*–soybean symbiosis and they do this by the means of exudates released into the growth medium. PGPR have been shown to produce many phytohormones and signal molecules, such as genistein. These molecules could be exuded from the bacterial cell and may have been present in large quantities in the supernatant. This might be the reason for improved morphological changes to the root hair compared with inoculating the PGPR directly into their site of action (Saharan and Nehra, 2011).

In addition, most of the PGPR inoculum prepared for direct application was obtained by spinning down the cells into a pellet and discarding the supernatant. It is possible that since most of the secondary metabolites produced by PGPR are extracellular, these might have been lost along with the supernatant. Following inoculation time will elapse before the PGPR can synthesize more of the plant stimulating molecules. On the other hand, when only the supernatant was added to the roots at the time of inoculation, the roots were exposed to the growth stimulating substances sooner than when they were inoculated directly with the PGPR cells.

However, this is a transient experience. Hence, the plants have to be frequently watered with the supernatant to ensure constant exposure to the growth promoting exudates. The observation that the supernatant of the PGPR cells not previously exposed to the plant tissues are capable of producing growth stimulating substances, is indicative of the fact that unlike the signal exchange process between the legumes and their nitrogen fixing symbionts, the production of bacteria to plant effector molecules in the PGPR system does not require a signal from the plant for initiation (Dashti, 1996).

Addition of genistein alleviates the effect of suboptimal root zone temperatures

Genistein, the most integral plant-to-bacteria signal in the soybean and *B. japonicum* symbiosis, is most active in the earliest phases of nodulation. Zhang et al. (1996) showed that roots of the plants germinated and grown at lower RZTs have lower genistein concentrations than do plants at higher RZT. The beneficial effects of genistein increased with decreasing RZT. At suboptimal RZTs (17.5 and 15°C), the most effective concentrations are in the 15–20 μM range, whereas at an optimal (25°C) RZT, 5 μM is the most effective concentration (Zhang and Smith, 1995). This is indicative of the fact that the effects of genistein increase with decreasing RZT (Zhang et al., 1996).

Studies conducted by Zhang and Smith (1995) showed that under controlled environmental and field conditions, preincubation of *B. japonicum* with genistein hastened the nitrogen fixation rate, increased nodulation, enlarged nodule size, and promoted plant growth. Zhang and Smith (1995) also showed that genistein stimulation of photosynthesis was only observed following the onset of nitrogen fixation. That is, the beneficial effect of genistein is most felt through promotion of nitrogen fixation.

Addition of the isoflavone genistein was also shown to partially overcome the effects of the suboptimal RZT on the growth and nodulation of soybean plants (Miransari and Smith, 2007, 2008). Addition of genistein significantly increased leaf area and shoot dry weight. The results obtained by Zhang and Smith (1995) were confirmed by Miransari and Smith (2008).

Suboptimal RZT was found to delay the stages involved in the formation of nodules on soybean roots. Inoculation of genistein (20 μM) was found to shorten this process. It can therefore be concluded that genistein compensated for the delay in the onset of nitrogen fixation in the nodules by shortening this stage. The increased availability of the fixed nitrogen due to this, promoted better soybean growth at suboptimal RZT (Miransari and Smith, 2008). Dashti et al. (2000), showed that at suboptimal RZT 17.5°C, a 15 μM concentration of genistein, preinoculated with *B. japonicum*, increased nodule number, nodule dry weight per plant, nodule size, and ratio of nodule weight to plant weight relative to the nongenistein inoculated controls. At 15°C RZT, *B. japonicum* coinoculated with 20 μM genistein showed increased nodule number, nodule dry weight, nodule size, and ratio of nodule weight to plant weight (Dashti et al., 2000).

Effects of coinoculation of *B. japonicum* with both genistein and PGPR

A controlled environmental experiment was conducted to investigate the combined effects of both PGPR and genistein to alleviate the stress of low RZT on soybean nodulation and nitrogen fixation (Dashti et al., 2000). Two PGPR strains were used, namely *S. liquefaciens* (2-68) and *S. proteamaculens* (1-102). Each of the PGPR strains was coinoculated with *B. japonicum* (either strain USDA110 or strain 532C), preincubated with different concentrations of genistein (0, 15, and 20 μM).

The resulting inocula were added to soybean rooting medium to test their ability to reduce the negative effects of suboptimal RZT on soybean growth and development by improving the physiological status of the plants. Three RZTs were tested: 25 (optimal), 17.5 (somewhat inhibitory), and 15°C (very inhibitory). At each temperature, PGPR strains and genistein increased the number of nodules formed and the amount of nitrogen fixed, but the most stimulatory combination of PGPR–genistein concentration and *B. japonicum* varied with temperature (Dashti et al., 2000).

The combinations that were most stimulatory were as follows: for 15°C RZT, the successful combination was *S. proteamaculans* (1-102), genistein concentration of 0 μM, and *B. japonicum* USDA110. For 17.5°C RZT, the best combination was again the *S. proteamaculans* strain with 15 μM genistein concentration and *B. japonicum* strain USDA110. At 25°C RZT, the best combination was the *S. proteamaculans* with a genistein concentration of 5 μM and *B. japonicum* strain USDA110. In at least some of the cases the stimulatory effects can be attributed to the cumulative effects of both PGPR and genistein in enhancing the number of nodules formed and the amount of nitrogen fixed by soybean plants.

The combinations of PGPR and genistein showed additive effects when compared with PGPR or genistein alone at higher RZT, while they show antagonistic effects at lower RZTs (Dashti et al., 2000). Preincubation of *B. japonicum* with either 5 μM of genistein or PGPR-268 alone, which was most effective at 25°C, showed antagonistic effects when both of them were added together. There was a marked decrease in the nodule dry weight per plant and nodule size. The same effect was observed at 15°C RZT.

In fact, the antagonistic effect between PGPR and genistein at this temperature was the highest compared with the other RZTs. On the other hand, some PGPR showed additive effects. The combination of PGPR 1-102, *B. japonicum* USDA110, and 5 μM genistein had an additive effect on the nodule dry weight per plant at 25°C RZT. At 25°C RZT many of the additive effects were approximately complete with the increases due to genistein addition being nearly the same in the presence and absence of PGPR.

At 17.5°C, the combination of PGPR 1-102, *B. japonicum* USDA110 and 15 μM genistein had an additive effect on the nodule number and nodule size (Dashti et al., 2000). However, level of additivity was not complete, with the increases due to addition of genistein being smaller or nonexistent in the presence of genistein than in its absence (Dashti et al., 2000). The replacement of PGPR 1-102 with PGPR 2-68; however, resulted in an antagonistic effect between the genistein and the PGPR. At 15°C, the combination of PGPR 1-102 and genistein showed antagonistic effects.

The additive effects observed at 25 and 17.5°C RZT could be explained in that genistein and PGPR may work by different mechanisms, stimulating different aspects of soybean physiology at optimum RZTs. The cause of the antagonistic effect displayed at suboptimal temperatures is still obscure and must be examined further. The frequent increases in the ratio of nodule weight to plant dry weight demonstrate that plants treated with either genistein or PGPR or both require more nodule mass to achieve each unit of accumulated weight (Dashti et al., 2000).

The efficiency with which additional nodule mass is able to support plant growth is less than for the nodule mass formed without the addition of these materials. These decreases in efficiency could be due to decreased relative efficiency for nitrogenase or greater restrictions of O_2 entry into the nodule (Hunt and Layzell, 1993). Based on the nitrogen distribution data (data not shown), it has been observed that applied PGPR–genistein treatments affected the nitrogen translocation from root nodules to shoot tissues.

The nitrogen concentrations of the plant shoots at 15°C were lower than at 17.5 and 25°C RZT. To summarize, the results of the aforementioned study concluded the following: (1) some PGPR strains combined with genistein can enhance soybean nodulation and nitrogen fixation; (2) PGPR 1-102, which performed poorly by itself at 25°C, proved to be extremely effective with the addition of 5 μM genistein; (3) the stimulations due to the PGPR–genistein combinations were probably due to the PGPR stimulation prior to the onset of nitrogen fixation; and (4) the combination of PGPR and genistein had a cumulative effect compared with their individual effects at the optimal RZT, a partial additive effect at 17.5°C RZT, and an antagonistic effect at 15°C RZT.

Dashti (1996) also demonstrated the effects of PGPR–genistein combination on the physiology and the development of the soybean plants at the suboptimal RZT. The experimental setup was similar to that in the study conducted to test the effects of PGPR–genistein combinations on nitrogen fixation. The results of the study indicated the following: (1) some PGPR strains combined with genistein can stimulate aspects of soybean growth and physiological development. (2) The genistein effect on photosynthesis was observed after the onset of nitrogen fixation while the effects of PGPR are felt prior to the onset of nitrogen (Zhang et al., 1996). (3) The variation of photosynthetic rate over time showed that plant photosynthesis was increased by some of the PGPR strains in the presence of genistein and *B. japonicum*. (4) As photosynthesis was increased by stimulatory strain combinations before the onset of nitrogen fixation, it can be concluded that the improvements in plant growth, development, and physiological activities must be due to the effects of PGPR on the overall plant physiology followed by the effect of genistein, which results in an improved nitrogen fixation (Dashti, 1996). (5) The stomatal conductance showed a linear change with the photosynthetic rate. When PGPR–genistein combinations caused increases in photosynthesis, the stomatal conductance also increased, while the internal CO_2 concentrations decreased. This increased photosynthesis allowed for decreased resistance to CO_2 entry and allowed increased CO_2 uptake.

Conclusions

Many biotic and abiotic factors disrupt the optimal growth and development of soybean, the most important of which is the suboptimal RZT. All stages of symbiotic establishment investigated to date, such as root hair curling, infection thread formation and penetration, and nodule development and function, are inhibited by suboptimal RZTs. In addition to this soybean growth, development, and physiological activities are also compromised. Studies have shown that coinoculation of *B. japonicum* either with genistein or PGPR or both together is capable of improving symbiotic nitrogen fixation, yield, and physiological parameters of the soybean even when RZT drops to 15°C. PGPR promote soybean growth by increasing plant growth, photosynthesis, amount of fixed N, and numbers of nodules formed. However, root zone temperature exerts a clear effect on the ability of PGPR to colonize soybean roots and this probably explains at least part of the differences in the colonization and plant growth stimulating abilities of PGPR. The ability of PGPR to colonize the root effectively is a prerequisite to their stimulatory effects. Genistein is an important bacterial signaling molecule and is integral in the early stages of nodulation. Coinoculation of genistein with *B. japonicum* has also been found to minimize the effects of low RZT. The combined effect of genistein and PGPR along with *B. japonicum,* varied with concentration of genistein used, the PGPR strain, and RZT. Some PGPR showed a cumulative effect with genistein and promoted growth, nodulation, nitrogen fixation, and physiological characteristics at certain RZTs while others manifested antagonistic effects in the presence of genistein, especially at lower RZTs. Addition of the PGPR supernatant results in a better symbiotic stimulation. This is, however, strain-specific and temperature-dependent as each PGPR probably releases a different growth stimulating substance. Given that they have similar effects on plant growth, they may be similar molecules.

References

Adesemoye, A.O., Ugoji, E.O., 2009. Evaluating *Pseudomonas aeruginosa* as plant growth promoting rhizobacteria in west Africa. Arch. Phytopathol. Plant Protect. 42, 188–200.

Ahemad, M., Kibret, M., 2014. Mechanisms and applications of plant growth promoting rhizobacteria: current perspective. J. King Saud Uni. Sci. 26, 1–20.

Anderson, J.J.B., Thomsen, K., Christiansen, C., 1987. High protein meals, insular hormones and urinary calcium excretion in human subjects. In: Christiansen, C., Johansen, J.S., Riis, B.J. (Eds.), Osteoporosis. Norhaven A/S, Viborg, Denmark, pp. 240–245.

Bakker, P.A.H.M., Pieterse, C.M.J., Van Loon, L.C., 2007. Induced systemic resistance by fluorescent *Pseudomonas* spp. Phytopathology 97 (Suppl. 2), 239–243.

Baon, J.B., Smith, S.E., Alston, A.M., 1994. Phosphorus uptake and growth of barley as affected by soil temperature and micorrhizal infection. J. Plant Nutri. 17, 479–491.

Beauchamp, C.J., Kloepper, J.W., Lemke, P.A., 1993. Luminometric analyses of plant root colonization by bioluminescent pseudomonads. Can. J. Microbiol. 39, 434–441.

Bellaloui, N., Reddy, K.N., Gillen, A.M., Fisher, D.K., Mengistu, A., 2011. Influence of planting date on seed protein, oil, sugars, minerals and nitrogen metabolism in soybean under irrigated and non-irrigated environments. Am. J. Plant Sci. 2, 702–715.

Dashti, N., 1996. Plant growth promoting rhizobacteria and soybean nodulation and nitrogen fixation under suboptimal root zone temperatures. Thesis PhD, McGill University, Montreal, Canada, 223 pp.

Dashti, N., Zhang, F., Hynes, R., Smith, D.L., 1997. Application of plant growth promoting rhizobacteria to soybean (*Glycine max* (L.) Merr.) increases protein and dry matter yield under short season conditions. Plant Soil 188, 33–41.

Dashti, N., Prithviraj, B., Hynes, R.K., Smith, D.L., 2000. Root and rhizosphere colonization of soybean [*Glycine max* (L.) Merr.] by plant growth promoting rhizobacteria at low root zone temperatures and under short-season conditions. J. Agron. Crop Sci. 185 (1), 15–22.

Dashti, N., Ali, N.Y., Cherian, V.M., Montasser, M.S., 2012. Application of plant growth-promoting rhizobacteria in combination with a mild strain of cucumber mosaic virus (CMV) associated with viral satellite RNAs to enhance growth and protection against a virulent strain of CMV in tomato. Can. J. Plant Pathol. 34 (2), 177–186.

Dashti, N., Cherian, V.M., Smith, D.L., 2013. PGPR to alleviate stress of suboptimal root zone temperature on leguminous plants. In: Miransari, M. (Ed.), Use of Microbes to Alleviate Soil Stresses, vol. 1. Springer Science and Business Media, New York, pp. 110–138.

Dashti, N., Montasser, M.S., Ali, N.Y.A., Cherian, V.M., 2014. Influence of plant-growth promoting rhizobacteria on fruit yield, pomological characteristics and chemical contents in cucumber mosaic virus-infected tomato plants. Kuwait J. Sci. 41, 205–220.

Dugje, I.Y., Omoigui, L.O., Ekeleme, F., Bandyopadhyay, R., Lava Kumar, P., Kamara, A.Y., 2009. Farmer's Guide to Soybean Production in Northern Nigeria. International Institute of Tropical Agriculture, Ibadan, Nigeria, 21 pp.

Duke, S.H., Schrader, L.E., Henson, C.A., Servaites, J.C., Robert, D., 1979. Low root temperatures effects on soybean nitrogen metabolism and photosynthesis. Plant Physiol. 63, 956–962.

Duzan, H.M., Zhou, X., Souleimanov, A., Smith, D.L., 2004. Perception of the *Bradyrhizobium japonicum* Nod factor by soybean [*Glycine max* (L.) Merr.] root hairs under abiotic conditions. J. Exp. Bot. 55 (408), 2641–2648.

Egamberdieva, D., Berg, G., Lindstorm, K., Rasanen, L.A., 2010. Co-inoculation of *Pseudomonas* spp. with *Rhizobium* improves growth and symbiotic performance of fodder galega (*Galega orientalis* Lam). Eur. J. Soil Biol. 46, 269–272.

Engels, C., Marschner, H., 1996. Effects of sub-optimal root zone temperatures and shoot demand on net translocation of micronutrients from the roots to the shoots of maize. Plant Soil 186, 311–320.

Hunt, H., Layzell, D., 1993. Gas exchange of legume nodules and regulation of nitrogenase activity. Annu. Rev. Plant Physiol. Plant Mol. Biol. 44, 483–511.

Hymowitz, T., Newell, C.A., 1981. Taxonomy of the genus *Glycine* domestication and uses of soybean. Econ. Bot. 35, 272–288.

Joseph, B., Ranjan, P.R., Lawrence, R., 2007. Characterization of plant growth promoting rhizobacteria associated with chickpea. Int. J. Plant Prod. 2, 141–152.

Liu, A., Wang, B., Hamel, C., 2004. Arbuscular mycorrhiza colonization and development at sub-optimal root zone temperatures. Mycorrhiza 14, 93–101.

Messina, M.J., 1999. Legumes and soybeans: overview of their nutritional profile and health effects. Am. J. Clin. Nutr. 70, 439S–450S.

Miransari, M., 2012. Microbial products and soil stresses. In: Maheswari, D.K. (Ed.), Bacteria in Agrobiology: Stress Management. Elsevier Science, Amsterdam, pp. 65–76.

Miransari, M., Smith, D.L., 2007. Overcoming the stressful effects of salinity and acidity on soybean [*Glycine max* (L.) Merr.] nodulation and yields using signal molecule genistein under field conditions. J. Plant Nutr. 30, 1967–1992.

Miransari, M., Smith, D.L., 2008. Using signal molecule genistein to alleviate stress of suboptimal root zone temperature on soybean-*Bradyrhizobium* symbiosis under different soil textures. J. Plant Interact. 3 (4), 287–295.

Miransari, M., Abrishamchi, A., Khoshbhakt, K., Niknam, V., 2013. Plant hormones as signals in arbuscular micorrhizal symbiosis. Crit. Rev. Biotechnol. 34 (2), 123–133.

Nelson, L.M., 2004. Plant growth promoting rhizobacteria: prospects for new inoculants. Crop Manage. 10, 310–305. Online doi: 10.1094/CM-2004-0301-05 RV http://plantmanagementnetwork.org/pub/cm/review/2004/rhizobacteria.

O'Sullivan, D.J., O'Gara, F., 1992. Traits of florescent *Pseudomonas* spp. involved in suppression of plant root pathogens. Microbiol. Rev. 56, 662–676.

Panwar, J.D.S., Laxmi, Vijay, 2005. Biological nitrogen fixation in pulses and cereals. In: Bose, B., Hemantaranjan, A. (Eds.), Developments in Physiology, Biochemistry and Molecular Biology of Plants, vol.1. New India Publishing Agency, Delhi, pp. 125–158.

Saharan, B.S., Nehra, V., 2011. Plant growth promoting rhizobacteria: a critical review. Life Sci. Med. Res., 2011: LMSR-21, 1–30.

Sindhu, S.S., Gupta, S.K., Suneja, S., Dadarwal, K.R., 2002. Enhancement of green gram nodulation and growth by *Bacillus* species. Biol. Plant. 45, 117–120.

Smit, G., Kijne, J.W., Lugtenberg, B.J.J., 1987. Involvement of both cellulose fibrils and a Ca^{2+} dependent adhesion in the attachment of *Rhizobium leguminosarum* to pea root hair tips. J. Bacteriol. 169, 4294–4301.

Smith, S.E., Bowen, J.D., 1979. Soil temperature, mycorrhizal infection and nodulation of *Medicago trunculata* and *Trifolium subterraneum*. Soil Biol. Biochem. 11, 469–473.

Smith, D.L., Hume, D.J., 1987. Comparison of assay methods of nitrogen fixation utilizing white bean and soybean. Can. J. Plant Sci. 67, 11–19.

Thomashow, M.F., Karlinsey, J.E., Marks, J.R., Hurlbert, R.E., 1987. Identification of a new virulence locus in *Agrobacterium tumefaciens* that affects polysaccharide composition and plant cell attachment. J. Bacteriol. 170, 3499–3508.

Vacheron, J., Desbrosses, G., Bouffaud, M.L., Touraine, B., Moenne-Loccoz, Y., Muller, D., Legendre, L., Wisniewski-Dye, F., Prigent-Combaret, C., 2013. Plant growth promoting rhizobacterium and root system functioning. Front. Plant Sci. 4, 1–19.

Van Der Ploeg, A., Heuvelink, E., 2005. Influence of suboptimal root zone temperature on tomato growth and yield: a review. J. Hort. Sci. Biotechnol. 80 (6), 652–659.

Volkmar, K.M., Woodbury, W., 1989. Effect of soil temperature and depth on colonization and root and shoot growth of barley inoculated with vesicular-arbuscular mycorrhizae indigenous to Canadian prairie soil. Can. J. Bot. 80, 571–576.

Weller, D.M., 1986. Effects of wheat genotype on root colonization by a take-all suppressive strain of *Pseudomonas fluorescens*. Phytopathology 76, 1059.

Yadegari, M., Rahmani, A., Noormohammadi, G., Ayneband, A., 2010. Plant growth promoting rhizobacteria increase growth, yield and nitrogen fixation in *Phaseolus vulgaris*. J. Plant Nutr. 33, 1733–1743.

Younesi, O., Moradi, A., Chaichi, M.R., 2013. Effects of different rhizobacteria on nodulation and nitrogen-fixation in alfalfa (*Medicago satvia*) at suboptimal root zone temperatures. Am. Eurasian J. Agric. Environ. Sci. 13 (10), 1370–1374.

Zhang, F., Smith, D.L., 1995. Pre-incubation of *Bradyrhizobium japonicum* with genistein accelerates nodule development of soybean [*Glycine max* (L.) Merr.] at suboptimal root zone temperatures. Plant Physiol. 108, 961–968.

Zhang, F., Dashti, N.H., Hynes, R.K., Smith, D., 1996. Plant growth promoting rhizobacteria and soyabean [*Glycine max* (L.) Merr.] nodulation and nitrogen fixation at suboptimal root zone temperatures. Ann. Bot. 77, 453–459.

Zhang, F., Dashti, N.H., Hynes, R.K., Smith, D.L., 1997. Plant growth promoting rhizobacteria and soybean [*Glycine max* (L.) Merr.] growth and physiology at suboptimal root zone temperatures. Ann. Bot. 79, 243–249.

Zhang, F., Smith, D.L., 2002. Interoganismal signaling in suboptimum environments: the legume-rhizobia symbiosis. Adv. Agron. 76, 125–161.

Soybean production and N fertilization

Mohammad Miransari
Department of Book & Article, AbtinBerkeh Ltd. Company, Isfahan, Iran

Introduction

Nitrogen (N) is the most important nutrient for plant growth and crop production. It is the most abundant element in the atmosphere at 79%, and is absorbable by plants as ammonia and nitrate (Broadbent, 1984; Crawford, 1995). Hence, to make it available to the plant, N must be reduced by the following processes: (1) production of chemical N fertilization; and (2) biological N fixation. Production of N-chemical fertilizer is via the Haber–Bosch in which atmospheric N is combined by hydrogen under high pressure (Galloway and Cowling, 2002; Wagner, 2011).

Biological N fixation consists in symbiotic N fixation by legumes and their symbiotic N fixing bacteria, rhizobia, and it is highly recommendable economically and environmentally. The economical benefits of using soybean inocula have been great (Galloway and Cowling, 2002; Wagner, 2011). Use of N-chemical fertilization is the quickest method of providing plants with their essential N; however, due to the properties of N compounds, especially nitrate, leaching occurs and hence this is not advisable, especially in large amounts (Miransari and Mackenzie, 2011, 2014).

Soybean (*Glycine max* (L.) Merr.) is among the most important crops for feeding people; it has a high protein content of 40% (Ferguson et al., 2010; Jensen et al., 2012). However, very high levels of N are essential for soybean growth and production (Smil, 2002). For example, according to Brazilian researchers, the amount of N essential for the production of 2737 kg ha^{-1} soybean is equal to 220 kg ha^{-1}. Under such conditions, soybean fields, which are not high in nitrogen, with an average of 15–30 kg N ha^{-1}, must have high amounts of N fertilizer applied. The efficiency of N fertilizer is 60% at its highest, hence an amount of at least 330 kg N ha^{-1} is essential (Hungria et al., 2006).

The process of biological N fixation was first indicated by Hellriegel and Wilfarth in 1888 and *Bradyrhizobium japonicum* was isolated 7 years later. In 1895, Nobbe and Hiltner received their first patent on the use of pure cultures of *Rhizobium* for treating soybean seeds. Before this, it had been indicated that using small amounts of the soil from soybean fields could result in the nodulation of soybean plants in new fields. The beginning of inoculation practices in soybean fields occurred during the early 1900s (Giller, 2001).

Soybean is mainly produced by the USA, Brazil, Argentina, and China. The yield of vegetable oil from soybean is 29.7% of the total vegetable oil produced globally.

Abiotic and Biotic Stresses in Soybean Production. http://dx.doi.org/10.1016/B978-0-12-801536-0.00011-6

Another important property of soybean is its isoflavone compounds, which are of both economical and health significance. In locations of soybean production, bacterial strains for use in commercial inocula are selected according to a set of field experiments in the main areas of soybean production (Graham, 1991; Hungria et al., 2005; Sensoz and Kaynar, 2006).

In tropical areas, 80% of the protein requirements for people is supplied by plants, and this will not change in the future. In the past, only 10% of the carbon produced by plants was used by humans, while it is now 40%. Accordingly, suitable techniques and strategies must be used to provide people with their food in the future. About 80% of biological N fixation occurs via the symbiotic association between *Rhizobium*, *Bradyrhizobium*, *Sinorhizobium*, *Allorhizobium*, *Mesorhizobium*, and *Azorhizobium* and the host plant. The remaining 20% is by actinorhizal (such as *Frankia*) and *Anabaena–Azollae* symbiotic association (Long, 2001; Ohyama, 2014).

Since N fertilizer is expensive, biological fertilization, specifically rhizobial inocula, is a suitable method economically and environmentally, as they are used in different parts of the world including the USA and Brazil. It is therefore important to select the rhizobial strains and plant varieties which allow the most efficient process of N symbiosis. As a result soybean plants will be able to acquire the highest level of their N requirement via the process of N fixation (Hungria et al., 2006; Mpepereki et al., 2000). Whether soybean plants, especially the newly released varieties with the potential to produce high yield, are able to acquire their N by the process of biological N fixation alone is a question yet to be investigated (Hungria et al., 2006; Yang et al., 2010).

During the process of N fixation, symbiotic bacteria are able to fix atmospheric N and reduce it to ammonia for use by the host plant. Significant amounts of N are fixed by the process of N fixation between the legumes and their symbiotic N-fixing bacteria. Legumes including soybean are able to acquire most of their essential N by the process of N fixation and the remaining part will be available to the plant by chemical fertilization. Accordingly, the process of N fixation is of great environmental and economical significance (Sinclair et al., 2003).

The important point about using N-chemical fertilizer for the production of soybean is its adverse effect on the process of N fixation, especially in excessive amounts. Under such conditions, activity of the N-fixing bacteria and hence the amount of fixed N by the process of symbiosis between the bacteria and the host plant decrease. As a result, the rate of N fertilization for soybean production must be determined so that the plant can establish an efficient symbiotic association with rhizobia for N fixation (Cassman et al., 2002; Stewart et al., 2005).

Using N as a starter, when the plant and the symbiotic rhizobia have not yet established the symbiotic association, is essential for the suitable growth of soybean plants at the beginning of the season. As a result the plants will be able to grow and absorb nutrients essential for their growth. However, before the initiation of the symbiotic process the amounts of N-chemical fertilization must be at minimum so that the plant and the bacteria can establish the symbiotic process (Miransari, 2011a). In this chapter some of the most important points relating to the effects of N-chemical fertilizer on the growth and yield of soybean, as well as the related interactions of N-chemical fertilizer with the process of biological N fixation by soybean and rhizobia, are presented and analyzed.

Soybean and N fertilizer

Similarly to the other nutrients, N is also essential for soybean growth and yield production. The quickest and most convenient method of providing N to the plant is the use of N-chemical fertilizer, although this is not recommended economically and environmentally, if used in large amounts. Under such conditions less organic matter is used and hence the fertility of the soil as well as the population and activities of soil microbes decrease (Adesemoye et al., 2009; Fageria and Baligar, 2005; Tilman et al., 2002).

Due to the low price of N fertilizers, they are used in large quantities and most is subjected to leaching, resulting in their increased concentration in the water and hence the process of eutrophication. Accordingly, the quality of water for human use is affected by N fertilizer. The activities of soil bacteria on chemical N fertilizer also result in the production of greenhouse gases such as N_2O and if the fertilizer is spread on the soil, a significant amount is volatilized (Conley et al., 2009; Fageria and Baligar, 2005).

However, if N fertilizer is used efficiently, most of the previously mentioned problems can be controlled. The essential nutrients for plant growth and yield production must be used at the right time, and in the right amount and place (not spread). The use of chemicals, which decrease the rate of nitrification, can also be an effective method for decreasing the rate of N leaching. Hence, the use of biological N fixation can be an effective tool to adjust the rate of N-chemical fertilizer. It is, however, important to indicate the right amount of N-chemical fertilizer when used with rhizobial inocula (Davidson, 2009; Dobbie and Smith, 2003).

Although a large set of data is available on the parameters affecting the yield of soybean with respect to its N requirements, including how soybean yield is affected by the processes of N-biological fixation and N fertilization, there have been no precise analyses done related to such processes and responses. Accordingly, using a dataset of 637, Salvagiotti et al. (2008) interpreted and analyzed such parameters. Such analysis can be used for the development of techniques and strategies which result in a higher efficiency of using biological N fixation and N fertilization.

Yield production of soybean has steadily increased since planting soybean in different parts of the world, which has been due to the use of new techniques and strategies, including the use of more efficient inocula and soybean varieties. The increase has been equal to 31 kg ha^{-1} in the United States (Specht et al., 1999) and 28 kg ha^{-1} worldwide (Wilcox, 2004). The new techniques and strategies have effectively increased the yield of soybean as, for example, in the corn belt of the USA, the average rate of soybean yield is in the range of 6–8 T ha^{-1} (Specht et al., 1999). The varieties of soybean with the ability of producing high yield must be able to photosynthesize at a high rate and have a high level of N in the grains (Lupwayi and Kennedy, 2007).

The nitrogen compounds in soybean leaf are present as certain sugar biophosphate products and the content of N in soybean leaf is strongly correlated with the rate of photosynthesis. Accordingly, the canopy of soybean plants must be able to absorb a high level of light so that a high rate of photosynthesis is achieved. The N source of the plant is by the process of biological N fixation and the N from the soil. However, it must be mentioned that the adverse interactions between the N from biological N fixation and the N from the soil may decrease the rate of N uptake by soybean plants (Marino et al., 2007; Masclaux-Daubresse et al., 2008; Vitousek et al., 2002; Witte, 2011).

The maximum rate of N fixation is between the R3 and R5 growth stage of soybean and the remaining part of N demand by soybean must be met from the soil or from N-chemical fertilization (Zapata et al., 1987). If the amounts of N are less than the requirement of plants, plant N will be directed to the grains and as a result of N deficiency, plant growth and yield decreases. Under conditions of high soybean yield the N-fixing potential of plant increases, which is due to the higher activation of nodule nitrogenase (Mengel, 1994).

Although soybean N demand which has not been met by biological N fertilization may be met by N-chemical fertilization, it has yet to be investigated how N-chemical fertilization may contribute to the N demand of soybean while not compromising the potential of N fixation in soybean plants. In the case of N-chemical fertilization, the plant is more apt to acquire its N demand via the fertilizer rather than from the process of biological N fixation (Herridge et al., 1984).

According to the analyses of Salvagiotti et al. (2008), who analyzed a dataset of 637 soybean fields with an average yield of higher than 4.5 ton ha^{-1}, the average total uptake of N by soybean plants was equal to 219 kg ha^{-1}. Accordingly, for the production of 1 kg of grain, a total of 12.7 kg (with a minimum and maximum of 6.4 and 18.8, respectively) of N was absorbed. Analyses of data sets indicated that 52% of grain N uptake resulted from the process of N fixation; however, for the fields, which received less than 10 kg ha^{-1} N fertilizer, 58% of the grain uptake was related to the process of N fixation.

The average rate of N fixation was in the range of 0–337 kg ha^{-1} and the minimum amount was related to those fields which had a high rate of nitrate in their root zone or where some kind of stress existed. Several authors have indicated higher rates of N fixation ranging from 360 to 450 kg ha^{-1} (Fageria, 2009; Giller, 2001). The analyses of the data sets indicated that under conditions of using less than 85 kg N ha^{-1} N fertilizer, and N uptake of more than 350 kg ha^{-1}, the amount of N provided via the process of biological N fixation was in the range of 35–86% as affected by soil N. However, the higher rates were related to the most suitable inocula being used under conditions of soil with lower amounts of N (Howarth et al., 2002; Hungria et al., 2006).

In the study by Salvagiotti et al. (2008), the amounts of N-fixed biologically and the rates of N fertilization (surface spread to a depth of 20 cm) were negatively and exponentially related. The important point about N fertilization, especially for high-producing soybean, is the level of yield at which the rate of biological N fixation may not be sufficient and hence the use of N-chemical fertilization is essential. By clarification of this aspect it is possible to ascertain the rate of optimum N fertilization, which is of environmental and economical significance.

However, future research must indicate the contribution of each of the N components for plant growth and yield production, including N fertilizer, biological N fixation, and N from the soil. Accordingly, the rate of N uptake by plant must be determined at each growth stage to determine the efficiency of the techniques and strategies which have been used and which may not decrease the rate of N fixation. The ability of rhizobia to fix N is determined by the properties of plants and bacterial strains as well as the conditions of environment such as the presence of stress (Howarth et al., 2002; Salvagiotti et al., 2008).

Importance of N fixation for soybean production

The process of biological N fixation includes some intimate interactions and molecular dialog between the host plant and rhizobia. Two types of nodules are produced on the host plant roots including the determinate (alfalfa, pea, clover) and indeterminate (soybean and bean). The bacteria are attracted to the root hairs as a result of signal communication between the two symbionts. Within minutes of attraction the bacteria are perpendicularly attached to the root hairs of the host plant. The penetration of rhizobia into the root hair cells results in the curling of the root hairs, within about 6–18 h following the inoculation. The rhizobial shifting in the root hair cells toward the root cortex takes place via the production of infection thread. The division of cells in the root cortex, before approaching the infection thread, results in the production of nodule cells (Garg, 2007; Liu et al., 2010).

The growing and shifting of infection thread into the nodule cells result in the release of rhizobia into the host cortex by the process of endocytosis. In the host cytoplasm, a prebacteroid membrane is produced around the rhizobia and hence the bacteria will be able to act as a symbiosome. With time the nodules grow and they can be seen 6–18 days after inoculation. Although the highest number of nodules are found on the top of the roots, some nodules are also produced on the lateral roots while the early top nodules begin senescence (Long, 2001; Wagner, 2011).

Nodule number and size are controlled by: (1) the genotype of the host plant and rhizobia; (2) the existing nodules; (3) the efficiency of the symbiotic process; and (4) environmental factors such as soil N and water (Giller, 2001; Long, 2001). Molecular research has indicated the genetic combination of *Rhizobium* and *Bradyrhizobium*; in rhizobia the related gene is found on the large symbiotic plasmids and in bradyrhizobia the gene is located on the chromosome (Lindstrom et al., 2010).

Two important proteins in rhizobia are essential for the process of N fixation including *nif* and *fix* gene. The *nif* gene regulates the activity of the nitrogenase enzyme of the N-fixation process. Although most nif proteins have a plasmid origin, they are placed on the chromosome in the bradyrhizobia. The N-fixation enzyme, which is activated by the *nifDK* and *nifH* genes has two components: (1) the molybdenum-iron protein (MoFe); and (2) iron, which contains protein. The subunits of MoFe are activated by *nifK* and *nifD* and the FeMo-cofactor regulates the activity of MoFe protein. The assemblage of this cofactor is due to the activity of *nifB, nifV, nifN, nifH,* and *nifE* genes. The *nifH* gene activates the Fe protein subunit (Ohyama, 2014). Although there is a wide range of *nif* genes in microbes, in most of them the regulation of all nif proteins is via NifA (a positive regulator of transcription) and NifL (a negative regulator of transcription).

Different parameters can regulate the activity of the *nif* gene including O_2 and the level of fixed N. For example, the elevated level of soil ammonia (NH_3 or NH_4^+) may cause the NifL to negatively regulate the process of gene expression by preventing the activity of NifA as an activator. The elevated level of O_2 can also prevent NifL from activating NifJ and hence the level of NifA remains constant. Since NifA regulates the activity of all nif proteins, elevated levels of O_2 decrease the production of the N-fixation enzyme and hence the process of N fixation (Fischer, 1994; Herridge and Rose, 2000; Monson et al., 1995).

The fix proteins are also important in regulating the process of biological N fixation. The other proteins in the microsymbiont regulate the activities of glutamine synthase (Carlson et al., 1987), decarboxylate (Jiang et al., 1989), hydrogen uptake (Baginsky et al., 2002), exopolysaccharide (Leigh and Walker, 1994), β-1,2-glucans (Miller et al., 1994), lipopolysaccharides (Carlson et al., 1987), and the efficiency of nodulation (Dowling and Broughton, 1986).

The contribution of N from the soil or from the process of N fixation is determined by the concentration of N in plants. Under high concentrations of N in the soil, the rate of symbiotic N fixation decreases and plants acquire most of their essential N from the soil. However, the lower the N content of the soil, the higher is the rate of N fixation. Establishing a suitable balance between such sources of N is important for efficiently providing N to the host plant. Although under suitable conditions, the process of N fixation is able to provide the host plant with most of its essential N, determining the conditions under which the rate of N fixation is at the highest is of great importance (Galloway and Cowling, 2002; Graham and Vance, 2003; Vitousek et al., 2002).

In brief, during the process of biological N fixation, the following steps result in the eventual production of root nodules, which are the location of rhizobial fixation of atmospheric N. (1) The legume host plants, including soybean, produce some biochemicals, such as genistein, which activate the bacterial *nod* genes in the rhizosphere and their subsequent activities. (2) In response, rhizobia produce biochemicals called lipochitooligosaccharides (*nod* factors) which are able to induce morphological and physiological alteration of root hair cells including the curling and bulging of the root hairs. (3) The production of infection thread results in the entrance of rhizobia into the root cortical cells. (4) Certain morphological and physiological changes in the root cortical cells result in the production of root nodules. (5) At this stage the cellular membrane is formed around the bacterial cells in the nodules, which are now called bacteroids and are able to fix N. (6) Rhizobia reduce atmospheric N to ammonium, which is available for the use of the legume host plant (Long, 2001; Miransari and Smith, 2007, 2008, 2009; Wagner, 2011).

As a result of the symbiosis between rhizobia and the host plant, significant amounts of N fixed by the bacteria are available for the use of the plant. The legume host plant provide the bacteria with essential carbon and the bacteria provide the host plant with most of its essential N. The symbiotic efficiency of biological N fixation is determined by the properties of the host plant and rhizobia, the properties of the environment, and the climate (Liu et al., 2010; Zhou et al., 2006).

Selecting the right plant variety and rhizobial strains may result in a higher efficiency of N fixation. For example, under stress conditions, if the rhizobial strains have been previously isolated from the stress conditions, they can more efficiently fix N for the use of their host plant, relative to the strains which have been isolated from nonstressed conditions. It is also recommendable to use plant varieties which have been produced for use under stress conditions. Some of the properties of the environment including pH and salinity may be adjusted using appropriate techniques and strategies. Accordingly, higher N symbiotic efficiency may be resulted (Alexandre and Oliveira, 2011; Miransari and Smith, 2007, 2008, 2009; Valentine et al., 2011).

In an interesting experiment, Chen et al. (1992) investigated the effects of soybean varieties, population densities, and chemical fertilization on the growth and yield production of soybean under the climatic conditions of Quebec, Canada. N-chemical fertilization decreased the process of biological N fixation by soybean; however, it enhanced soybean growth and yield under the conditions of low N in the soil. High plant population did not significantly affect the nodulation of soybean varieties, although the fresh weight of nodules increased per unit area. The higher densities of plant population resulted in a higher soybean grain yield. There were no significant differences between the grain yields of the two soybean varieties.

Brown and Bethlenfalvay (1987) conducted an interesting 56-day-long experiment under growth chamber conditions using soybean, *B. japonicum* (USDA 136) and arbuscular mycorrhiza (AM) fungi (*Glomus mosseae* (Nicol. & Gerd) Gerd. and Trappe). The experiment was done under deficient conditions of soil N or P and different combinations were tested including the use of both microbes, each of them, or none of them. Nonmycorrhizal and nonnodulated plants were supplemented with P and N, respectively, to create similar growth conditions. These researchers found the following interesting results related to the tripartite symbiosis.

The weight (dry) of plants related to all four treatments was not different at harvest; however, mycorrhizal plants resulted in a higher CO_2 exchange rate and lower P concentration in the leaf. The N content in plant leaf was lower in nodulated plants as compared to plants fertilized with N, although the starch concentration was higher. Plant N and starch content was significantly and negatively correlated. According to statistical analysis, some plant properties (CO_2 exchange, P content, and leaf area) were correlated with P nutrition (or AM fungal presence) and some (fresh leaf weight, fresh root weight) were correlated with N (or rhizobial presence) nutrition. Related to the single symbiosis between soybean and AM fungi or rhizobia, the tripartite symbiosis resulted in lower activity and development of symbiotic organelles and nodules (Brown and Bethlenfalvay, 1987).

Seneviratne et al. (2000) tested the effects of rhizobial inoculation, including a control treatment (without biological and N-chemical fertilization), on the growth and yield of soybean. At different growth times, plant parameters were determined. At flowering, the weight (dry) of nodule and plant biomass as well as plant density were determined. At harvest, grain yield and seed N content were analyzed. The authors found that under tropical conditions, if a higher soybean population with inocula is used, a higher yield will be produced.

However, such results were not related to the use of N-chemical fertilizer. Under high soybean density, the rate of nodulation decreased, which was due to the lower numbers of rhizobia in the soil. Using a total of 46 kg ha^{-1} urea at seeding and flowering did not adversely affect the process of biological N fixation. The researchers accordingly considered that the proper rate of plant density with respect to the use of inocula must be determined to result in the highest soybean yield (Seneviratne et al., 2000).

In relation to the conditions at which the plant and the bacteria become active and grow, it is advantageous to determine the optimal properties for the highest rate of N fixation, because both the host plant and the rhizobia must be under optimal conditions to be able to act efficiently. However, even under stress conditions, it is expedient to alleviate any stress so that the host plant and the symbiotic rhizobia can activate more efficiently (Miransari et al., 2013; Miransari, 2014a).

Using N fertilization is a quick and almost convenient method of providing the plant with its essential N. However, some important details related to N-chemical fertilization render the use of such sources of N not economically and environmentally recommended as indicated in the following (Miransari and Mackenzie, 2011, 2014). Accordingly, the use of microbial inocula that can enable the host plant to establish a symbiotic association with rhizobia is highly beneficial (Botha et al., 2004; Miransari, 2011a). (1) The price of N fertilizer is high and is increasing year on year, related to the use of microbial inocula essential for the process of symbiotic N fixation; (2) the efficiency of N fixation is much higher than using N-chemical fertilization because most of the N-fertilizer compounds are subject to leaching as such products are highly soluble; (3) as a result, the sources of N-chemical fertilization contaminate the environment, which is of both environmental and economical significance (Fageria and Baligar, 2005; Salvagiotti et al., 2008).

According to the previously mentioned details, the optimum rate of fertilization including N-chemical fertilization and biological N fixation must be determined and used. Such a determination is dependent on the properties of the plant, soil, climate, and rhizobial strains as the presence of stress can also affect the efficiency of fertilization. The optimum combinations of such factors make the optimum fertilization of the host plant likely (Miransari, 2011a).

Plant properties including the variety and the related morphological and physiological properties can affect the use of N-chemical fertilization by the host plant as well as the rate of biological N fixation. Some varieties are able to establish a more efficient symbiotic association with rhizobia, which is, in particular, determined by plant morphological and physiological properties. A more extensive root network and production of higher rate of biochemicals can result in a higher attraction of rhizobia to the host plant roots and hence a higher rate of N-symbiotic association is possible. The morphological and physiological properties of the roots are the most important ones determining the rate of N-symbiotic association with rhizobia (Downie, 2010; Gage, 2004; Subramanian et al., 2007).

Plant physiological properties can affect: (1) the production of biochemicals from the plant roots; (2) the actions and interactions of the host plant with the soil microbes; (3) the attraction of soil microbes toward the plant roots; (4) the activity of soil microbes including rhizobia; (5) the interactions between the soil microbes; and (6) the properties of the rhizosphere including pH, the presence of organic products, etc. (Manavalan et al., 2009; Miransari, 2011b; Zhu et al., 2005).

The abundance of root hairs in the rhizosphere and their production of biochemicals which attract the rhizobia toward the plant roots, the presence of other microbes in the rhizosphere, the environmental properties such as pH, heat, water, organic matter, etc. are among the most important factors indicating the efficiency of symbiotic establishment and development. Accordingly, some plant varieties are able to establish a more efficient symbiosis with rhizobia. The actions and interactions of the host plant and rhizobia can, importantly, affect the process of biological N fixation (Hirel et al., 2007; Marino et al., 2007; Manavalan et al., 2009).

If the host plant is able to produce higher amounts of biochemicals, the rate of rhizobial attraction toward the plant roots increases and hence the rate of N-symbiotic fixation can increase. However, under situations such as the presence of stress, as the morphology and physiology of the host plant is adversely affected, plant activity including the production of such biochemicals decreases. Interestingly, research has indicated that it is possible to alleviate such stresses so that the host plant and the rhizobia are able to act and interact more efficiently (Dimkpa et al., 2009; Miransari and Smith, 2007, 2008, 2009; Yoon et al., 2009).

The interactions between the soil microbes in the rhizosphere can also affect the efficiency of the symbiotic rhizobia. For example, symbiotic AM fungi are able to favorably affect the activities of rhizobia in the rhizosphere and inside plant cells and hence increases the rate of symbiotic association between the bacteria and the host plant. The presence of AM fungi and rhizobia in the rhizosphere can result in the establishment of a tripartite symbiosis between the host plant, rhizobia, and AM fungi. The soil microbes may interact by: (1) producing biochemicals; (2) residing on each other; (3) affecting the activities of plant roots; (4) affecting the properties of the rhizosphere, etc. (Kuiper et al., 2004; Miransari, 2011b; Pozo et al., 2005).

The properties of the soil also affect is the efficiency of the symbiotic N fixation. The more suitable the conditions of soil, the higher is the efficiency of biological symbiotic association. Under stress conditions such as unfavorable pH, salinity, drought, heavy metals, etc., the activities of both the host plant and rhizobia decrease (Caliskan et al., 2008; Lindstrom et al., 2010). Providing more suitable conditions for the host plant and the bacteria may result in a higher symbiotic efficiency. It is accordingly possible to alleviate the soil stresses by using appropriate techniques and strategies. For example, the adverse effects of unfavorable soil pH, salinity, drought, and heavy metals on the process of symbiotic N fixation can be controlled resulting in a higher efficiency of symbiotic N fixation (Filoso et al., 2006; Marino et al., 2007; Miransari and Smith, 2007, 2008, 2009).

Plant genotype, N fertilizer, and N fixation

Proper use of N fertilizer for its efficient use by plant and the sustainability of agriculture is of great significance. With respect to the properties of N products, the loss of such products by volatilization, leaching, and denitrification decreases their efficiency. Accordingly, such N products must be managed efficiently so that the highest quantity can be absorbed and used by plant. The loss of N fertilizer products increases the expense of crop production and results in the pollution of the environment. As a result, the efficiency of using N fertilizer must be improved so that the highest amount of N fertilizer can be absorbed by the plant (Fageria and Baligar, 2005).

Plant properties are greatly affected by plant genotype or the genetic combination of plant. Under different conditions, including stress, plants behave according to the sequence of their proteins. Each plant genotype may behave more efficiently under specific conditions. For example, under stress, the genotypes which have the appropriate related genes and proteins can resist the stress and hence grow and produce yield. The response of plants to the environment is determined by such proteins. If the related proteins can enable the plant to tolerate the stress then the plant can complete its life cycle, instead the plant not being able to survive the stress (Desclaux et al., 2000; Schauer and Fernie, 2006; Tollenaar and Lee, 2002).

Fabre and Planchon (2000) analyzed the properties of two varieties of soybean resulting from the cross of two soybean genotypes with different origins for their N-fixing ability, assimilation of nitrate, and the traits of seed. They evaluated the effects of both sources of N, including biological N fixation and N fertilizer on the growth and yield of soybean. They indicated the importance of biological N fixation during the stage of reproduction until the late stage of growth to the protein content of seed. The production of yield was more affected by N assimilation during the early stage of reproduction and efficiency of N fixation during the late stage of reproduction.

The response of plants to the nutrients is also a function of its genotype or genetic combination. Some genotypes may be more efficient in absorbing higher amounts of nutrients due to the following: (1) a more extensive root network; (2) the production of different biochemicals by plant roots, which enhances the availability and hence the uptake of nutrients by plant roots; (3) the symbiotic and nonsymbiotic association of the host plant by the soil microbes in the rhizosphere; and (4) the interactions between the soil microbes affecting the growth of the host plant and hence the uptake of nutrients (McCulley et al., 2004; Miransari, 2013, 2014 a, b, c; Wardle et al., 2004).

Plant roots may be among the most important plant properties affecting plant survival under stress. The extensive root network is able to grow more deeply into the soil and absorb higher amounts of water and nutrients under different conditions including stress. Plants can also tolerate stress by altering the morphology and physiology of roots. For example, under water stress, the plant roots may be able to grow deeper in the soil and absorb water for the growth of the plant. In response to the availability of soil nutrients such as phosphorus, root morphology may be altered so that the plant can absorb higher rates of phosphorus (Jobbagy and Jackson, 2001; Lopez-Bucio et al., 2003; Steudle, 2000).

With respect to the properties of plants, the production of different biochemicals by plant roots can also affect the availability and hence uptake of nutrients by the host plant. For example, production of higher amounts of H^+ and organic products by plant roots may enhance the solubility and hence the availability of nutrients to the host plant. Such abilities are affected by the properties of soil and under suitable conditions the plant roots are able to absorb higher rate of nutrients (Downie, 2010; Rodriguez-Navarro et al., 2011).

The symbiotic association with soil microbes can also increase the uptake of nutrients by the host plant. As previously mentioned, N-fixing bacteria can provide the host plant with most of its essential N. Mycorrhizal fungi can enhance the uptake of different nutrients by the host plant, including phosphorus, under different conditions including stress. The nonsymbiotic bacteria including plant growth promoting rhizobacteria (PGPR) can also increase the uptake of different nutrients, including N, by the host plant (Miransari, 2014a).

The presence and actions and interactions between the soil microbes, can also determine the availability and hence uptake of nutrients by the host plant. In the case of mycorrhizal fungi and rhizobia, the tripartite association between the host plant and the microbes can increase the uptake of N and P by the host plant. Such associations may also make the host plant more tolerant to stress. However, there may also be some negative interactions between the soil microbes, adversely affecting their activities (Powell et al., 2007; Wang et al., 2011). For example, at the time of using commercial inocula there may be some competition between the indigenous and the commercial strains of rhizobia affecting their inoculating and N-fixing abilities. The stronger the commercial strains the more efficient they can act and fix atmospheric N (Miransari, 2011a, b).

Soybean is able to absorb N from the soil and from the process of biological N fixation. N is a mobile nutrient and accordingly can quickly be absorbed by the host plant or be leached into the ground water. Plant genotype is among the most important factors affecting N uptake by the host plant. Some plant genotypes are more efficient and can absorb a higher rate of nutrients under different conditions including stress. This is especially of significance under stress and can enable the plant to survive under such conditions (Fageria, 2009; Liu et al., 2008).

Soil microbes including PGPR, rhizobia, and mycorrhizal fungi may also make the plant more tolerant to the stress by enhancing plant nutrient uptake. However, it must be mentioned that some soil microbes such as rhizobia may not be tolerant under stress, especially if they have not been isolated from stress conditions. Other strains and species of soil microbes such as mycorrhizal fungi can be more tolerant under stress and help the host plant to grow under such conditions. Such a capability is heavily determined by the properties of the host plant and microbes (Dimkpa et al., 2009; Egamberdieva et al., 2013).

If the host plant and the symbiotic and nonsymbiotic microbes are naturally tolerant (with respect to their genetic combination) they can survive the stress. However, using the new techniques it is possible to produce plant species and microbial strains which are more tolerant under stress conditions. The use of soybean genotypes which are resistant under stress in combination with the tolerant rhizobial strains, can enable the host plant to grow under stress and produce a reasonable yield (Atkinson and Urwin, 2012; Egamberdieva et al., 2013; Manavalan et al., 2009).

B. japonicum and N fertilization

The use of rhizobial inocula for the production of soybean is of economical and environmental significance. However, in almost all cases the use of N fertilizer for soybean growth and yield is unavoidable. N fertilizer can be used at different times of plant growth stage including seeding as the starter. Although N fertilizer can enhance seedling growth, especially at the beginning of the growing season, higher amounts of N fertilizer can adversely affect the activity of *B. japonicum* and hence the process of N fixation (Chen et al., 1992; Seneviratne et al., 2000).

The production of inocula is important for the use of rhizobia in the field. Using peat or liquid inocula as microbial carriers is among the most suitable methods of inoculation under field conditions. The most efficient strains of rhizobia are selected and grown using appropriate media. The growth of bacteria under septic conditions results in the production of a high bacterial population, which can be used for the inoculation of seeds in the field. Using efficient and populous inocula which have been produced under sterilized conditions are essential for the production of inocula with high quality (Lindstrom et al., 2010; Mpepereki et al., 2000; Seneviratne et al., 2000).

The most suitable method of using inocula is inoculating the seeds. The seeds can be inoculated at seeding or before seeding under shading and 15 min later the seeds can be planted. For an efficient inoculation the seeds must have a high population of bacteria, so that during the growth of the seedling the bacteria can inoculate the seedling roots and develop a symbiotic association with the host plant (Temprano et al., 2002; Wan et al., 2005).

If the contribution of N components including rhizobial inocula, chemical N fertilizer, and N from the soil is determined, it is likely to enhance the efficiency of such components and hence the uptake of N by the host plant. At high rates of N-chemical fertilizer and N from the soil the activity of *B. japonicum* for the biological fixation of N decreases, which is mostly due to decreased activity of the related N-fixing enzyme. The N-fixing bacterium *B. japonicum* must have been properly produced and stored under suitable conditions to be able to fix high amounts of N in symbiosis with the host plant (Stewart et al., 2005).

Since the process of biological N fixation is expensive to the host plant, the plant will tend to use the other sources of N including chemical N fertilizer and N from the soil unless such sources of N are available at too low a rate. If the host plant properties are improved in such a way that it is more prone to develop a symbiotic association with the N-fixing bacteria, this will be more likely to alleviate the adverse effects of N sources such as chemical N fertilizer and N from the soil on the process of N fixation, especially if such sources are available at high rates.

B. japonicum, or N fertilization, or both?

Soybean inocula including *B. japonicum* can significantly contribute to the production of soybean yield under field conditions by the fixation of atmospheric N (Miller et al., 1994; Wan et al., 2005). However, much research has been done to determine whether the use of inocula alone can fulfill the N requirements of soybean plants under field conditions. The research so far has indicated that rhizobial inocula can provide the soybean host plant with most of its N requirements and that the remaining requirement must be supplied using N fertilizer (Caliskan et al., 2008; Lindstrom et al., 2010). Hence, an important point that must be addressed in research work is how the rate of N fixation by rhizobial inocula can be increased so that there will be no need for the use of N fertilizer.

A high rate of N fertilization significantly decreases the activity of rhizobia and hence the rate of biological N fixation. However, at appropriate amounts, N fertilization can enable the soybean plants to become established at the beginning of the season until the initiation of biological N fixation by rhizobia. During the season, sources of N-chemical fertilizer provide the remaining part of soybean N requirements that have not been supplied by biological N fixation (Chen et al., 1992; Ferguson et al., 2010; Seneviratne et al., 2000).

Different factors determine the appropriate combination of rhizobial inocula and N fertilizer including: (1) bacterial strains; (2) plant species; and (3) environment, including climate. The higher the efficiency of rhizobial strains in the establishment of symbiotic association with the host plant, the higher is the rate of N fixation. For a higher rate of N fixation, the rhizobial inocula must be competitive enough with the native bacteria to establish their symbiosis with their host plant and fix N. Soybean plant species must be able to develop an efficient symbiosis with rhizobia. Different soybean varieties and rhizobial strains may differ in their potential for the establishment of a symbiotic association. The more efficient the combination of the two symbionts the higher is the rate of N fixation (Ferguson and Mathesius, 2003; Martínez-Romero, 2009; Oldroyd and Downie, 2008; Souleimanov et al., 2002; Yang et al., 2010).

The effects of the environment on the process of biological N fixation are significant. Under optimum conditions the rate of N fixation is at its highest. However, under stress both the growth and the activity of the host plant and the rhizobia are adversely affected. Depending on the stress tolerance of the two symbionts the level of N-biological fixation may differ. The more tolerant the strains of rhizobia and the species of soybean, the higher is the rate of N fixation under stress. It is also possible to enhance the stress tolerance of rhizobia and soybean by using different techniques and strategies (Gresshoff, 2003; Miransari, 2014a, b, c; Zhang and Smith, 2002).

Because of the previously mentioned details and the following, the use of both rhizobial inocula and chemical N fertilizer is essential for the production of soybean yield under different conditions including stress. Using rhizobial inocula is a healthy method of providing the plant with its essential N. (1) So far, the sole use of rhizobia has not been able to provide the host soybean with its complete essential N; (2) at the beginning of the growing season before the initiation of the symbiotic process between soybean and rhizobia, the use of N-chemical fertilizer is essential for the establishment and growth of soybean seedlings; (3) the appropriate rate of N-chemical fertilizer is not suppressive of the process of biological N fixation and can help the plant to acquire the optimum rate of N and produce high yield; (4) if the rhizobial inocula are not efficient enough, N-chemical fertilizer must be used to provide soybean with its essential N (Miransari, 2011a; Seneviratne et al., 2000).

N fixation and its environmental and economical significance

Biological N fixation is a very useful way of providing legumes with their essential N, environmentally and economically. Rhizobia are able to efficiently fix atmospheric N and reduce it so that the host plant can assimilate it and use it as a source of N for their growth and yield production. Biologically fixed N is immediately used by the host plant and is not subjected to environmental factors such as leaching. The price of commercial inocula is much less than that of N-chemical fertilizer and hence it is also of great economical significance.

Under optimum conditions, rhizobia are able to fix large amounts of N for use by their host plant. The use of legumes is also a suitable way of providing the essential N for the use of the next crop plant as legumes are high in N. Although it has been indicated that biological N fixation can only supply most of the N essential for the use of the host plant, such a contribution is also of great environmental and economical significance. N fertilizer can quickly provide the host plant with its essential N; however, related to the process of biological N fixation, it is not efficient enough to provide the host plant with its essential N and most of it may not be used by the host plant (Chen et al., 1992).

If the proper rate of biological N fixation and N-chemical fertilization are combined, they can provide the legume host plant with its essential N, and is environmentally and economically recommendable. Research must focus more on the techniques and strategies, which may result in a higher rate of biological N fixation under different conditions including stress. Use of rhizobial strains which can act well enough under the environmental and climatic conditions of the region, as well as the soybean varieties which have been specifically produced for such conditions, are among the most important factors for the efficient use of symbiotic N fixation (Egamberdieva et al., 2013; Wan et al., 2005).

Although microbial inocula are used in different parts of the world, they must be used more widely by farmers, so that they can contribute more to the health of the environment. With respect to the genetic combination of rhizobia and the soybean host plant, efficient commercial inocula and soybean species must be produced and used worldwide so that related climatic effects can also benefit humans healthwise.

Soybean N fixation and stress

Stress can have adverse effects on the growth and activities of both rhizobia and the host plant. Depending on the rate and time of stress, the unfavorable effects of stress on the process of N-symbiotic fixation differ. For example, drought and salinity are among the most widespread stresses worldwide and significantly decrease the rate of biological N fixation and hence the growth and the yield of soybean. Tolerant plant species and rhizobial strains can utilize mechanisms which may alleviate the negative effects of stress on the survival and activities of the two symbionts and hence the process of biological N fixation (Miransari, 2014a, b, c).

Different mechanisms may be used by rhizobia and the host plant to alleviate the adverse effects of stress on their growth and activities including the process of biological N fixation. Alteration of morphological and physiological properties can help the bacteria and the host plant to tolerate stress. Accordingly, the bacteria and the host plant may: (1) produce different organic products; (2) enhance their rate of signaling communication; (3) become inactive; (4) be more interactive with the other microbes; and (5) alter the properties of soil (Adesemoye et al., 2009; Egamberdieva et al., 2013).

Production of different products by bacteria and the host plant can result in: (1) the adjustment of the osmotic potential; (2) the uptake of nutrients by the plant; (3) the alteration of rhizosphere properties; and (4) the behavior of soil microbes in the rhizosphere. Higher production of signals under stress is an important factor affecting the establishment and progress of the symbiotic association. Some soil bacteria may become inactive under unfavorable conditions such as stress by producing spores. They can become reactivated when the conditions are favorable (Dimkpa et al., 2009; Zhang and Smith, 2002).

Many studies have indicated that it is possible to mitigate the unfavorable effects of stress on the growth and activities of rhizobia and soybean host plant. For the alleviation of stress, the most sensitive stages of the N-symbiotic association must be determined and, accordingly, the appropriate techniques used. For example, Miransari and Smith (2007, 2008, 2009) hypothesized that under salinity, unfavorable soil pH, and suboptimal root-zone degree, the initial stages of N-symbiotic fixation including the signaling communication between the symbionts is adversely affected and hence the bacteria and the host plant are not able to detect the presence of each other for the development of a symbiotic N fixation. They proved that the pretreatment of *B. japonicum* with the signal molecule genistein is able to activate the bacterial *nod* genes and hence make the bacteria proceed with the next stages of the N-symbiotic process. Interestingly, it was indicated that the adverse effects of such stresses on the growth and activates of bacteria and soybean host plant can be alleviated.

Conclusions and future perspectives

N is among the most important nutrients for the growth and yield of soybean. The plant can absorb its essential N by the process of biological N fixation, from N-chemical fertilizer and from the soil. Because at extra amounts of N-chemical fertilizer, the process of biological N fixation is adversely affected, the appropriate amount of fertilizer must be applied with respect to the contribution of each N component. Although *B. japonicum* is able to provide most of the host plant's essential N, the remaining requirement must be supplied using N fertilizer. Soybean N demand is affected by different factors such as bacterial strains, soybean varieties, and the environment, including climate. The proper combination of rhizobial strain and soybean variety may result in the most efficient N fixation by rhizobia and soybean. With respect to the economic and environmental benefits of biological N fixation, numerous studies have been conducted to increase the rate of N fixation under different conditions including stress. For example, under stress, if the bacterial strains are isolated from the stress environment and a tolerant soybean variety is used, the highest N fixation efficiency may result. Research has indicated that it is possible to alleviate the adverse effects of stress on the process of N fixation by rhizobia and soybean. Future research may focus on the production of highly efficient rhizobial strains and high-producing soybean varieties that are able to fix high amounts of biological N. As a result, the use of N-chemical fertilizer may be reduced to a minimum level. Under any condition, the appropriate contribution of each component must be determined and accordingly an appropriate amount of rhizobial inocula and N fertilizer should be used.

References

Adesemoye, O., Torbert, H., Kloepper, J., 2009. Plant growth-promoting rhizobacteria allow reduced application rates of chemical fertilizers. Microb. Ecol. 58, 921–929.

Alexandre, A., Oliveira, S., 2011. Most heat-tolerant rhizobia show high induction of major chaperone genes upon stress. FEMS Microbiol. Ecol. 75, 28–36.

Atkinson, N., Urwin, P., 2012. The interaction of plant biotic and abiotic stresses: from genes to the field. J. Exp. Bot. 63, 3523–3543.

Baginsky, C., Brito, B., Imperial, J., Palacios, J.-M., Ruiz-Argüeso, T., 2002. Diversity and evolution of hydrogenase systems in rhizobia. Appl. Environ. Microbiol. 68, 4915–4924.

Botha, W., Jaftha, J., Bloem, J., Habig, J., Law, I., 2004. Effect of soil bradyrhizobia on the success of soybean inoculant strain CB 1809. Microbiol. Res. 159, 219–231.

Broadbent, F.E., 1984. Plant use of soil nitrogen. In: Hauck, R. (Ed.), Nitrogen in Crop Production. ASA-CSSA-SSSA, Madison, WI. pp. 171–182.

Brown, M., Bethlenfalvay, G., 1987. Glycine-glomus-rhizobium symbiosis. VI. Photosynthesis in nodulated, mycorrhizal, or N- and P-fertilized soybean plants. Plant Physiol. 85, 120–123.

Caliskan, S., Ozkaya, I., Caliskan, M.E., Arslan, M., 2008. The effects of nitrogen and iron fertilization on growth, yield and fertilizer use efficiency of soybean in a Mediterranean-type soil. Field Crops Res. 108, 126–132.

Carlson, R.W., Kalembasa, S., Turowski, D., Pachori, P., Noel, K.D., 1987. Characterization of the lipopolysaccharide from a *Rhizobium phaseoli* mutant that is defective in infection thread development. J. Bacteriol. 169, 4923–4928.

Cassman, K., Dobermann, A., Walters, D., 2002. Agroecosystems, nitrogen-use efficiency, and nitrogen management. Ambio 31, 132–140.

Chen, Z., MacKenzie, A.F., Fanous, M.A., 1992. Soybean nodulation and grain yield as influenced by N-fertilizer rate, plant population density and cultivar in southern Quebec. Can. J. Plant Sci. 72, 1049–1056.

Conley, D., Paerl, H., Howarth, R., Boesch, D., Seitzinger, S., Havens, K., Lancelot, C., Likens, G., 2009. Controlling eutrophication: nitrogen and phosphorus. Science 323, 1014–1015.

Crawford, N.M., 1995. Nitrate: nutrient and signal for plant growth. Plant Cell 7, 859–868.

Davidson, E., 2009. The contribution of manure and fertilizer nitrogen to atmospheric nitrous oxide since 1860. Nat. Geosci. 2, 659–662.

Desclaux, D., Huynh, T., Roumet, P., 2000. Identification of soybean plant characteristics that indicate the timing of drought stress. Crop Sci. 40, 716–722.

Dimkpa, C., Weinand, T., Asch, F., 2009. Plant–rhizobacteria interactions alleviate abiotic stress conditions. Plant Cell Environ. 32, 1682–1694.

Dobbie, K., Smith, K., 2003. Impact of different forms of N fertilizer on N_2O emissions from intensive grassland. Nutr. Cycling Agroecosyst. 67, 37–46.

Dowling, D.N., Broughton, W.J., 1986. Competition for nodulation of legumes. Annu. Rev. Microbiol. 40, 131–157.

Downie, J., 2010. The roles of extracellular proteins, polysaccharides and signals in the interactions of rhizobia with legume roots. FEMS Microbiol. Rev. 34, 150–170.

Egamberdieva, D., Jabborova, D., Wirth, S., 2013. Alleviation of salt stress in legumes by co-inoculation with *Pseudomonas* and *Rhizobium*. In: Arora, N. (Ed.), Plant Microbe Symbiosis: Fundamentals and Advances. Springer, India, pp. 291–303.

Fabre, F., Planchon, C., 2000. Nitrogen nutrition, yield and protein content in soybean. Plant Sci. 152, 51–58.

Fageria, N., 2009. The Use of Nutrients in Crop Plants. CRC Press, Boca Raton, 448 pp.

Fageria, N., Baligar, V., 2005. Enhancing nitrogen use efficiency in crop plants. Adv. Agron. 88, 97–185.

Ferguson, B.J., Mathesius, U., 2003. Signaling interactions during nodule development. J. Plant Growth Regul. 22, 42–72.

Ferguson, B., Indrasumunar, A., Hayashi, S., Lin, M., Lin, Y., Reid, D., Gresshoff, P., 2010. Molecular analysis of legume nodule development and autoregulation. J. Integr. Plant Biol. 52, 61–76.

Filoso, S., Martinelli, L., Howarth, R., Boyer, E., Dentener, F., 2006. Human activities changing the nitrogen cycle in Brazil. Biogeochemistry 79, 61–89.

Fischer, H.M., 1994. Genetic regulation of nitrogen fixation in rhizobia. Microbiol. Rev. 58, 352–386.

Gage, D., 2004. Infection and invasion of roots by symbiotic, nitrogen-fixing rhizobia during nodulation of temperate legumes. Microbiol. Mol. Biol. Rev. 68, 280–300.

Galloway, J., Cowling, E.B., 2002. Reactive nitrogen and the world: 200 years of change. Ambio 31, 64–67.

Garg, N., 2007. Symbiotic nitrogen fixation in legume nodules: process and signaling. A review. Agron. Sustain. Dev. 27, 59–68.

Giller, K., 2001. Nitrogen Fixation in Tropical Cropping Systems. CABI, Wallingford, UK.

Graham, T., 1991. Flavonoid and isoflavonoid distribution in developing soybean seedling tissues and in seed and root exudates. Plant Physiol. 95, 594–603.

Graham, P., Vance, C., 2003. Legumes: importance and constraints to greater use. Plant Physiol. 131, 872–877.

Gresshoff, P.M., 2003. Post-genomic insights into plant nodulation symbioses. Genome Biol. 4, 201.

Hellriegel, H., Wilfarth, H., 1888. Untersuchungen Über die stickstoffnährung der Gramineen und Leguminosen. Beil. Ztscher. Ver. Dt. ZuckInd, 234 pp.

Herridge, D., Rose, I., 2000. Breeding for enhanced nitrogen fixation in crop legumes. Field Crops Res. 65, 229–248.

Herridge, D., Roughley, R., Brockwell, J., 1984. Effect of rhizobia and soil nitrate on the establishment and functioning of the soybean symbiosis in the field. Aust. J. Agric. Res. 35, 149–161.

Hirel, B., Le Gouis, J., Ney, B., Gallais, A., 2007. The challenge of improving nitrogen use efficiency in crop plants: towards a more central role for genetic variability and quantitative genetics within integrated approaches. J. Exp. Bot. 58, 2369–2387.

Howarth, R., Boyer, E., Pabich, W., Galloway, J., 2002. Nitrogen use in the United States from 1961–2000 and potential future trends. Ambio 31, 88–96.

Hungria, M., Franchini, J.C., Campo, R.J., Graham, P.H., 2005. The importance of nitrogen fixation to soybean cropping in South America. In: Werner, D., Newton, W.E. (Eds.), Nitrogen Fixation in Agriculture, Forestry, Ecology, and the Environment. Springer, Dordrecht.

Hungria, M., Campo, R.J., Mendes, I.C., Graham, P.H., 2006. Contribution of biological nitrogen fixation to the N nutrition of grain crops in the tropics: the success of soybean (*Glycine max* L. Merr.) in South America. In: Singh, R.P., Shankar, N., Jaiwal, P.K. (Eds.), Nitrogen Nutrition in Plant Productivity. Stadium Press/LLC, Houston, TX, pp. 43–93.

Jensen, E., Peoples, M., Boddey, R., Gresshoff, P., Hauggaard-Nielsen, H., Alves, B., Morrison, M., 2012. Legumes for mitigation of climate change and the provision of feedstock for biofuels and biorefineries. A review. Agron. Sustain. Dev. 32, 329–364.

Jiang, J., Gu, B.H., Albright, L.M., Nixon, B.T., 1989. Conservation between coding and regulatory elements of *Rhizobium meliloti* and *Rhizobium leguminosarum dct* genes. J. Bacteriol. 171, 5244–5253.

Jobbagy, E., Jackson, R., 2001. The distribution of soil nutrients with depth: global patterns and the imprint of plants. Biogeochemistry 53, 51–77.

Kuiper, I., Lagendijk, E., Bloemberg, G., Lugtenberg, B., 2004. Rhizoremediation: a beneficial plant–microbe interaction. Mol. Plant Microbe Interact. 17, 6–15.

Leigh, J.A., Walker, G.C., 1994. Exopolysaccharides of *Rhizobium*: synthesis, regulation and symbiotic function. Trends Genet. 10, 63–67.

Lindstrom, K., Murwira, M., Willems, A., Altier, N., 2010. The biodiversity of beneficial microbe-host mutualism: the case of rhizobia. Res. Microbiol. 161, 453–463.

Liu, X., Jin, J., Wang, G., Herbert, S., 2008. Soybean yield physiology and development of high-yielding practices in Northeast China. Field Crops Res. 105, 157–171.

Liu, Y., Wu, L., Baddeley, J., Watson, C., 2010. Models of biological nitrogen fixation of legumes. Agron. Sustain. Dev. 31, 155–172.

Long, S., 2001. Genes and signals in the *rhizobium*-legume symbiosis. Plant Physiol. 125, 69–72.

Lopez-Bucio, J., Cruz-Ramirez, A., Herrera-Estrella, L., 2003. The role of nutrient availability in regulating root architecture. Curr. Opin. Plant Biol. 6, 280–287.

Lupwayi, N., Kennedy, A., 2007. Grain legumes in Northern Great Plains. Agron. J. 99, 1700–1709.

Manavalan, L., Guttikonda, S., Tran, L., Nguyen, H., 2009. Physiological and molecular approaches to improve drought resistance in soybean. Plant Cell Physiol. 50, 1260–1276.

Marino, D., Frendo, P., Ladrera, R., Zabalza, A., Puppo, A., Arrese-Igor, C., Gonzalez, E., 2007. Nitrogen fixation control under drought stress. Localized or systemic? Plant Physiol. 143, 1968–1974.

Martínez-Romero, E., 2009. Coevolution in *Rhizobium*–legume symbiosis? DNA Cell Biol. 28, 361–370.

Masclaux-Daubresse, C., Reisdorf-Cren, M., Orsel, M., 2008. Leaf nitrogen remobilisation for plant development and grain filling. Plant Biol. 10, 23–36.

McCulley, R., Jobbágy, E., Pockman, W., Jackson, R., 2004. Nutrient uptake as a contributing explanation for deep rooting in arid and semi-arid ecosystems. Oecologia 141, 620–628.

Mengel, K., 1994. Symbiotic dinitrogen fixation – its dependence on plant nutrition and its ecophysiological impact. Zeitschrift Pflanzenernährung Bodenkunde 157, 233–241.

Miller, K.J., Hadley, J.A., Gustine, D.L., 1994. Cyclic [beta]-1,6-1,3-glucans of *Bradyrhizobium japonicum* USDA 110 elicit isoflavonoid production in the soybean (*Glycine max*) host. Plant Physiol. 104, 917–923.

Miransari, M., 2011a. Soil microbes and plant fertilization. Appl. Microbiol. Biotechnol. 92, 875–885.

Miransari, M., 2011b. Interactions between arbuscular mycorrhizal fungi and soil bacteria. Appl. Microbiol. Biotechnol. 89, 917–930.

Miransari, M., 2013. Soil microbes and the availability of soil nutrients. Acta Physiol. Plant. 35, 3075–3084.

Miransari, M., 2014a. Plant growth promoting rhizobacteria. J. Plant Nutr. 37, 2227–2235.

Miransari, M., 2014b. Use of Microbes for the Alleviation of Soil Stresses, vol. 1. Springer, New York.

Miransari, M., 2014c. Use of microbes for the alleviation of soil stresses. In: Alleviation of Soil Stress by PGPR and Mycorrhizal Fungi, vol. 2, Springer, New York.

Miransari, M., Mackenzie, A.F., 2011. Development of a soil N test for fertilizer requirements for wheat. J. Plant Nutr. 34, 762–777.

Miransari, M., Mackenzie, A.F., 2014. Optimal N fertilization, using total and mineral N, affecting corn (Zea mays L.) grain N uptake. J. Plant Nutr. 37, 232–243.

Miransari, M., Smith, D.L., 2007. Overcoming the stressful effects of salinity and acidity on soybean [Glycine max (L.) Merr.] nodulation and yields using signal molecule genistein under field conditions. J. Plant Nutr. 30, 1967–1992.

Miransari, M., Smith, D.L., 2008. Using signal molecule genistein to alleviate the stress of suboptimal root zone temperature on soybean-Bradyrhizobium symbiosis under different soil textures. J. Plant Interact. 3, 287–295.

Miransari, M., Smith, D., 2009. Alleviating salt stress on soybean (Glycine max (L.) Merr.) – Bradyrhizobium japonicum symbiosis, using signal molecule genistein. Eur. J. Soil Biol. 45, 146–152.

Miransari, M., Riahi, H., Eftekhar, F., Minaie, A., Smith, D.L., 2013. Improving soybean (Glycine max L.) N$_2$-fixation under stress. J. Plant Growth Regul. 32, 909–921.

Monson, E., Ditta, G., Helinski, D., 1995. The oxygen sensor protein, FixL, of Rhizobium meliloti. Role of histidine residues in heme binding, phosphorylation, and signal transduction. J. Biol. Chem. 270, 5243–5250.

Mpepereki, S., Javaherib, F., Davis, P., Giller, K., 2000. Soybeans and sustainable agriculture: promiscuous soybeans in southern Africa. Field Crops Res. 65, 137–149.

Ohyama, T., 2014. Advances in Biology and Ecology of Nitrogen Fixation. InTech, p. 282. DOI: 10.5772/56990.

Oldroyd, G., Downie, J., 2008. Coordinating nodule morphogenesis with rhizobial infection in legumes. Annu. Rev. Plant Biol. 59, 519–546.

Powell, J., Klironomos, J., Gulden, R., Hart, M., Campbell, R., Levy-Booth, D., Dunfield, K., Pauls, K., Swanton, C., Trevors, J., 2007. Mycorrhizal and rhizobial colonization of genetically-modified and conventional soybeans. Appl. Environ. Microbiol. 73, 4365–4367.

Pozo, M., Van Loon, L., Pieterse, C., 2005. Jasmonates – signals in plant-microbe interactions. J. Plant Growth Regul. 23, 211–222.

Rodriguez-Navarro, D., Oliver, M., Contreras, A., Ruiz-Sainz, J., 2011. Soybean interactions with soil microbes, agronomical and molecular aspects. Agron. Sustain. Dev. 31, 173–190.

Salvagiotti, F., Cassman, K., Specht, J., Walters, D., Weiss, A., Dobermann, A., 2008. Nitrogen uptake, fixation and response to fertilizer N in soybeans: a review. Field Crops Res. 108, 1–13.

Schauer, N., Fernie, A., 2006. Plant metabolomics: towards biological function and mechanism. Trends Plant Sci. 11, 508–516.

Seneviratne, G., Van Holm, L., Ekanayake, E., 2000. Agronomic benefits of rhizobial inoculant use over nitrogen fertilizer application in tropical soybean. Field Crops Res. 68, 199–203.

Sensoz, S., Kaynar, I., 2006. Bio-oil production from soybean (Glycine max L.); fuel properties of bio-oil. Ind. Crops Prod. 23, 99–105.

Sinclair, T.R., Farias, J.R., Neumaier, N., Nepomuceno, A.L., 2003. Modeling nitrogen accumulation and use by soybean. Field Crops Res. 81, 149–158.

Smil, V., 2002. Nitrogen and food production: proteins for human diets. Ambio 31, 126–131.

Souleimanov, A., Prithiviraj, B., Smith, D.L., 2002. The major Nod factor of Bradyrhizobium japonicum promotes early growth of soybean and corn. J. Exp. Bot. 53, 1929–1934.

Specht, J.E., Hume, D.J., Kumudini, S.V., 1999. Soybean yield potential – a genetic and physiological perspective. Crop Sci. 39, 1560–1570.

Steudle, E., 2000. Water uptake by roots: effects of water deficit. J. Exp. Bot. 51, 1531–1542.

Stewart, W., Dibb, D., Johnston, A., Smyth, T., 2005. The contribution of commercial fertilizer nutrients to food production. Agron. J. 97, 1–6.

Subramanian, S., Stacey, G., Yu, O., 2007. Distinct, crucial roles of flavonoids during legume nodulation. Trends Plant Sci. 12, 282–285.

Temprano, F., Albareda, M., Camacho, M., Santamarí, A., Rodríguez-Navarro, D., 2002. Survival of several *Rhizobium/Bradyrhizobium* strains on different inoculant formulations and inoculated seeds. Int. Microbiol. 5, 81–86.

Tilman, D., Cassman, K., Matson, P., Naylor, R., Polasky, S., 2002. Agricultural sustainability and intensive production practices. Nature 418, 671–677.

Tollenaar, M., Lee, E., 2002. Yield potential, yield stability and stress tolerance in maize. Field Crops Res. 75, 161–169.

Valentine, A., Benedito, V., Kang, Y., 2011. Legume nitrogen fixation and soil abiotic stress: from physiology to genomics and beyond. Annu. Plant Rev. 42, 207–248.

Vitousek, P., Cassman, K., Cleveland, C., Crews, T., Field, C., Grimm, N., Howarth, R., Marino, R., Martinelli, L., Rasetter, E., Sprent, J., 2002. Towards an ecological understanding of biological nitrogen fixation. Biogeochemistry 57–58, 1–45.

Wagner, S.C., 2011. Biological nitrogen fixation. Nat. Educ. Knowledge 3, 15.

Wan, J., Torres, M., Ganapathy, A., Thelen, J., DaGue, B., Mooney, B., Xu, D., Stacey, G., 2005. Proteomic analysis of soybean root hairs after infection by *Bradyrhizobium japonicum*. Mol. Plant Microbe Interact. 18, 458–467.

Wang, X., Pan, Q., Chen, F., Yan, X., Liao, H., 2011. Effects of co-inoculation with arbuscular mycorrhizal fungi and rhizobia on soybean growth as related to root architecture and availability of N and P. Mycorrhiza 21, 173–181.

Wardle, D., Bardgett, R., Klironomos, J., Setala, H., van der Putten, W., Wall, D., 2004. Ecological linkages between aboveground and belowground biota. Science 304, 1629–1633.

Wilcox, J.R., 2004. World distribution and trade of soybean. In: Boerma, H.R., Specht, J.E. (Eds.), Soybeans: Improvement, Production and Uses. ASA, CSSA, ASSA, Madison, WI, pp. 1–13.

Witte, C., 2011. Urea metabolism in plants. Plant Sci. 180, 431–438.

Yang, S., Tang, F., Gao, M., Krishnan, H., Zhu, H., 2010. *R* gene-controlled host specificity in the legume–rhizobia symbiosis. Proc. Natl Acad. Sci. USA 107, 18735–18740.

Yoon, J., Hamayun, M., Lee, S., Lee, I., 2009. Methyl jasmonate alleviated salinity stress in soybean. J. Crop Sci. Biotechnol. 12, 63–68.

Zapata, F., Danso, S., Hardarson, G., Fried, M., 1987. Time course of nitrogen fixation in field-grown soybean using nitrogen-15 methodology. Agron. J. 79, 172–176.

Zhang, F., Smith, D.L., 2002. Interorganismal signaling in suboptimum environments: the legume-rhizobia symbiosis. Adv. Agron. 76, 125–161.

Zhou, X., Liang, Y., Chen, H., Shen, S., Jing, Y., 2006. Effects of rhizobia inoculation and nitrogen fertilization on photosynthetic physiology of soybean. Photosynthetica 44, 530–535.

Zhu, H., Choi, H., Cook, D., Shoemaker, R., 2005. Bridging model and crop legumes through comparative genomics. Plant Physiol. 137, 1189–1196.

Heat stress responses and thermotolerance in soybean

12

Kamrun Nahar*, **, Mirza Hasanuzzaman†, Masayuki Fujita*
*Laboratory of Plant Stress Responses, Department of Applied Biological Science, Faculty of Agriculture, Kagawa University, Miki-cho, Kita-gun, Kagawa, Japan; **Department of Agricultural Botany, Faculty of Agriculture, Sher-e-Bangla Agricultural University, Dhaka, Bangladesh; †Department of Agronomy, Faculty of Agriculture, Sher-e-Bangla Agricultural University, Dhaka, Bangladesh

Introduction

Temperature extremes are consequences of climatic change and plants encounter threats to their survival therefrom. The records of extremities of elevated temperatures are such that high temperature (HT) stress should be considered a central concern and deemed to be one of the top ranking abiotic stress exerting threatening the survival of living organisms. Satellite measurements of ground temperature recorded 70.7°C in 2005 in the Lut Desert, Iran (Mildrexler et al., 2011). A ground temperature of 84°C was recorded in Port Sudan, Sudan (Nicholson, 2011). The highest natural ground surface temperature ever evidenced was 93.9°C (201°F) in Furnace Creek, Death Valley, California, USA on July 15, 1972 (Kubecka, 2001).

The global climate change scenario has shown that global average temperatures are expected to increase from 1.8°C to 4.0°C or higher by 2100, compared with the average temperature of 1980–2000 (IPCC, 2007). Extreme temperatures can cause species extinction. In general, this rise in temperature severely affects plant productivity. A rise in growing season temperature by 1°C may decrease crop yield up to 17% (Lobell and Asner, 2003; Ozturk et al., 2015).

Soybean (*Glycine max* L.) is a globally important commercial crop widely cultivated throughout the world covering a 102,386,923 ha area of Africa, Americas, Asia, Europe, Oceania, and Canada. Globally, it yields 39,761,852 ton oil. Global soybean seed production is 6,983,352 ton used for various purposes (FAOSTAT, 2010; http://faostat.fao.org/). Soybean provides vitamins, minerals, fiber, and antioxidant compounds.

Soybean seed contains 40% protein and 20% oil. As a single crop, every year soybean provides more protein and vegetable oil than any other cultivated crop. Its nutritional value is high: 100 g soybean seed contains 446 kcal energy, 36.5 g protein, 30.2 g carbohydrate, 19.9 g fat, 9.3 g fiber, 377 mg calcium, 704 mg phosphorus, and 15.7 mg iron (Khedar et al., 2008). Recently, soybean has attracted attention because of high value-added products including functional nutraceuticals, such as phospholipids, saponins, isoflavones, oligosaccharides, and edible fibers.

The isoflavones of soybean exert anticancer, antioxidant, positive cardiovascular, and cerebrovascular effects (Lui, 2004). Soybean is used as a high protein

Abiotic and Biotic Stresses in Soybean Production. http://dx.doi.org/10.1016/B978-0-12-801536-0.00012-8

feed supplement for livestock. Soybean plays a vital role in improving soil fertility through biological nitrogen fixation. Soybean can meet 50–60% of its nitrogen demand through biological nitrogen fixation (Salvagiotti et al., 2008). Under suitable conditions, soybean–*Bradyrhizobium* symbiosis can fix about 300 kg N ha^{-1} (Keyser and Li, 1992). Use of soybean oil as a source of biodiesel reveals other avenues in the exploration of soybean products (Pestana-Calsa et al., 2012).

Soybean plants are often subjected to HT stress, especially that of tropical and semiarid tropical region. The distribution, growth, yield, and quality of soybean is highly influenced by temperature (Liu et al., 2008). HT stress imposes a direct effect on photosynthesis and transpiration, consequently affecting soybean yield (Puteh et al., 2013). Each degree increase in temperature decreases soybean yield by 16–17% (Kucharik and Serbin, 2008; Lobell and Asner 2003).

Schlenker and Roberts (2009) detected the negative effect of temperature with a threshold level of 30°C for soybeans. Increase in soybean yield has been reported at temperatures between 18/12 and 26/20°C (day/night). Yield started decreasing at temperatures higher than 26/20°C (Sionit et al., 1987) and if temperature rises to 29/20 to 34/20°C during the seed-filling stage, there is a significant decrease in soybean seed yield (Dornbos and Mullen, 1991). It was reported that a daytime soil and air temperature of 28 and 38°C during the start of flowering to maturity reduced pod yield by 50% (Prasad et al., 2000).

It has been predicted that by 2100, a 6°C increase in temperature could lead to a 49% decrease in soybean yield (Schlenker and Roberts, 2009). In considering the damaging effects of HT on soybean, this chapter has gathered some available information regarding the morphophysiology of the HT-affected soybean plant. HT is an unavoidable stress and groups of scientists are trying to investigate ways to improve performance of soybean plant and yield under such a stress. This chapter also presents some approaches related to HT-tolerance development from existing literature, which may explore new ways and means to develop HT-tolerant soybean plants.

Effects of high temperature on soybean

Morphophysiology of soybean

Germination

Studies of germination of soybean revealed that a certain range of temperature is suitable for improved germination and healthy seedling establishment. According to Liu et al. (2008), a suitable temperature for emergence of soybean is 15–22°C. Seeds of soybean were exposed to different HTs (50, 60, and 70°C for 10 h). Compared with controls, a gradual increase of temperature reduced the percentage germination, moisture content, and seedling vigor of soybean (Anto and Jayaram, 2010). At different temperature levels, such as 24.5, 28.5, 33.0, 36.5, and 40.0°C, germination performance of soybean seedlings was tested.

The number of germinated seeds were recorded at 2 h intervals at those temperature levels, until this parameter reached about 80%. Of these temperatures, the optimal temperature range for germination was 33.0–36.5°C. Beyond this range the longest period was required for germination, the lowest percentage of seeds were germinated, and seedlings showed abnormal appearance (Edwards, 1934). Soybean (*G. max* L., cv. Grant) seeds were germinated for 4 days and were evaluated for hypocotyl length at temperatures of 33 and 39°C. Seedling length dropped from 76.3 mm to 25.3 mm, respectively, at 33 and 39°C. Germination at 39°C was only 62%, compared to 29°C (Ndunguru and Sununerfield, 1975). Covell et al. (1986) reported the optimum temperature for 50% germination of soybean: base, optimum, the ceiling temperature for 50% germination of soybean were 4, 34.3, and 51°C, respectively.

Vegetative growth

Performance of soybean plant was examined considering three dates of sowing along with temperature increases during delayed sowing treatments. In delayed sowing treatment, growth duration declined significantly. In normal sowing, the growth duration was 114 d; in the first delayed treatment the growth duration was 106 d; and for later it was 78 d. Increase of day length along with increasing temperature in delayed sown treatment resulted in reduction of plant height and leaf area, biomass of leaves, stem, pods, roots, and total biomass (Kumar et al., 2008).

Soybean (*G. max* L. Merr.) plants were subjected to HT (10, 25, 50, and 110°C) at different growth stages. Germination was adversely affected among different genotypes. HT also disrupted events of phenology of soybean at different growth stages. The most susceptible growth stage was flowering (Sapra and Anaek, 1991). Seddigh and Jolliff (1984) studied the performance of soybean plant using three levels of temperature stress, i.e., 10, 16 ± 1, and 24 ± 2°C. High night temperature had no significant effect on morphological characteristics such as plant height, number of auxiliary branches, and number of nodes in soybean, compared with 10°C. But compared with 10°C, high night temperature promoted early vegetative growth and hastened physiological maturity.

In other study, different soybean genotypes (D 88-5320, D 90-9216, Stalwart III, PI 471938, DG 5630 RR, and DP 4933 RR) were exposed to two temperature regimes of 30/22 and 38/30°C. Compared with the normal temperature of 30/22°C, HT (38/30°C) decreased vegetative growth. Significant reductions in plant height, leaf area, and total biomass were noticed in HT-affected plants. Considering relative injury among different genotypes under HT, a total stress response index was determined. PI 471938 and D 88-5320 were selected as the most tolerant genotypes (Koti et al., 2007).

HT increased thicknesses of palisade and spongy layers and lower epidermis of soybean leaf. Increases in upper epidermis thickness, palisade layer I thickness, palisade layer II thickness, spongy cell thickness, lower epidermis thickness, and total leaf thickness were characteristic features of HT-affected leaves of soybean plants (Djanaguiraman et al., 2011b). Nitrogen fixation and ethylene production in soybean root was inhibited by an HT of 40°C. Nitrogen fixation activity decreased with increasing HT-exposure duration from 2 h to 6 h. After 2 h of HT, fixation rate decreased by 15%. The fixation rate decreased by 70% after 4 h to 6 h of HT exposure.

However, the fixation capacity recovered between 24 h and 144 h after the end of HT stress (Keerio et al., 2001). Sinclair and Weisz (1985) reported that in soybean, soil temperatures above 34°C negatively affected nitrogen fixation. According to Munevar and Wollum (1982) the optimum rate of N fixation in soybean occurs at 28°C and the lowest rate at 38°C. In another study, the detached soybean nodules indicated a maximum fixation rate up to 30°C, which then decreased rapidly at 35°C (Waughman, 1977).

Reproductive development

Survival and succession of seed crop plants are ensured by successful reproduction, and the reproductive phase is considered to be one of the most sensitive growth phases, which is affected by the surrounding environmental conditions (Nakamoto et al., 2001; Zheng et al., 2002). Soybean plants are highly sensitive to reproductive stage HT stress. HT stress hinders flowering and postflowering development and distorts reproductive development. HT of 32–38°C during reproductive periods reduces seed yield components of soybean (Dornbos and Mullen, 1991; Gibson and Mullen, 1996; Huxley et al., 1976).

The decreases induced by HT in photosynthesis, abscission and abortion of flowers, developing seeds, and young pods are among the most important reasons for seed yield reduction (Prasad et al., 2002, 2006, 2008). HT decreased pollen viability and stigma receptivity, which decreased seed set (Prasad et al., 2002, 2006; Snider et al., 2009). Lindey and Thomson (2012) reported that the optimum temperature of soybean ranges from 25°C to 29°C and pod setting is severely affected above 37°C.

Compared with the control temperature (30/22°C), HT of 38/30°C severely affected pollen development in soybean genotypes. Pollens, which were developed under HT were shriveled without apertures and disturbed exine ornamentation. Genotypes were classified based on total stress response index, which was calculated from pollen germination percentage, pollen tube length, pollen number anther, and flower length. Genotypes Deltagrow 5630RR and Delsoy 88-5320 were tolerant, genotypes Delsoy 90-9216 and 471938 were intermediate, and genotypes Stalwart III and Deltapine 4933RR were sensitive (Koti et al., 2005).

Different temperatures were imposed on soybean plants at different reproductive stages. Flowering and pod set, seed fill and maturation, and the entire reproductive period were exposed to 30/20, 30/30, 35/20, and 35/30°C day/night temperatures, respectively. High day temperature reduced seed formation. HT stress during flowering and pod-set stage decreased seed growth and seed fill. Seed growth reductions in high day temperature-affected plants were accompanied by decreased photosynthesis rates (Gibson and Mullen, 1996).

Djanaguiraman et al. (2011a) exposed soybean plant to HT (38/28°C) for 14 d at the flowering stage. Compared with the control (28/18°C), HT stress increased flower abscission, which was the reason for decreased pod-set percentage. Sionit et al. (1987) reported that reductions in pod number resulted from decreased effectiveness of pollination and fertilization under HT; as a consequence pod setting was poor (Prasad et al., 2001).

Moderate increases in daytime temperature (18–26°C) during seed filling have been reported as beneficial for increasing soybean yield (Sionit et al., 1987). Reduced pollen production by 34%, pollen germination by 56%, and pollen tube elongation by 33% were reported for 38/30°C, compared with 30/22°C (Salem et al., 2007). Elevated CO_2 with HT increased soybean flower production (Nakamoto et al., 2001; Zheng et al., 2002). Temperatures from 18/12°C to 26/20°C for long duration in the growing season is favorable for increasing weight per seed (Sionit et al., 1987).

Temperature higher than 26/20°C decreases weight per seed (Baker et al., 1989; Hesketh et al., 1973; Huxley et al., 1976). Flowering and pod-development stages exposed to temperatures higher than 30/25°C showed significantly reduced weight per seed (Egli and Wardlaw, 1980). Temperatures beyond 29/20°C during seed fill reduced weight per seed (Dornbos and Mullen, 1991). Kitano et al. (2006) reported several disorders caused by HT in soybean reproductive development. HT resulted in abortion of flowers, young pods, and developing seeds. Reduced pod setting in soybean was noticed due to lower pollen viability at HT of 37/27°C (day/night).

Hatfield et al. (2008) studied the performance of different soybean cultivars under HT. From averaging over many cultivars, cardinal, optimum, and failure point temperatures were 13.2, 30.2, and 47.2°C for pollen germination, and 12.1, 36.1, and 47.0°C for pollen tube growth, respectively (Hatfield et al., 2008). During anthesis time, the soybean has a cardinal or base temperature of 6°C and an optimum of 26°C (Boote et al., 1998).

The optimum temperature of postanthesis phase was 23°C (Baker et al., 1989; Boote et al., 2005; Egli and Wardlaw, 1980; Pan, 1996; Thomas, 2001). High day (39/20°C) and night temperature (30/29°C) decreased pollen viability by 37 and 39%, respectively. Such high day and night heat decreased the pollen germination by 35 and 28%, respectively, compared with control (30/20°C). Increasing nighttime temperature from 23°C to 29°C resulted in a decrease in pollen viability and pollen germination.

Pollen viability decreased by 19, 29, and 39%, and pollen germination decreased by 11, 20, and 28%, respectively, at 23, 26, and 29°C, compared with the optimum temperature (20°C). HT decreased saturated phospholipids and phosphatidic acid in pollen grains, which resulted in decreased pollen viability and germination (Djanaguiraman et al., 2013). For soybean plants, suitable temperature for flowering is 20–25°C, and for maturity is 15–22°C (Liu et al., 2008).

Seddigh and Jolliff (1984) studied soybean plants using three night temperature treatments at 10, 16 ± 1, and 24 ± 2°C. HTs reduced duration of developmental stages. Warmer night temperatures shortened the period between plant emergence and the appearance of first flower. HT also reduced duration essential for flower appearance at the node directly below the uppermost node. Physiological maturity was also hastened by higher night temperatures (Seddigh and Jolliff, 1984). Changes or reduction in photosynthate supply under HT stress during flowering and pod-set growth stages may result in abortion of flowers, young pods, and seeds (Egli and Bruening, 2006).

Photosynthesis

Photosynthesis is hampered by HT in different ways. Soybean plants were grown under HT (38/28°C) during the flowering stage and performance was compared with that of plants grown under normal growth temperature (28/18°C). Leaf photosynthesis rate and stomatal conductance were decreased by HT at 20.2 and 12.8%, respectively. HT also induced anatomical and structural changes in cells and cell organelles, especially in chloroplasts and mitochondria. Stomatal density and stomata diameter significantly decreased by 6 and 20%, respectively, in HT affected soybean leaf, compared with control.

HT-distorted plasma membrane, chloroplast, thylakoid and mitochondrial membranes, cristae, and matrix (Djanaguiraman et al., 2011b). Response of different soybean genotypes (D 88-5320, D 90-9216, Stalwart III, PI 471938, DG 5630 RR, and DP 4933 RR) was evaluated for 30/22 and 38/30°C. HT adversely affected vegetative and physiological parameters. HT significantly affected net photosynthesis, total chlorophyll (chl), phenolic and wax content, and vegetative growth. The studied physiological parameters were used for the determination of total stress response index, and the genotypes were classified as tolerant (PI 471938 and D 88-5320), intermediate (DG 5630 RR and D 90-9216), and sensitive (DP 4933 RR and Stalwart III) (Koti et al., 2007).

HT (45 and 48°C) decreased the efficiency of identical maximum Photosystem II (PSII) photochemistry (Fv/Fm ratio) in soybean seedlings. Leaves at 48°C showed high damage due to formation of higher complexes of oxygen evolving along with lower rates of O_2 evolution, compared to the leaves at 45°C. Donor side of PSII was damaged severely at 48°C (Li et al., 2009). HT (39/20°C) decreased leaf chl content, photosynthesis rate, photochemical quenching, and electron transport rate (compared with a normal temperature of 30/20°C).

HT increased leaf respiration rates. High night temperatures of 23 and 26°C increased thylakoid membrane damage by 10 and 83%, respectively, compared to optimum temperature (20°C). Nonphotochemical quenching increased with high day (39/20°C) or night temperature (30/29°C) compared with optimum temperature (30/20°C). HT decreased photosynthesis electron transport, which resulted in decreased photosynthesis rates. Decreased quantum yield of PSII (35%) and photochemical quenching (25%) due to high day temperature indicates that maximum inhibition of electron transport occurred at quantum yield of PSII compared with photochemical quenching (Djanaguiraman et al., 2013).

Djanaguiraman et al. (2011a) detected HT stress (38/28°C for 14 d) induced damage on chl and photosynthesis in soybean, compared to control temperature (28/18°C). Decreases of total chl content (by 17.8%), chl a content (by 7.0%), chl a/b ratio (by 2.5%), sucrose content (by 9.0%), and stomatal conductance (by 16.2%) were noticed under HT. HT also decreased PSII quantum efficiency and photosynthesis rate. PSII photochemistry (Fv/Fm ratio) decreased by 5.3% after 14 d of HT exposure. Photosynthesis rate decreased by 7.4, 15.6, 19.6, and 19.7% after 2, 6, 10, and 14 d of HT exposure, respectively (Djanaguiraman et al., 2011a).

HT caused dissociation of the oxygen-evolving complex (OEC) in soybean. This resulted in imbalance between electron flows from OEC to reaction center. This also restricted electron flow toward the acceptor side of PSII in the direction of PSI. Due to uncoupling of OEC, alternative internal electron donor-like proline (Pro) or ascorbate (AsA) takes part in donating electrons to PSII instead of H_2O. Due to the inactivation of the reaction center after HT exposure of 2 days, the antisense soybean plants were more susceptible to photoinhibition. HT resulted in an increased proportion of inactivated (non-QA reducing or heat sink center) reaction center, which decreased the rate of photosynthesis (de Ronde et al., 2004a).

Soybean plants were exposed to HT (38/28°C) for 14 days and their performance was compared with control (28/18°C) at the beginning of pod set. Leaf senescence occurred at later stages of reproductive development. Chlorophyll a content decreased significantly by 44% under HT stress, compared with control. Chlorophyll b content was also significantly affected by HT and chl b content decreased by 25.2% and the chl a/b ratio by 26%, compared with control. HT also resulted in a decrease in soluble protein content (by 22.4%). The parameters of chlorophyll a fluorescence such as Fv/Fm ratio and Fo/Fm ratio (ratio of basal florescence to maximum fluorescence indicating thylakoid membrane damage) were negatively affected by HT. The rate of thylakoid membrane damage was significantly increased by HT. Stomatal conductance was reduced.

Photosynthesis decreased by 12.7% under HT stress, compared with control. Reduction in photosynthesis ultimately decreased the photosynthetic product, which is evident from decreased sucrose content (by 21.5%) (Djanaguiraman and Prasad, 2010). HT and humidity stress often cause preharvest deterioration during soybean seed development and maturation in the field (Wang et al., 2012). Soybean plants were subjected to 40/30°C and 100/70% humidity (RH) for 10/14 h cycle (light/dark) for 7 days. Plant exposure to 30/20°C and 70% RH, and to 10/14 h (light/dark), was considered as control treatment.

Thinner cell wall, smaller chloroplasts and protein bodies, irregular cell nucleus, and nuclear core were noticed under HT. The cell also had bigger vacuoles and lipid bodies. HT decreased starch grains, and increased mitochondria and smooth endoplasmic reticulum in developing soybean seed. Compared with control (168 h), HT-affected seeds contained more saturated (palmitic and stearic) and monounsaturated (oleic) fatty acids. HT affected seeds contained lower oil and polyunsaturated (linoleic and linolenic acid) fatty acids. HT increased soluble protein content. Seeds also showed enhanced nitric oxide production (Wang et al., 2012).

Oxidative stress

Extreme temperature stress accelerates generation of reactive oxygen species (ROS) such as singlet oxygen ($^1O_2$), superoxide radical ($O_2^{\cdot-}$), hydrogen peroxide (H_2O_2), and hydroxyl radical (OH·), which induce oxidative stress (Mittler, 2002; Yin et al., 2008). HT of 38/28°C resulted in oxidative damage to soybean. Compared to control treatment, HT significantly increased $O_2^{\cdot-}$ (by 63%) and H_2O_2 (by 70.4%) content, which resulted in severe membrane damage (by 54.7%) (Djanaguiraman and Prasad, 2010). Electron transfer from PS II is extremely HT sensitive. Electrolyte leakage is a common phenomenon under HT.

Different soybean genotypes were subjected to various temperature regimes (10, 25, 50, 110°C). The genotypes PI 408.155, PI 423. 827B, PI 423.759, and Peshing were selected as HT tolerant considering the electrolyte leakage rate (Sapra and Anaek, 1991). According to Lin et al. (1984) oxidative damage resulted in enhanced membrane permeability and electrolyte leakage in soybean. The ability of the plasma membrane to retain solutes and water was also reduced by oxidative damage (Lin et al., 1984).

Soybean plant showed dissociation of OEC under HT (38/25 ± 2°C day/night, 2 days) stress, which was the reason to create imbalance between electron flow from the OEC to reaction center. This imbalance between electron flow may result in oxidative stress (de Ronde et al., 2004a). Djanaguiraman et al. (2011b) reported that the major generation sites of ROS are cell chloroplast and mitochondria. In HT-affected soybean plants, the thylakoid membrane was not properly stacked. Thylakoid dilation and membrane loss of the stromal and chloroplast envelope were evident due to loss of integrity, which rendered chloroplasts swollen. Discontinuous mitochondrial membrane was indicated under HT.

The damage to membranes of chloroplasts and mitochondria and cell plasma membrane indicated oxidative damage (Djanaguiraman et al., 2011b). Increased membrane lipid peroxidation under HT was noted by Tan et al. (2011) in terms of membrane injury in soybean. Heat-induced oxidative stress has been reported in soybean due to 38/28°C day/night temperature, compared with control (28/18°C). HT increased H_2O_2 content and membrane damage. Significant increases of $O_2^{\bullet-}$ and malondialdehyde (MDA, a lipid peroxidation product) contents by 85.6 and 174%, respectively, were noted in HT-affected soybean plant (Djanaguiraman et al., 2011a).

Martineau et al. (1979) isolated HT-tolerant soybean genotypes on the basis of membrane thermostability. The HT-sensitive genotypes showed higher electrolyte leakage with reduced membrane thermostability in heat-damaged leaf tissue cells under HT. At 40°C, accelerated aging (with 100% RH) in soybean seeds resulted, which increased respiration rate and ROS including $O_2^{\bullet-}$ and H_2O_2. Activities of ROS-scavenging enzymes decreased under HT, which also amplified ROS and caused lipid peroxidation.

The MDA content of soybean seeds significantly increased with accelerated aging (80°C, 34%), after the 15th and 20th days. Finally, the cells died with continuation of the stress (Tian et al., 2008). Developing seed showed higher accumulation of ROS and higher accumulation of H_2O_2, which increased lipid peroxidation under HT (40/30°C, 100/70% relative humidity under light/dark), compared with control (Wang et al., 2012). The effects of HT on the morphophysiology and yield of soybean are presented in Table 12.1.

Yield attributes and yield

Several research findings have common agreement that HT negatively affects growth and yield of soybean. But favorable temperature ranges for growing soybean were slightly different due to differences in growing season and region or soybean cultivar. Sionit et al. (1987) reported that seed yield of soybean increased when temperature

Table 12.1 **Effect of high temperature on the morphophysiology and yield of soybean**

Temperature and duration	Effects	References
39/20°C, 10 d	Decreased leaf chl content, photosynthetic rate, photochemical quenching, electron transport rate, pollen viability and germination, decreased pod set and seed weight per plant.	Djanaguiraman et al. (2013)
35°C, 14 d	Caused pod abortion; reproductive stage showed decrease in seed yield components such as the number of pods per plant, number of seeds per pod, and 100-seed weight.	Puteh et al. (2013)
38/28°C day/night temperature, 14 d	Increased ethylene production rates, oxidative damage; decreased antioxidant-enzyme activity, chlorophyll, sucrose, soluble protein, and PSII photochemistry; caused premature leaf senescence, increased flower abscission, and decreased pod-set percentage.	Djanaguiraman et al. (2011a)
45°C for 10, 25, 40, 60, 90, 120, 150, and 180 min	Decreased Fv/Fm, damaged the OEC, reduced O_2 evolution rate, damaged donor side of PSII.	Li et al. (2009)
38/28°C, 14 d	Increased oxidative damage. Decreased chl b content and chl a/b ratio, photochemical efficiency, photosynthetic rate, sucrose content, activities of SOD, CAT, and POX seed set percentage, seed size, and seed yield per plant.	Djanaguiraman and Prasad (2010)
38/28°C, 14 d	Decreased leaf photosynthesis rate and stomatal conductance, increased the thicknesses of palisade and spongy layers and lower epidermis. Damaged plasma membrane, chloroplasts, thylakoid and mitochondrial membranes, cristae, and matrix.	Djanaguiraman et al. (2011b)
40°C with 100% relative humidity; 0, 5, 10, 15, and 20 d	Decreased activities of SOD, APX (ascorbate peroxidase), CAT, and GR (glutathione reductase). Increase in $O_2^{\cdot-}$ and H_2O_2 and lipid oxidation and cell death.	Tian et al. (2008)
35/25°C (day/night temperature), 10 d	Increased H_2O_2 content, lipid hydroperoxide, lipid peroxidation, Pro, and AsA. Decreased RWC, activity of GR and GST (glutathione S-transferase).	Kocsy et al. (2005)

(Continued)

Table 12.1 **Effect of high temperature on the morphophysiology and yield of soybean** (*cont.*)

Temperature and duration	Effects	References
32–42°C, 2 d	NADP$^+$ levels decreased. Activity of *P5CR* (L-Δ^1-pyrroline-5-carboxylate reductase) enzyme, Pro content, and OEC complex increased. Reduced photosynthetic activity and e-feeding, damaged PSII.	de Ronde et al. (2004a)
40/30°C (with 100/70% relative humidity), 7 d	Increased ROS production and lipid peroxidation. Reduced CAT activity. Resulted in thinner cell wall, smaller chloroplasts and protein bodies, irregular cell nucleus and nuclear core. Altered seed protein and fatty acid metabolism.	Wang et al. (2012)
45°C, 2 h	Decreased primary root, hypocotyl and seedling length. Distorted membrane integrity, increased electrolyte leakage, and MDA content.	Amooaghaie and Moghym (2011)
38/30°C, vegetative to 10d after flowering	Decreased net photosynthesis, chl, phenolic and wax content. Decreased plant height, leaf area and total biomass.	Koti et al. (2007)
30°C	Reduction in photosynthate supply to reproductive organ caused abortion of flowers, young pods, and seeds.	Egli and Bruening (2006)

increased between 18/12°C (day/night) and 26/20°C, and seed yield decreased at temperatures higher than 26/20°C. In other findings, it was reported that increasing temperature from 29/20 to 34/20°C during the stage of seed filling significantly reduced soybean seed yield (Dornbos and Mullen, 1991). Among seed yield components, the seeds per pod parameter is the least affected (Baker et al., 1989; Huxley et al., 1976; Sionit et al., 1987). Another research finding reported that temperatures above 30/25°C (day/night) during flowering and pod development adversely affected soybean physiology and reduced weight per seed (Egli and Wardlaw, 1980).

Koti et al. (2005) gave different reasons for yield reduction in soybean under HT. At the stage of flowering, HT increased the abscission of flowers and decreased the percentages of pod set as well as the individual seed weight in comparison with optimum temperature. HT also reduced pollen viability, which decreased pod-set percentage. Reductions in the number of pods and individual seed weight were recognized as the damage effects of HT. All these combined acted to reduce yield (Koti et al., 2005).

High daytime (39/20°C) and night time temperature (30/29°C) significantly reduced the pod set and seed weight, compared to optimum temperature (30/20°C). Pod set was more sensitive to HT. Pod-set decreased at night time temperatures

>23°C. On the contrary, seed weight decreased at night time temperatures >26°C (Djanaguiraman et al., 2013). Performance of three soybean varieties, namely Dieng (small seeded), Willis (medium seeded), and AGS190 (large seeded) was observed under HT.

Reproductive stage exposed to HT showed a decrease in seed yield components including the number of pods per plant, number of seeds per pod, and 100-seed weight, indicating that HT during flowering and pod set can have its highest effect on the soybean seed yield. Seed yield components were slightly affected at 30°C and severely affected at 35°C (Puteh et al., 2013). A reduction of yield about 27% was reported in soybean under 35°C, 10 h (Gibson and Mullen, 1996).

According to Dornbos and Mullen (1991) temperature increased to between 18/12°C (day/night) and 26/20°C is favorable to increase soybean yield. Temperature higher than 26/20°C decreases soybean yield (Huxley et al., 1976; Sionit et al., 1987). A temperature increase from 29/20 to 34/20°C at the seed-fill stage decreased seed yield (Dornbos and Mullen, 1991). Gibson and Mullen (1996) studied the effect of reproductive stage temperature on yield of soybean. Soybean plants at their different growth stages were subjected to different day/night temperature regimes: 30/20, 30/30, 35/20, and 35/30°C were imposed during flowering, pod set, seed fill and maturation, and during the entire reproductive period, respectively.

The effect of HT was significant to impede reproductive development, which resulted in yield reduction in some growth stages. Significant reduction of seed weight per plant during flowering and pod set was recorded due to high day and night temperature. The largest yield reduction was 27%, which was due to exposure at 35°C, 10 h per day from flowering to seed maturity. High night temperature at any reproductive growth phase did not significantly affect the yield (Gibson and Mullen, 1996).

Kumar et al. (2008) reported that soybean seed sown in delayed growing season was affected by HT, which reduced growth duration, and negatively affected growth and yield attributes. Number of pods per plant, number of seeds per pod, dry matter accumulation, stover yield, and grain yield were correlated with the degree days of growing, and with heliothermal and photo thermal units. Due to HT, yield components of soybean were significantly and adversely affected as found by Djanaguiraman and Prasad (2010).

Decreases in the number of filled pods per plant (by 50.9%), seed set (by 18.6%), number of filled seeds per plant (by 30.5%), and seed size (by 64.5%) were recognized under HT, compared with control. The reduction in yield components was enough to cause severe yield loss in HT-affected soybean. Reduction in seed yield was 71.50% and reduction in harvest index was 78.2% in soybean under HT, compared with control (Djanaguiraman and Prasad, 2010).

Compared to 10°C night temperature, 16 ± 1°C and 24 ± 2°C temperatures increased seed weight per plant. This parameter increased by 27 and 37% for 16°C, and increased by 20 and 23% for 24°C treatments in 1981 and 1982, respectively. Increased seed size (1000-seed weight) accounted for the increase of seed yield under increased night temperature. Final numbers of seeds per plant and pods per plant were not significantly affected by temperature changes. But the number of seeds per pod for the 24°C treatment was significantly higher relative to 10°C (Seddigh and

Jolliff, 1984). Seed yields at harvest maturity decreased by 29% due to HT (Ferris et al., 1999). Baker et al. (1989) and Boote et al. (2005) concluded that the seed harvest index was reduced above 23–27°C. Pan (1996) and Thomas (2001) concluded that a mean temperature of 39°C is lethal to soybean yield production.

Approaches to develop high-temperature stress tolerance in soybean

Roles of osmoprotectants or compatible solutes

Stress-induced change of osmoprotectants within plants is a common event and several reports described the beneficial roles of endogenously upregulated or exogenously applied osmoprotectants under HT stress. Genetic manipulation of Pro synthesis can affect Pro level and other antioxidants. The *P5CR* gene is involved in Pro biosynthesis. Wild type soybean (*G. max* cv. Ibis) and transgenic soybean containing *P5CR* (antisense and sense transformants, respectively) were tested in drought and HT (35/25°C day/night temperature) conditions.

After the period of recovery, antisense produced the highest H_2O_2 and lipid hydroperoxide concentrations, and showed the greatest HT injury. In contrast, sense transformants showed the lowest H_2O_2 content and the lowest injury percentage. Sense transformants showed the highest Pro and AsA accumulation. Antisense transformants had the highest glutathione (GSH) and reduced homoglutathione (hGSH) contents, AsA/dehydroascorbate (DHA) and (h)GSH/(h) oxidized homoglutathione (GSSG) ratios, and APX and GR activities (Kocsy et al., 2005).

de Ronde et al. (2000, 2001, 2004a, b) conducted several experiments with transgenic (with *P5CR*) soybean and examined the roles of Pro in the adaptive response under HT stress. Soybean plants showed improved drought and HT tolerances when overexpressing Arabidopsis *P5CR* (*AtP5R*) (de Ronde et al., 2000, 2004a, b). The overexpression of soybean with *P5CR* in the sense and antisense orientation resulted in a heat shock cassette, containing an inducible heat shock promoter (de Ronde et al., 2000). Control plants (without expression of the *P5CR* gene) showed reduced protein synthesis and Pro shortage, compared to soybean plant overexpressing the *P5CR* gene.

However, transgenic lines showed a lower rate of seed production, which means that the antisense *P5CR* gene adversely influenced seed production. Decreased Pro content due to the activated antisense *P5CR* gene illustrated a greater sensitivity to osmotic and HT stress indicating the Pro-induced adaptive response in soybean plants (de Ronde et al., 2001). de Ronde et al. (2004a, b) found that under field conditions the *P5CR* transgenic soybean lines were tolerant to the stresses of drought and HT, compared with wild type cultivars (de Ronde et al., 2004a, b).

de Ronde et al. (2004a) investigated the biochemical basis of heat and drought tolerances in antisense and sense transgenic soybean plants by comparing the response of soybean plants, which are antisense and sense transgenic, and overexpress *Arabidopsis P5CR*. Both types of plants were subjected to simultaneous drought and HT stress (38/25°C day/night heat treatment, 2 d) and then subjected to rewatering at a normal temperature of 25°C. Sense plants had the highest ability to accumulate Pro during stress and to metabolize Pro after rewatering.

At the period of recovery, sense plants exhibited mild symptoms of stress and antisense plants indicated severe damage effects resulting from HT stress. Decrease in nicotinamide adenine dinucleotide phosphate (NADP$^+$) levels was noted in antisense plants. The sense plants showed an increase in these levels at the period of recovery in which OEC dissociation was bypassed by Pro feeding electrons into PSII, so prohibiting further damage (de Ronde et al., 2004).

The seedlings of transgenic (with *P5CR*) soybean (*G. max* (L.) Merr. cv. Ibis) were subjected to stress at 35/25°C day/night temperature and at the same time water stress was imposed by withholding water for 10 d. Overexpression of *P5CR* not only regulated accumulation of Pro but also influenced the concentration of other amino acids. It was suggested that overexpression of *P5CR* has roles in regulation of distinct metabolic pathways related to other amino acids. Related to the wild type, the transgenic soybean (*G. max* cv. Ibis) transformed with *P5CR* in sense showed a rapid increase in Pro content. The sense *P5CR* soybean transformed also showed least water loss. The level of the Pro precursors such as Glu (glutamine) and Arg (arginine) was also higher in sense transformants, compared with antisense and wild type plants during exposure to stress (Simon-Sarkadi et al., 2005).

In a study Salem et al. (2005) reported the positive effects of glycine betaine (GB) to provide protection in soybean under extreme temperature conditions. The temperature treatments consisted of 15 and 45°C in different soybean genotypes, which were imposed *in vitro* or *in vivo* with or without GB application. HT reduced the pollen germination percentage significantly. Soybean plants had GB applied at 2 kg per ha as a foliar spray before flowering, which was considered to be *in vivo*.

For *in vitro* treatment, pollens were collected from flowers and were germinated on a pollen germination medium supplemented by GB. Pollen germination was found to be 15 and 21% higher under *in vivo* and *in vitro* GB treatments under HT, respectively, when compared to the control. Considering the response of soybean plant with GB application under HT, the tolerance of soybean genotypes was determined. Extremely responsive genotypes were Hutcheson, D 68-0102, and Stress land; the non-responsive genotype was DG 5630RR, the less-responsive genotypes were P 9594, Maverick, and DARE, and the more-responsive genotypes were DP 4690RR, DK 3964RR, Williams 82, and Stalwart IIIand. It was also concluded that manipulation of the biosynthetic pathway of GB into soybean would be a strong tool in developing soybean plants under environmental stress conditions (Salem et al., 2005).

Polyamine

Polyamines (PAs) are aliphatic amines and organic polycations with a low molecular mass, which have important roles in improving abiotic stress tolerance including HT tolerance. Moghym et al. (2010) studied the effects of exogenous PA in alleviating the adverse effects of HT stress. They reported that although soybean (*G. max* (L.) Merr.) is a thermophilic plant, a temperature of up to 40°C inhibits its seed germination.

Exogenous PAs (putrescine, Put; spermidine, Spd; and spermine, Spm) were applied to investigate the heat-shock protection of soybean seedlings cv. Sahar.

The effect of pretreatment with PA inhibitor on seedling growth was examined. Heat-shock decreased growth of soybean. In contrast, application of PA including putrescine and spermidine increased thermotolerance and growth recovery. Furthermore, lysine, an amino acid, was applied at equivalent concentration to demonstrate that heat-shock protection is in actuality a PA phenomenon and not simply a growth effect from a source of nitrogen. PAs including Put, Spd, Spm did not act merely as a source of nitrogen when stimulating growth. PAs had HT-stress adaptive roles (Moghym et al., 2010).

Soybean seeds (*G. max* (L.) Merr. cv. Sahar) were subjected to heat-shock at 45°C (for 2 h), which resulted in a 47% decrease in primary root length and a 50% decrease in hypocotyl length. In other treatments, PAs: Put (1 mM), Spd (1 mM), and Spm (1 mM) were applied as a pretreatment (at the normal temperature of 28°C for 2 h before subjecting the seedlings to heat shock). Application of PAs resulted in the protection of membrane integrity, and decreased electrolyte leakage and MDA content from different tissue sections. Seedling growth was improved by PAs. Pretreatment with PAs significantly enhanced thermoprotection capability of soybean seedlings. Among these PAs, Put was the most effective in conferring thermotolerance, followed by Spm (Amooaghaie and Moghym, 2011).

Effect of temperature on the germination and seedling development of soybean, and on the PA titers in soybean seeds, was investigated. Three germination temperatures, 25, 30, and 36°C, were considered in order to evaluate the influence of PA concentrations on soybean seeds germinated at 76 and 90 h. Compared to the germination at 25°C, HT of 36°C reduced total germination from 97.2 to 92.5% and HT increased the frequency of abnormal seedling growth. The PAs including cadaverine (Cad), Put, Spd, agmatine (Agm), and Spm in soybean seed were measured. Cad, Put, Agm, and Spd contents declined with increasing the germination temperature from 25°C to 36°C. Decrease in PAs titers has a correlation with decrease in germination and abnormal seedling generation under HT. In contrast, Spm increased considerably with the rise in temperature (Mejia, 1999).

Roles of antioxidant components

Both nonenzymatic and enzymatic components of antioxidant defense system are vital to impart HT-induced oxidative stress tolerance. The available research findings explore the roles of antioxidant components in improving the HT tolerance of soybean. According to D'Souza (2013) HT-induced enhanced activities of peroxidase (POX), GR, and APX, which prevented oxidative damage in soybean plants. Soybean plants exposed to 38/28°C day/night temperature increased oxidative damage significantly and decreased SOD, POX, and CAT enzyme activity (by 16.8, 30.1, and 103.1%, respectively), compared with optimum temperature, which was related to HT-induced ethylene response.

However, application of the ethylene perception inhibitor 1-methylcyclopropene (1-MCP) minimized the oxidative damage effects in soybean. Application of 1-MCP significantly increased the activities of CAT, SOD, and POX (by 14.4, 50.7, and

23.1%, respectively), compared with the untreated control, which significantly reduced H_2O_2 content, $O_2^{\bullet-}$ production, MDA content, and membrane damage (by 8.8, 4.1, 4.0, and 17%, respectively), compared with the untreated control (Djanaguiraman et al., 2011a).

Effect of HT stress resulted in the production of ethylene, and its effects on antioxidant system of soybean were studied. HT significantly increased ethylene production rate. Decreases in the activities of antioxidant enzymes were noted under HT. Superoxide dismutase activity decreased by 13.3%, CAT activity decreased by 44.6%, and POX activity decreased by 42.9% under HT. Decreases in the antioxidant system were corroborated by significant increases in $O_2^{\bullet-}$ and H_2O_2 content and membrane damage. However, foliar spray of 1-MCP inhibited the production of ethylene, increased antioxidative capacity, and decreased oxidative stress to some extent (Djanaguiraman and Prasad, 2010). Tian et al. (2008) reported that the reduction in the activities of antioxidant enzymes was correlated with increased oxidative damage in soybean seed.

Freshly harvested soybean seeds were subjected to accelerated aging at 40°C and 100% relative humidity for 0, 5, 10, 15, and 20 days. Activities of SOD, GR, and APX of soybean seed decreased gradually until 10 days of accelerated aging, and then increased by the 15th day. The activities finally decreased after 20 days of stress. The activity of CAT decreased by 77% after accelerated aging for 20 days. The significant decrease in GR activities was noted up to day 10 of accelerated aging (Tian et al., 2008).

The interaction effects of long-term temperature ($25 \pm 2°C$ and $35 \pm 2°C$) and salt stress (NaCl with 0, −0.1, −0.4, and −0.7 MPa) were studied in different soybean cultivars. All the cultivars of soybean (G. max L. Merr., A 3935, CX-415, Mitchell, Nazlıcan, SA 88 and Türksoy) showed increased POX activity under 35°C, compared with 25°C (at the same salinity levels). Higher GR activity resulted in all cultivars, except for Mitchell and Nazlıcan at 35°C, compared with 25°C. The APX activity enhanced in Mitchell, Nazlıcan, SA 88 cultivars at 35°C, compared with 25°C (Çiçek and Çakırlar, 2008). Biochemical changes during seed aging are important for maintaining seed quality and seed longevity.

Soybean seeds were exposed to 42°C and relative humidity of 100%. Decrease in SOD and POX activities was significant on treatment involving applied accelerated aging at 42°C and relative humidity of 100%, which was associated with enhanced lipid peroxidation, and loss of seed vigor and seed viability (Balešević-Tubić et al., 2011).

If the production of proline synthesis is genetically modified, the production of Pro level and other antioxidants can also be affected. The wild-type soybean and transgenic soybean (G. max cv. Ibis) containing P5CR (antisense and sense transformants) were subjected to drought and HT (35/25°C day/night temperature) conditions. After the period of recovery, the antisense had the highest H_2O_2 and lipid hydroperoxide concentrations and showed the greatest HT injury.

In contrast, sense transformants showed the lowest H_2O_2 content and the lowest injury percentage. Transgenic soybean showed higher endogenous Pro and AsA levels under HT (35/25°C) compared with optimum temperature (25/15°C). This transgenic cultivar also regulated the GSH and hGSH contents, AsA/DHA, and (h)GSH/(h)GSSG ratios, and APX and glutathione GR activity. The transgenic soybean thus showed reduced oxidative damage and HT injury (Kocsy et al., 2005).

The activity of soybean seed CAT, as one of the important antioxidant enzymes, increased at the early HT (40/30°C, 100/70% relative humidity under light/dark) stress stage (0–24 h), compared with the control temperature (30/20°C, 70% RH, and 10/14 h under light/dark). After 24 h of HT stress, a sharp decrease in CAT activity was noticed over time, which continued up to 168 h of HT. Decrease pattern of CAT activity was correlated with increased oxidative damage to seed (Wang et al., 2012).

Seeds of four soybean cultivars were subjected to HT (41°C, 48 h) and humidity (artificial aging). The content of α-tocopherol in soybean seeds increased gradually with increasing storage time. The initial levels of α-tocopherol protect seeds from oxidative damage by reducing the levels of ROS and by inhibiting lipid peroxidation. Long-term exposure of HT resulted in degradation of α-tocopherol, which consequently increased lipid peroxidation and caused cell death (Giurizatto et al., 2012).

Effect of growth temperature on tocopherol composition in developing seeds of different genotypes was studied under different temperature regimes (15.5, 19.5, 23.5, and 27.5°C). Total tocopherol content increased gradually with increase in temperature from 15.5°C to 27.5°C. But α-tocopherol and δ-tocopherol decreased; whereas, γ content increased with the increase of the temperature in a similar pattern (Almonor et al., 1998).

Chennupati et al. (2011) studied HT-induced modulation of tocopherol concentrations in soybean plant. Tocopherol content increased greatly at the seed formation stage, compared with the other stages, resulting in a 752% increase of α-tocopherol (Chennupati et al., 2011). Increases in α-tocopherol and corresponding decreases in δ-tocopherol and γ-tocopherol were documented in soybean plant when the temperature was increased from 23°C to 28°C (Britz and Kremer, 2002). In other study soybean seeds were subjected to accelerated aging stress by exposure to 40°C and 100% relative humidity. Tocopherol and organic free radical contents were measured. Slight or no significant increase in tocopherol homologs (α, γ, δ) were noted, which was correlated with the relatively stable organic contents of free radicals (Priestley et al., 1980).

Phytohormones

Roles of plant growth regulators under HT have been little studied but some existing research has explored the antagonistic effect of plant growth regulators under HT.

Ethylene is considered a stress hormone produced in stress conditions the overproduction of which results in the early senescence and abscission of vegetative and reproductive organs. Effects of HT (38/28°C) stress during flowering of soybean plants were examined. HT stress significantly increased ethylene production. HT caused premature leaf senescence, increased flower abscission, and decreased pod-set percentage.

HT-affected plants were also characterized by increase of oxidative damage and decrease of antioxidant enzyme activity. But application of 1-MCP reduced ethylene and ROS production, enhanced antioxidant enzyme activity, increased the stability of membrane, delayed senescence of leaf, decreased abscission of flower, and increased percentage of pod set (Djanaguiraman et al., 2011a). According to Tan et al. (1988) 1-amino-cyclopropane-1-carboxylic acid (ACC, a precursor of ethylene biosynthesis) accumulated in soybean at 40°C. With increasing temperature up to 40°C, ethylene production in soybean hypocotyls increased through conversion of ACC into ethylene. Ethylene production inhibited the growth of soybean hypocotyls at 45°C (Tan et al., 1988).

Djanaguiraman and Prasad (2010) showed that the increased rate of ethylene production under HT was the main reason for the premature senescence of leaf, whereas 1-MCP application reduced the premature senescence of leaf by inhibiting ethylene production. Application of 1-MCP also restored the antioxidant defense system by enhancing the activities of antioxidant enzymes, which reduced oxidative damage. Foliar spray of 1-MCP also increased the percent of seed set and seed size of HT-affected soybean plant, which resulted in a higher yield, compared with control without spraying 1-MCP (Djanaguiraman and Prasad, 2010).

Soybean seedlings treated with salicylic acid improved seedling emergence and vigor, evaluated under HT of 41°C (48 h) (Brunes et al., 2014). Advantageous roles of $CaCl_2$ in HT-affected soybean plants were reported by Amooaghaie and Moghym (2011) who reported that exposure of soybean to 45°C (2 h) resulted in reduced primary root length and hypocotyl length and membrane damage. Compared with the heat-shock treatment, $CaCl_2$ application with HT treatment reduced the leakage of electrolyte and peroxidation of lipid from the sections of root and hypocotyl tissue. Again, in contrast to $CaCl_2$ application, application of EGTA (a chelator of calcium), resulted in stress injury and growth inhibition, and increased electrolyte leakage and MDA content of roots. These results suggests the protective role of $CaCl_2$ in recovery from HT injury (Amooaghaie and Moghym, 2011).

Heat-induced proteins

Specific heat-shock proteins (HSPs) have been identified in different plant species and their roles have proved HT tolerance in plants. Induction of HSPs may occur if plants experience a sudden or gradual rise in temperature (Nakamoto and Hiyama, 1999; Schöffl et al., 1999). Plants of arid and semiarid regions are often documented to synthesize and accumulate sizeable contents of HSPs (Hopf et al., 1992). Major HSPs are very much homologous among distinct organisms.

In higher plants, HSPs can be synthesized at any stage of development on exposure to HT (Vierling, 1991). There are several examples of HSP induction in soybean. The expression of HSP68 localizing in mitochondria increases under HT in stressed cells of soybean (Neumann et al., 1993). Soybean (G. max var. Wayne) seedlings showed dramatic change in protein synthesis when they were transferred from 28°C up to 40°C. This heat shock significantly reduced the synthesis of normal protein and generated a new set of proteins known as HSPs, which enhance thermotolerance.

The synthesis of HSPs also resulted by the exposure of seedlings to 45°C for 10 min followed by incubation at 28°C. However, prolonged exposure to 45°C for

1–2 h greatly impaired protein synthesis and resulted in the malfunctioning of seedlings. A pretreatment at 40°C for 2 h or a brief pulse treatment at 45°C for 10 min followed by a 28°C incubation provided thermal tolerance to a subsequent exposure at 45°C (Lin et al., 1984).

Lee et al. (1994) reported that the mRNA for HSP101 was not detected in soybean seedlings grown at 28°C. At elevated temperature levels, HSP101 was induced in soybean seedlings, which has roles in acquiring thermotolerance. The chaperones of Hsp100/Clp family can be developmentally expressed in plants. The expression of this family of chaperons is often highly induced by different environmental stresses including= HT stress. Hsp100/Clp proteins were induced in soybean and other plants such as *Arabidopsis*, tobacco, rice, maize, wheat (Adam et al., 2001; Agarwal et al., 2001; Keeler et al., 2000).

Without HSP, induction of heat-inducible genes may be attributed to heat-shock elements (HSEs). HSEs have been identified as locating in the TATA box proximal 5′ flanking regions of heat shock genes (Schöffl et al., 1999). Through deletion analysis of soybean heat-shock genes in sunflower, the TATA box was demonstrated (Czarnecka et al., 1989). Expression of other protein, termed ubiquitin, may be stimulated upon HT stress (Sun and Callis, 1997).

In soybean, biosynthesis of ubiquitin and conjugated ubiquitin for the first 30 min of exposure to HT has been reported as the most important mechanism of HT tolerance (Ortiz and Cardemil, 2001). Absence of heat-shock transcription factor (HSF) A1B may render soybean more sensitive to HT. In contrast, another class A HSF may enhance HT tolerance (Kotak et al., 2004; Li et al., 2004; Sung et al., 2003). Plant HSFs may act as H_2O_2 sensors and they have roles in redox regulation. Under prolonged HT or heat shock and conditions of recovery, HSFA2 controls the expression of characters.

HSFA2 expression can result from high luminosity and exposure to H_2O_2, which are important for stress-induced responses. In soybean the characteristics of HSFA4A and HSFA8 were fairly similar and these may act as sensors of ROS (Miller and Mittler, 2006). The HSFs also mediate cross talk between signaling cascades in soybean, which can be a vital issue for heat shock and development of soybean genotypes sustaining tolerance to HT and other abiotic stresses (Soares-Cavalcanti et al., 2012).

Conclusions and future perspectives

One of the predicted effects of global warming is a negative effect of HT on plants. Different plant studies have already provided the notion that even a couple of degrees of HT above the optimum may result in serious consequences for plant growth, metabolism, and productivity. Unlike other abiotic stresses, developing thermotolerance in plants is difficult due to their complex nature. Although numerous research results have been published on thermotolerance of plants, a clear-cut mechanism of thermotolerance has still to be elucidated. In spite of soybean being an HT-sensitive plant, relatively few studies have been conducted to cover the complex issue of HT stress and thermotolerance. Most of the existing studies focused on the responses of soybean to HT stress and approaches to thermotolerance have only sporadically been addressed. Therefore, the future outlook toward the development of soybean varieties which are HT-stress tolerant may include the exploitation of physiological and molecular mechanisms of heat stress tolerance and finding a way to tailor tolerant traits by using genetic engineering. Agronomic practices to avoid heat stress episodes may be explored to grow soybean in tropical areas where HT stress is common in summer. Exogenous protectants such as osmoregulators, antioxidants, phytohormones, and signaling molecules, which can positively alter the physiology of plants, must be explored so that they can be exploited in mitigating HT stress in soybean.

References

Adam, Z., Adamska, I., Nakabayashi, K., Ostersetzer, O., Haussuhl, K., Manuell, A., Zheng, B., Vallon, O., Rodermel, S.R., Shinozaki, K., Clarke, A.K., 2001. Chloroplast and mitochondrial proteases in *Arabidopsis*. A proposed nomenclature. Plant Physiol. 125, 1912–1918.

Agarwal, M., Katiyar-Agarwal, S., Sahi, C., Gallie, D.R., Grover, A., 2001. *Arabidopsis thaliana* Hsp100 proteins: kith and kin. Cell Stress Chaperones 6, 219–224.

Almonor, G.O., Fenner, G.P., Wilson, R.F., 1998. Temperature effects on tocopherol composition in soybeans with genetically improved oil quality. J. Am. Oil Chem. Soc. 75, 591–596.

Amooaghaie, R., Moghym, S., 2011. Effect of polyamines on thermotolerance and membrane stability of soybean seedling. Afr. J. Biotechnol. 10, 9673–9679.

Anto, K.B., Jayaram, K.M., 2010. Effect of temperature treatment on seed water content and viability of green pea (*Pisum sativum* L.) and soybean (*Glycine max* L. Merr.) seeds. Int. J. Bot. 6, 122–126.

Baker, J.T., Allen, L.H., Boote, K.J., 1989. Response of soybean to air temperature and carbon dioxide concentration. Crop Sci. 29, 98–105.

Baleševic̄-Tubic̄, S., Tatic̄, M., Dordevic̄, V., Nikolic̄, Z., Subic̄, J., Dukic̄, V., 2011. Changes in soybean seeds as affected by accelerated and natural aging. Rom. Biotechnol. Lett. 16, 6740–6747.

Boote, K.J., Jones, J.W., Hoogenboom, G., 1998. Simulation of crop growth: CROPGRO model. In: Peart, R.M., Curry, R.B. (Eds.), Agricultural Systems Modeling and Simulation. Marcel Dekker, New York, pp. 651–692, Chapter 18.

Boote, K.J., Allen, L.H., Prasad, P.V.V., Baker, J.T., Gesch, R.W., Snyder, A.M., Pan, D., Thomas, J.M.G., 2005. Elevated temperature and CO_2 impacts on pollination, reproductive growth, and yield of several globally important crops. J. Agric. Meteorol. 60, 469–474.

Britz, S.J., Kremer, D.F., 2002. Warm temperatures or drought during seed maturation increase free α-tocopherol in seeds of soybean (*Glycine max* [L.] Merr.). J. Agric. Food Chem. 50, 6058–6063.

Brunes, A.P., Lemes, E.S., Dias, L.W., Gehling, V.M., Villela, F.A., 2014. Performance of soybean seeds treated with salicylic acid doses. Centro Científico Conhecer Goiânia 10, 1467–1474.

Chennupati, P., Seguin, P., Liu, W., 2011. Effects of high temperature stress at different development stages on soybean isoflavone and tocopherol concentrations. J. Agric. Food Chem. 59, 13081–13088.

Çiçek, N., Çakırlar, H., 2008. Changes in some antioxidant enzyme activities in six soybean cultivars in response to long-term salinity at two different temperatures. Gen. Appl. Plant Physiol. 34, 267–280.

Covell, S., Ellis, R.H., Roberts, E.H., Summerfield, R.J., 1986. The influence of temperature on seed germination rate in grain legumes. I. A comparison of chickpea, lentil, soyabean and cowpea at constant temperatures. J. Exp. Bot. 37, 705–715.

Czarnecka, E., Key, J.L., Gurley, W.B., 1989. Regulatory domains of the Gmhsp 17.5-E heat shock promoter of soybean: a mutational analysis. Mol. Cell. Biol. 9, 3457–3463.

D'Souza, M.R., 2013. Effect of traditional processing methods on nutritional quality of field bean. Adv. Biores. 4, 29–33.

de Ronde, J.A., Spreeth, M.H., Cres's, W.A., 2000. Effect of antisense 1-D1-pyrroline-5-carboxylate reductase transgenic soybean plants subjected to osmotic and drought stress. Plant Growth Regul. 32, 13–26.

de Ronde, J.A., Cress, W.A., Van Staden, J., 2001. Interaction of osmotic and temperature stress on transgenic soybean. S. Afr. J. Bot. 67, 655–660.

de Ronde, J.A., Cress, W.A., Krüger, G.H.J., Strasser, R.J., Van Staden, J., 2004a. Photosynthetic response of transgenic soybean plants, containing an *Arabidopsis P5CR* gene, during heat and drought stress. J. Plant Physiol. 161, 1211–1224.

de Ronde, J.A., Laurie, R.N., Caetano, T., Gray Ling, M.M., Kerepesi, I., 2004b. Comparative study between transgenic and non-transgenic soybean lines proved transgenic lines to be more drought tolerant. Euphytica 138, 123–132.

Djanaguiraman, M., Prasad, P.V.V., 2010. Ethylene production under high temperature stress causes premature leaf senescence in soybean. Funct. Plant Biol. 37, 1071–1084.

Djanaguiraman, M., Prasad, P.V.V., Al-Khatib, K., 2011a. Ethylene perception inhibitor 1-MCP decreases oxidative damage of leaves through enhanced antioxidant defense mechanisms in soybean plants grown under high temperature stress. Environ. Exp. Bot. 71, 215–223.

Djanaguiraman, M., Prasad, P.V.V., Boyle, D.L., Schapaugh, W.T., 2011b. High temperature stress and soybean leaves: leaf anatomy and photosynthesis. Crop Sci. 51, 2125–2131.

Djanaguiraman, M., Prasad, P.V.V., Schapaugh, W.T., 2013. High day- or nighttime temperature alters leaf assimilation, reproductive success, and phosphatidic acid of pollen grain in soybean [*Glycine max* (L.) Merr.]. Crop Sci. 53, 1594–1604.

Dornbos, D.L., Mullen, R.E., 1991. Influence of stress during soybean seed fill on seed weight, germination, and seedling growth rate. J. Plant Sci. 71, 373–383.

Edwards, T.I., 1934. Relations of germinating soybeans to temperature and length of incubation time. Plant Physiol. 9, 1–30.

Egli, D.B., Bruening, W.P., 2006. Fruit development and reproductive survival in soybean: position and age effects. Field Crops Res. 98, 195–202.

Egli, D.B., Wardlaw, I.F., 1980. Temperature response of seed growth characteristics of soybeans. Agron. J. 72, 560–564.

FAOSTAT, 2010. http://faostat.fao.org/default.aspx.

Ferris, R., Wheeler, T.R., Ellis, R.H., Hadley, P., 1999. Seed yield after environmental stress in soybean grown under elevated CO_2. Crop Sci. 39, 710–718.

Gibson, L.R., Mullen, R.E., 1996. Influence of day and night temperature on soybean seed yield. Crop Sci. 36, 98–104.

Giurizatto, M.I.K., Ferrarese-Filho, O., Ferrarese, M.D.L., Robaina, A.D., Gonçalves, M.C., Cardoso, C.A.L., 2012. α-Tocopherol levels in natural and artificial aging of soybean seeds. Acta Scient. Agron. 34, 339–343.

Hatfield, J., Boote, K., Fay, P., Hahn, L., Izaurralde, C., Kimball, B.A., Mader, T., Morgan, J., Ort, D., Polley, W., Thomson, A., Wolf, D., 2008. Agriculture. In: The effects of climate change on agriculture, land resources, water resources, and biodiversity in the United States. A Report by the U.S. Climate Change Science Program and the Subcommitee on Global Change Research. Washington, DC, USA, pp. 362.

Hesketh, J.D., Myhre, D.L., Willey, C.R., 1973. Temperature control of time intervals between vegetative and reproductive events in soybeans. Crop Sci. 13, 250–254.

Hopf, N., Plesofskv-Vig, N., Brambl, R., 1992. The heat response of pollen and other tissues of maize. Plant Mol. Biol. 19, 623–630.

Huxley, P.A., Summerfied, R.J., Hughes, P., 1976. Growth and development of soybean CV-TK5 as affected by tropical day lengths, day/night temperatures and nitrogen nutrition. Ann. Appl. Biol. 82, 117–133.

IPCC 2007, Climate Change. The physical science basis. Summary for Policymakers. Contribution of Working Group I to the Fourth Assessment Report of the Intergovernmental Panel on Climate Change.

Keeler, S.J., Boettger, C.M., Haynes, J.G., Kuches, K.A., Johnson, M.M., Thureen, D.L., Keeler, Jr, C.L., Kitto, S.L., 2000. Acquired thermotolerance and expression of the HSP100/ ClpB genes of lima bean. Plant Physiol. 123, 1121–1132.

Keerio, M.I., Chang, S.Y., Mirjat, M.A., Lakho, M.H., Bhatti, I.P., 2001. The rate of nitrogen fixation in soybean root nodules after heat stress and recovery period. Int. J. Agric. Biol. 3, 512–514.

Keyser, H.H., Li, F., 1992. Potential for increasing biological nitrogen fixation in soybean. Dev. Plant Soil Sci. 141, 119–135.

Khedar, O.P., Singh, R.V., Shrimali, M., Singh, N.P., 2008. Pulses: Status and Cultivation Technology. Aavishkar Publishers, Jaipur.

Kitano, M., Saitoh, K., Kuroda, T., 2006. Effect of high temperature on flowering and pod set in soybean. Scientific Report of the Faculty of Agriculture, Okayama University, vol. 95, pp. 49–55.

Kocsy, G., Laurie, R., Szalai, G., Szilágyi, V., Simon-Sarkadi, L., Galiba, G., de Ronde, J.A., 2005. Genetic manipulation of proline levels affects antioxidants in soybean subjected to simultaneous drought and heat stresses. Physiol. Plant. 124, 227–235.

Kotak, S., Port, M., Ganguli, A., Bicker, F., von Koskull-Doring, P., 2004. Characterization of C-terminal domains of *Arabidopsis* heat stress transcription factors (Hsfs) and identification of a new signature combination of plant class a Hsfs with AHA and NES motifs essential for activator function and intracellular localization. Plant J. 39, 98–112.

Koti, S., Reddy, K.R., Reddy, V.R., Kakani, V.G., Zhao, D., 2005. Interactive effects of carbon dioxide, temperature, and ultraviolet-B radiation on soybean (*Glycine max* L.) flower and pollen morphology, pollen production, germination, and tube lengths. J. Exp. Bot. 56, 725–736.

Koti, S.K., Reddya, V.R., Kakani, G., Zhaob, A.D., Gaoc, W., 2007. Effects of carbon dioxide, temperature and ultraviolet-B radiation and their interactions on soybean (*Glycine max* L.) growth and development. Environ. Exp. Bot. 60, 1–10.

Kubecka, P., 2001. A possible world record maximum natural ground surface temperature. Weather 56, 218–221.

Kucharik, C.J., Serbin, S.P., 2008. Impacts of recent climate change on wisconsin corn and soybean yield trends. Environ. Res. Lett. 3, 034003.

Kumar, A., Pandey, V., Shekh, A.M., Kumar, M., 2008. Growth and yield response of soybean (*Glycine max* L.) in relation to temperature, photoperiod and sunshine duration at Anand, Gujarat, India. Am. Eur. J. Agron. 1, 45–50.

Lee, Y.R.J., Nagao, R.T., Key, J.L., 1994. A soybean 101-kD heat shock protein complements a yeast HSP 104 deletion mutant in acquiring thermotolerance. Plant Cell 6, 1889–1897.

Li, H.-Y., Chang, C.-S., Lu, L.-S., Liu, C.-A., Chan, M.-T., Charng, Y.-Y., 2004. Over-expression of *Arabidopsis thaliana* heat shock factor gene (*AtHsfA1b*) enhances chilling tolerance in transgenic tomato. Bot. Bull. Acad. Sin. 44, 129–140.

Li, P., Cheng, L., Gao, H., Jiang, C., Peng, T., 2009. Heterogeneous behavior of PSII in soybean (*Glycine max*) leaves with identical PSII photochemistry efficiency under different high temperature treatments. J. Plant Physiol. 166, 1607–1615.

Lin, C.Y., Roberts, J.K., Key, J.L., 1984. Acquisition of thermotolerance in soybean seedlings. Plant Physiol. 74, 152–160.

Lindey, L., Thomson, P., 2012. High temperature effects on corn and soybean. Crop Observ. Recommend. Netw. Newsl. 2012, 23–26.

Liu, X., Jian, J., Guanghua, W., Herbert, S.J., 2008. Soybean yield physiology and development of high-yielding practices in Northeast China. Field Crops Res. 105, 157–171.

Lobell, D.B., Asner, G.P., 2003. Climate and management contributions to recent trends in U.S. agricultural yields. Science 299, 1032.

Lui, K., 2004. Soybeans as a powerhouse of nutrients and phytochemicals. In: Lui, K. (Ed.), Soybeans as Functional Foods and Ingredients. AOCS Press, Champaign, IL, pp. 1–53.

Martineau, J.R., Specht, J.E., Williams, J.H., Sullivan, C.Y., 1979. Temperature tolerance in soybeans. I. Evaluation of a technique for assessing cellular membrane thermostability. Crop Sci. 19, 75–78.

Mejia, R.P., 1999. Effects of stress temperatures of germination on polyamine titers of soybean seeds. A dissertation submitted to the graduate faculty in partial fulfillment of the requirements for the degree of Doctor of Philosophy. Iowa State University, Ames, Iowa.

Mildrexler, D.J., Zhao, M., Running, S.W., 2011. Satellite finds highest land skin temperatures on earth. Bull. Am. Meteorol. Soc. 92, 855–860.

Miller, G., Mittler, R., 2006. Could heat shock transcription factors function as hydrogen peroxide sensors in plant? Ann. Bot. 98, 279–288.

Mittler, R., 2002. Oxidative stress, antioxidants and stress tolerance. Trends Plant Sci. 7, 405–410.

Moghym, S., Amooaghaie, R., Shareghi, B., 2010. Protective role of polyamines against heat shock in soybean seedlings. Iran. J. Plant Biol. 2, 31–39.

Munevar, F., Wollum, A.G., 1982. Response of soybean plants to high root temperature as affected by plant cultivar and *Rhizobium* strain. Agron. J. 74, 138–142.

Nakamoto, H., Hiyama, T., 1999. Heat-shock proteins and temperature stress. In: Pessarakli, M. (Ed.), Handbook of Plant and Crop Stress. Marcel Dekker, New York, pp. 399–416.

Nakamoto, H., Zheng, S., Furuya, T., Tanaka, K., Yamazaki, A., Fukuyama, M., 2001. Effects of long-term exposure to atmospheric carbon dioxide enrichment on flowering and poding in soybean. J. Fac. Agric. Kyushu Univ. 46, 23–29.

Ndunguru, B.J., Sununerfield, R.J., 1975. Comparative laboratory studies of cowpea *(Vigna unguiculata)* and soybean *(Glycine max)* under tropical temperature conditions. I. Germination and hypocotyl conditions. East Afr. Agric. Forest. J. 41, 58–64.

Neumann, D.M., Emmermann, M., Thierfelder, J.M., Zur Nieden, U., Clericus, M., Braun, H.P., Nover, L., Schmitz, U.K., 1993. HSP68 – a DNAK-like heat-stress protein of plant mitochondria. Planta 190, 32–43.

Nicholson, S.E., 2011. Dryland Climatology. Cambridge University Press, Cambridge, p. 158.

Ortiz, C., Cardemil, L., 2001. Heat-shock responses in two leguminous plants: a comparative study. J. Exp. Bot. 52, 1711–1719.

Ozturk, M., Hakeem, K.R., Faridah-Hanum, I., Efe, R. (Eds.), 2015. Climate Change Impacts on High-Altitude Ecosystems. Springer Science + Business Media, New York, 736 pp.

Pan, D., 1996. Soybean responses to elevated temperature and doubled CO_2. PhD dissertation. University of Florida, Gainesville, 227 pp.

Pestana-Calsa, M.C., Pacheco, C.M., de Castro, R.C., de Almeida, R.R., de Lira, N.P., Junior, T.C., 2012. Cell wall, lignin and fatty acid-related transcriptome in soybean: achieving gene expression patterns for bioenergy legume. Genet. Mol. Biol. 35, 322–330.

Prasad, P.V.V., Craufurd, P.Q., Summerfield, R.J., 2000. Effect of high air and soil temperature on dry matter production, pod yield and yield components of groundnut. Plant Soil 222, 231–239.

Prasad, P.V.V., Craufurd, P.Q., Kakani, V.G., 2001. Influence of high temperature during pre- and post-anthesis stages of floral development on fruit-set and pollen germination in peanut. Aust. J. Plant Physiol. 28, 233–240.

Prasad, P.V.V., Boote, K., Allen, L.H., Jean, M.G., 2002. Effects of elevated temperature and carbon dioxide on seed-set and yield of kidney bean. Global Change Biol. 8, 710–721.

Prasad, P.V.V., Boote, K.J., Allen, L.H., Sheehy, J.E., Thomas, J.M.G., 2006. Species, ecotype and cultivar differences in spikelet fertility and harvest index of rice in response to high temperature stress. Field Crops Res. 95, 398–411.

Prasad, P.V.V., Pisipati, S.R., Mutava, R.N., Tuinstra, M.R., 2008. Sensitivity of grain sorghum to high temperature stress during reproductive development. Crop Sci. 48, 1911–1917.

Priestley, D.A., Mcbride, M.B., Leopold, C., 1980. Tocopherol and organic free radical levels in soybean seeds during natural and accelerated aging. Plant Physiol. 66, 715–719.

Puteh, A.B., ThuZar, M., Mondal, M.M.A., Abdullah, N.A.P.B., Halim, M.R.A., 2013. Soybean [Glycine max (L.) Merrill] seed yield response to high temperature stress during reproductive growth stages. Aust. J. Crop Sci. 7, 1472–1479.

Salem, M.S., Koti, S., Kakani, V.G., Reddy, K.R., 2005. Glycinebetaine and its role in heat tolerance as studied by pollen germination techniques in soybean genotypes. The ASA-CSSA-SSSA International Annual Meetings, November 6–10, 2005, Salt Lake City, Utah, USA.

Salem, M.A., Kakani, V.G., Koti, S., Reddy, K.R., 2007. Pollen-based screening of soybean genotypes for high temperature. Crop Sci. 47, 219–231.

Salvagiotti, F., Cassman, K.G., Specht, J.E., Walters, D.T., Weiss, A., Dobermann, A., 2008. Nitrogen uptake, fixation and response to fertilizer N in soybeans: a review. Field Crops Res 108, 1–13.

Sapra, V.T., Anaek, A.O., 1991. Screening soya bean genotypes for drought and heat tolerance. J. Agron. Crop Sci. 167, 96–102.

Schlenker, W., Roberts, M.J., 2009. Nonlinear temperature effects indicate severe damages to U.S. crop yields under climate change. Proc. Natl Acad. Sci. USA 106, 15594–15598.

Schöffl, F., Prandl, R., Reindl, A., 1999. Molecular responses to heat stress. In: Shinozaki, K., Yamaguchi-Shinozaki, K. (Eds.), Molecular Responses to Cold, Drought, Heat and Salt Stress in Higher Plants. R.G. Landes Co., Austin, TX, pp. 81–98.

Seddigh, M., Jolliff, G.D., 1984. Night temperature effects on morphology, phenology, yield and yield components of indeterminate field-grown soybean. Agron. J. 76, 824–828.

Simon-Sarkadi, L., Kocsy, G., Várhegyi, A., Galiba, G., de Ronde, J.A., 2005. Genetic manipulation of proline accumulation influences the concentrations of other aminoacids in soybean subjected to simultaneous drought and heat stress. J. Agric. Food Chem. 53, 7512–7517.

Sinclair, T.R., Weisz, P.R., 1985. Response to soil temperature of dinitrogen fixation (acetylene reduction) rates by field grown soybean. Agron. J. 77, 685–688.

Sionit, N., Strain, B.R., Flint, E.P., 1987. Interaction of temperature and CO_2 enrichment on soybean: growth and dry matter partitioning. Can. J. Plant Sci. 67, 59–67.

Snider, J.L., Oosterhuis, D.M., Skulman, B.W., Kawakami, E.M., 2009. Heat-induced limitations to reproductive success in Gossypium hirsutum. Physiol. Plant. 137, 125–138.

Soares-Cavalcanti, N.M., Belarmino, L.C., Kido, E.A., Pandolfi, V., Marcelino-Guimarães, F.C., Rodrigues, F.A., Pereira, G.A.G., Benko-Iseppon, A.M., 2012. Overall picture of expressed heat shock factors in Glycine max, Lotus japonicus and Medicago truncatula. Genet. Mol. Biol. 35, 247–259.

Sun, C.W., Callis, J., 1997. Independent modulation of *Arabidopsis thaliana* polyubiquitin mRNAs in different organs of and in response to environmental changes. Plant J. 11, 1017–1027.

Sung, D.Y., Kaplan, F., Lee, K.-J., Guy, C., 2003. Acquired tolerance to temperature extremes. Trends Plant Sci. 8, 179–187.

Tan, C., Yu, Z.W., Yang, H.D., Yu, S.W., 1988. Effect of high temperature on ethylene production in two plant tissues. Acta Phytophysiol. Sin. 14, 373–379.

Tan, W., Meng, Q.W., Brestic, M., Olsovska, K., Yang, X., 2011. Photosynthesis is improved by exogenous calcium in heat-stressed tobacco plants. J. Plant Physiol. 168, 2063–2071.

Thomas, J.M.G., 2001. Impact of elevated temperature and carbon dioxide on development and composition of soybean seed. PhD Dissertation. University of Florida. Gainesville, Florida, USA, 185 pp.

Tian, X., Song, S., Lei, Y., 2008. Cell death and reactive oxygen species metabolism during accelerated ageing of soybean axes. Russ. J. Plant Physiol. 55, 33–40.

Vierling, E., 1991. The role of heat shock proteins in plants. Annu. Rev. Plant Physiol. Plant Mol. Biol. 42, 579–620.

Wang, L., Ma, H., Song, L., Shu, Y., Gu, W., 2012. Comparative proteomics analysis reveals the mechanism of pre-harvest seed deterioration of soybean under high temperature and humidity stress. J. Proteom. 75, 2109–2127.

Waughman, G.J., 1977. The effect of temperature on nitrogenase activity. J. Exp. Bot. 28, 949–960.

Yin, H., Chen, Q.M., Yi, M., 2008. Effects of short-term heat stress on oxidative damage and responses of antioxidant system in *Lilium longiflorum*. Plant Growth Regul. 54, 45–54.

Zheng, S., Nakamoto, H., Yoshikawa, K., Furuya, T., Fukuyama, M., 2002. Influences of high night temperature on flowering and pod setting in soybean. Plant Prod. Sci. 5, 215–218.

Strategies, challenges, and future perspectives for soybean production under stress

Mohammad Miransari

Department of Book & Article, AbtinBerkeh Ltd. Company, Isfahan, Iran

Introduction

It is important to plan research and development for the next 50 years so that it will be likely to provide the world's ever-increasing population with their essential food. Although the growth of science has been astonishingly rapid resulting in a great contribution to the improved life of human beings, some challenges may lessen the rate of progress. Research should look for techniques that are able to enhance human health and the environment, while contributing to increased food production.

Feeding the world is among the most important goals of research and development activities as the world population will be equal to 9.2 billion by the year 2050. Accordingly, the production of food must increase by 50–70% to provide people with their essential food (FAO, 2009). However, it is a difficult task as the resources of soil and water are decreasing. Accordingly, by the use of research the most efficient techniques for the production of crop must be determined and used so that plants can produce suitable amounts of yield under different conditions including stress (Miransari, 2014a, b).

A useful method to investigate the effects of different parameters on the production of crop plants, cycling of nutrients, and the environmental effects of agriculture is the use of long-term data of more than 20 years. Such data are suitable for evaluating the biogeochemical, biological, and environmental aspects of agriculture, which enable future predictions of the globe and validation of the related models likely (Rasmussen et al., 1998).

Climate change is also another problem, which may adversely affect the production of food, worldwide. Although the use of techniques which are suitable for the alleviation of stresses caused by climate change, including the reduction of greenhouse gases (Table 13.1), may be expensive, such techniques are essential. Such problems will also result in an increased rate of stresses, worldwide (Craine et al., 2012; Mir et al., 2012).

Accordingly, the most important research areas regarding soybean must address the following: (1) alleviation of stresses on the growth and yield of soybean; (2) improving the quality of yield including seed fortification; (3) enhancing plant efficiency in utilization of resources such as water, nutrients, etc.; (4) efficient use of biological fertilization including inocula and other sources of biological fertilization; and

Abiotic and Biotic Stresses in Soybean Production. http://dx.doi.org/10.1016/B978-0-12-801536-0.00013-X

Table 13.1 **Some examples related to the emission of N_2O from grass pastures and crops fertilized with N, legumes in the field, and unfertilized fields of Europe, North and South America, South and East Asia, New Zealand, and Australia**

Conditions and species	Number of site years	Total amount of N_2O emitted in the season or in the year (kg N_2O–N h^{-1})	
		Range	Mean
Legume			
White clover	3	0.50–0.90	0.79
Alfalfa	14	0.67–4.57	1.99
Mixed pasture*			
Grass clover	8	0.10–1.30	0.54
Legume crops			
Lupin	1	–	0.05
Faba bean	1	–	0.41
Soybean	33	0.29–7.09	1.58
Chickpea	5	0.03–0.16	0.06
Field pea	6	0.38–1.73	0.65
Means of all legumes	71		1.29
Pasture fertilized with N**			
Grass	19	0.3–8.16	4.49
Crops fertilized with N			
Maize	22	0.16–12.67	2.72
Wheat	18	0.09–8.57	2.73
Canola	8	0.13–8.60	2.65
[†]Mean of all N fertilized fields	67		3.22
Fields without N fertilization of legumes	33	0.03–4.80	1.20

* N fertilizer was not used or legumes were fertilized with just 5 kg fertilizer N ha^{-1} as starter N at seeding with some exceptions.
** N fertilizer was used.
[†] Unplanted fields or nonlegume species where N fertilizer was not used.
From Aulakh et al. (2001), Chen et al. (2008), Hénault et al. (1998), Jensen et al. (2012), Mahmood et al. (1998), Ruz-Jerez et al. (1994) .

(5) organic production of soybean for its healthy production (de Ponti et al., 2012; Galili and Amir, 2013; Lavania et al., 2015; Miransari, 2011a; Miransari et al., 2013; Naveed et al., 2015).

Improving plant growth and crop production under different conditions including stress is among the most important research subjects. In the past 20 years use of plant biotechnology has been an effective method for the modern production of crop yield. The commercialization of genetically modified food started in 1996 and, ever since, high amounts of biotech crop has been produced globally, which in 2009 was equal to a coverage of 134 million hectares (Blumel et al., 2015; James, 2009). By 2008, this

Table 13.2 **The components (%) of some crop plants including soybean (Jensen et al., 2012)**

Crop and crop residue	Starch	Other carbohydrate	Fat	Protein	Lignin
Soybean	15	14	21	41	6
Maize	71	14	5	9	2
Wheat	66	17	3	13	2
Lupin	22	23	5	45	16
Pea	55	18	2	25	6
Faba bean	42	21	1	31	9
Pea residue	0	81	2	7	41
Wheat stubble	0	92	2	3	45
Alfalfa	2	72	3	20	31
Grass clover	2	62	4	22	20

increased by 9 million hectares indicating the significance of biotechnology and the related benefits for the production of food (James, 2009).

Soybean is the seventh most important food product after milk, rice, different types of meat, and wheat (FAOSTAT, 2014). Soybean is among the most important crop plants for the production of grain (40% protein) and oil (20%) (higher than other crop plants in the world, Table 13.2) and is able to feed a large number of people across the world. Soybean can establish a symbiotic association with the N-fixing bacteria, rhizobia, to obtain most of the nitrogen required for its growth and production. Photosynthesis is the most important biological process in nature followed by biological N fixation (Jensen et al., 2012; Lambin and Meyfroidt, 2011; Priester et al., 2012; Silvente et al., 2012).

Although there is a high level of N in the atmosphere, in most cases it is not available to the plant due to its chemical properties. Just a few prokaryotes are able to fix atmospheric N and some of them are also able to turn it into ammonia and make it available to their host plant. The process of N fixation is by the production of signals by the host plant roots, which attract the N-fixing bacteria toward the host plant roots. Accordingly, the bacterial *nod* genes are activated and in response, the bacteria are able to produce some biochemicals called nod factors inducing some morphological and physiological changes in the host plant roots (Esfahani and Mostajeran, 2011; Graham and Vance, 2000; Long, 2001; Wagner, 2011).

Soybean originated in China more than 5000 years ago and it is over 200 years since soybean was planted in North America, with the highest rate of yield production at the time (Qiu and Chang, 2010). Although soybean is mainly planted for its grain oil, it also contains some products with high values including saponins, oligosaccharides, isoflavonoids, phospholipids, and edible fiber. Soybean also contains the signaling molecules genistein and daidzein, which are important for the establishment of symbiosis between the plant and the bacteria, and can also attenuate osteoporosis; they also have some interesting effects on the health of people. Soybean can also be used for producing biodiesel (Du et al., 2003).

With respect to the above discussion and the importance of soybean as a food and oil crop, some of the most important considerations relating to the strategies and challenges related to the alleviation of soybean growth and yield production under stress are presented in greater detail below. Some conclusions and future perspectives are also offered which make the production of soybean more likely under stress conditions.

Soybean (*Glycine max* (L.) Merr.)

Soybean, an N fixer, has been used as a model plant for the evaluation of N-fixing process. It is able to establish symbiotic association with the N-fixing bacteria, rhizobia, and acquire most of its nitrogen for growth and production thereby. The efficiency of biological N fixation is a function of climatic factors such as heat and photoperiod. The efficiency of the process of N fixation is also determined by the interaction of the soybean genotype with the soil nutrients and moisture (Jung et al., 2008). There are also other factors affecting the efficiency of the process of N fixation, the most important of which are the properties of the bacterial inocula relating to the native soil bacteria. Accordingly, the most-efficient strains must be selected for the process of N fixation. For example, under stress conditions the strains that have been isolated from stress environments must be used (Abi-Ghanem et al., 2012; Jensen et al., 2012; Salvagiotti et al., 2008; Wang et al., 2012a).

In a symbiotic association, eventually root nodules are produced, which are the sita of rhizobial bacteria, specifically, *Bradyrhizobium japonicum*, for the fixation of atmospheric N. The bacteria are able to turn the atmospheric N into ammonia and hence make it available for the use of the host plant. A high rate of atmospheric N is fixed by this important process so providing the greater part of atmospheric N used by the host plant for growth and yield production (Manavalan et al., 2009; Miransari and Smith, 2007, 2008, 2009).

Due to the importance of biological N fixation, a significant amount of research has been conducted in this respect to increase soybean growth and yield under different conditions including stress. The process of biological N fixation by the host plant, soybean, and the activity of symbiotic the N-fixing bacterium, *B. japonicum*, has been investigated in great detail. Accordingly, appropriate strategies and techniques for the regulation and adjustment of the process of biological N fixation have been examined, approved, and used under different conditions including stress (Adesemoye and Klopper, 2009; Lindström et al., 2010; Miransari et al., 2013).

B. japonicum

B. japonicum are Gram-negative bacterium (rod shaped) that is able to develop a symbiotic association with its specific host plant, soybean, and fix atmospheric N. The strain that is widely used is USDA110, originally from Florida, USA. It is an important strain and is able to fix high amounts of N in association with its host plant, relative to other strains. The bacterial genome (circular) sequence has been identified, with a length of 9.11 bp, 8373 predicted gene and average GC content of 64.1% (Kaneko et al., 2002a, b; Li et al., 2011; Sachs et al., 2011).

The bacteria are attracted to the roots of the host plant after the production of signals by the host's roots and its activated proteins result in the production of lipochitooligosaccharide, which is able to initiate the formation of root nodules. The bacteria then reside in the nodules and initiate the fixation of atmospheric N, which becomes available to the host plant in the form of ammonia. In most cases *B. japonicum* is able to increase soybean seed yield. Carbohydrate for the use of *B. japonicum* is trehalose, which is converted to CO_2 (Müller et al., 2001).

Interestingly, and similarly, jasmonate can also induce morphological and physiological changes in plant roots (Mabood et al., 2006; Munoz et al., 2014; Wang et al., 2012b). The direct and enhancing effects of lipochitooligosaccharide on the growth and yield of legumes and nonleguminous plant have also been elucidated, initially by Denarie and Cullimore (1993). Such Nod factors can also have similar activities to auxin and cytokinin in promoting and developing the growth of embryo and plant.

Process of N fixation

As previously mentioned, during the process of N fixation between rhizobia and legumes atmospheric N is fixed by the bacteria and can be used as a useful source of N for the host plant. Legumes including soybean are able to acquire most of their N by the process of N fixation and hence the use of chemical fertilization can be significantly decreased, which is of environmental and economical significance. Large amounts of N are fixed by legumes worldwide (Kaneko et al., 2011; Wagner, 2011).

The process of N fixation, which is done by some prokaryotes including *Rhizobium* and *Bradyrhizobium*, was first discovered by Beijerinck (1901). By using the enzyme nitrogenase, the bacteria are able to reduce the atmospheric N to ammonia, which is easily used and assimilated by the host plants. Although N is the most dominant element, it can only be used by plants if reduced. For the process of N fixation the bacteria require 16 molecules of ATP, supplied by the oxidation of organic products for the reduction of each molecule of N. Symbiotic and nonsymbiotic N-fixing bacteria obtain such products from the rhizosphere of their host plant (National Research Council, 1994).

By the production of biochemicals such as flavonoids by the plant, the bacteria are attracted toward the host plant roots where they become attached to the root hairs. Such an attachment consists of a two-step process including attachment via a Ca^{2+} binding protein, rhicadhesin, followed by a firmer attachment of products by plant and bacteria, respectively, including lectins, cellulose firbrils, and fimbriae (Wagner, 2011).

The Nod factors, which are produced by the bacteria, are sensed by the host plant resulting in the alteration of root hairs and hence subsequent root hair bulging and curling. Subsequently, the bacteria enter the root hairs and form the infection thread (tubular structure). At this time the bacteria are able to induce cortical cell division resulting in the production of root nodules. In the meantime, a plant-derived membrane is formed around the bacteria, which are released into the nodules and bacteroids are produced. They (the bacteroids) are the changed form of rhizobia, where the cell wall is not present, with a high rate of morphological changes resulting in the production of branching cells with an irregular shape (Long, 2001; Miransari et al., 2013; Wagner, 2011).

Such interactions are so specific as just one strain of *Rhizobium* is able to nodulate its host plant, which is mostly determined by the production of organic molecules (flavonoids) such as genistein by the host plant and the Nod factors produced by rhizobia. For, example, *B. japonicum* is able to nodulate its specific host plant, soybean. The other important interactive process between the host plant and the rhizobia is the production and function of an organic molecule called leghemoglobin, which is similar to human hemoglobin. The molecule is able to bind to the oxygen present and hence adjust its level for the use of rhizobia. This is because the N-fixing enzyme, nitrogenase, is not able to act in the presence of high levels of oxygen in the cytoplasm of nodule cells. The heme protein is produced by the bacteria (producing the heme) and the legume host plant (producing the apoprotein) (Bergersen, 1971; Graham and Vance, 2003; Schubert, 1986; Sinclair and Goudriaan, 1981).

Soybean and stress

Research has indicated that although soybean is not resistant under stress conditions, it is possible to enhance its tolerance under stress using different strategies and techniques. This finding is accordingly likely to increase the production of soybean crops under stress in different parts of the world. Use of: (1) tolerant varieties; (2) molecular techniques; (3) biochemicals; (4) tolerant strains of rhizobia, etc., presented in greater detail in the following sections, are among the most important techniques and strategies that can improve soybean growth and production under different conditions including stress (Clement et al., 2008; Kohler et al., 2008; Silvente et al., 2012; Zhang et al., 2010; Zilli et al., 2008).

Using varieties that have been produced to tolerate the unfavorable effects of stress is one method that can alleviate stress. Tolerant plants are produced by using the stress genes that have been identified under stress. Such proteins are expressed under stress and make the plant resistant under such conditions. However, before producing genetically modified (GM) plants all the related aspects must be evaluated (Clement et al., 2008; Sinclair, 2011; You et al., 2011).

Molecular techniques related to the use of different molecular products that are usually present in the plant under stress, but not at high enough levels, can help the plant to survive stress if present in greater quantity. For example, it has been hypothesized, tested, and proved that if *B. japonicum* inocula are treated with flavonoids such as genistein before inoculation under the stress, they are able to mitigate the effects of the stress on the growth and yield of soybean (Miransari and Smith, 2007, 2008, 2009; Silvente et al., 2012).

Some biochemicals such as plant hormones, and hydrogen sulfide are able to enhance plant resistance under stress by adjusting the metabolic activities of the plant, for example by increasing the activity of antioxidant enzymes including superoxide dismutase (EC 1.1.5.1.1) and catalase (EC1.11.1.6). Under stress the increased production of such enzymes can enhance plant tolerance to stress by scavenging the production of reactive oxygen species (Liao et al., 2008; Sajedi et al., 2010, 2011; Zhang et al., 2010).

Production of tolerant rhizobial strains is also another method for the alleviation of stress on bacterial activities and plant growth and yield. Such strains are able to act more efficiently under stress and hence are able to establish a symbiotic association with their host plant. It has been indicated that if such strains are collected from stressed locations, they can be more effective under stress. It is also possible to produce strains that have been genetically modified and are able to be more responsive under stress (Alexandre and Oliveira, 2011; Gil-Quintana et al., 2013; Indrasumuna et al., 2011; Laranjo and Oliveira, 2011).

Alleviating strategies

It is important to use strategies that enable the plants to survive under different conditions including stress. The strategies must be selected with respect to conditions such as the properties of plant, climate, and soil. Among such strategies the most important ones are the use of: (1) naturally tolerant plants, (2) genetically modified plants, (3) water efficient plants, (4) balanced fertilization, (5) microbial inocula, (6) organic farming, and (7) organic products (Jensen et al., 2012; Silvente et al., 2012; Zhang et al., 2010; Zilli et al., 2008).

Some plants are resistant to stress as they are able to use mechanisms that enable them to survive by altering their morphological and physiological properties. Different strategies are used by plants to tolerate stress. For example, there are plants that can adjust the angle and the position of their leaf under stress so that less water is evaporated and the plant is able to use the water more efficiently (Nilsen and Orcutt, 1996).

In some plants, the growth of their roots is changed under stress and the roots are able to grow more deeply under drought stress and absorb more water. Some of them are able to establish a more extensive root network under stress and as a result due to the longer and greater number of root hairs the plant will be able to survive. Some plants can adjust the growth of their aerial parts via the physiological processes of their roots including the production of certain biochemicals (Nilsen and Orcutt, 1996).

It is also possible to render plants resistant to stress by genetically modifying them. Different techniques can be used to insert the related stress genes, which are expressed under stress and hence make the plant tolerant. The use of GM plants has resulted in a higher yield under stress with some favorable properties. However, it must be mentioned that the use of such plants must be beneficial to the health of humans without any side-effects (Sinclair, 2011).

Another important point about planting under stress is the use of water efficiently by plant under stresses such as drought and salinity. If the plants are able to use water more efficiently, they will be able to produce greater yields under stress and hence the unfavorable effects of stress will be alleviated. The use of suitable watering techniques is also of great importance under stress. The right amount of water must be supplied to the plant at the right time. Accordingly, an efficient method of watering must be selected so that the plant can produce a higher yield under different conditions including stress (Bacon, 2009; Sinclair et al., 1984).

Plants are also able to use other techniques to alleviate stress; for example, under drought and salinity stress, plants are able to produce certain organic products, such as proteins, which are able to adjust plant water potential and hence make the plant absorb water at a higher rate and so survive. Such products are also able to regulate water potential at the cellular level and hence enable the water to circulate at different cellular parts (Showalter, 1993).

Providing plants with sufficient amounts of nutrients can also make plant growth more efficient under different conditions such as stress. If the balanced rate of nutrients is used for the growth of the plant, it can produce higher yield and survive the stress more efficiently. The important point about balanced fertilization is the combined use of organic and chemical fertilization. Using only chemical fertilization is not recommendable environmentally and economically. Hence, organic fertilization, including the use of organic products such as compost, vermincompost, plant residue, and microbial (biological) fertilization is a suitable method to provide plants with their nutrients. Organic fertilization can increase plant growth by: (1) providing nutrients, (2) enhancing the properties of soil, (3) increasing the microbial population, (4) being suitable for the environment, etc. However, the appropriate rate of combined organic and chemical fertilization must be determined (Miransari, 2011a).

Use of microbial inocula is also an efficient method for providing plants with their nutrients. Microbial inocula can be used as a source of biological fertilization for plant growth and yield production. Microbial inocula include the use of soil microbes such as arbuscular mycorrhiza (AM) fungi and plant growth-promoting rhizobacteria (PGPR). Such microbes are able to establish a symbiotic or nonsymbiotic association with their host plants and increase the uptake of water and nutrients by their host (Miransari, 2011a).

With organic farming, healthier and more nutritious crop yield is produced, and the method is also friendly to the environment. In such a method, plant nutrients are supplied by using organic fertilization, and chemicals are either not used or are used at a minimum level. For example, weeds are controlled by tillage under organic farming. However, due to the use of such techniques, the prices of organic products are higher than those of conventionally produced products (Rigby and Caceres, 2001). More details are presented in the next section.

Some biochemicals are able to affect plant behavior under different conditions including stress. For example, Miransari and Smith (2007, 2008, 2009) found that under stress the early stages of symbiotic association between the legume host plant, soybean, and the related rhizobial bacterium, *B. japonicum*, are adversely affected including the production of biochemicals such as genistein by plant roots. Such a molecule is able to induce genetic modification in *B. japonicum* and hence enable the bacteria to detect the presence of the host plant and establish a symbiotic association with it. Under stress the production of such molecules and the rate of symbiosis decreases. To alleviate such a stress Miransari and Smith (2007, 2008, 2009) preincubated the inocula of *B. japonicum* with genistein and found that such treatment is able to alleviate the stress and hence increase the rate of N fixation and soybean yield production under greenhouse and field conditions.

Soybean organic production

The two important issues related to organic farming are: (1) the benefits of farming, and (2) the environment related issues. Farmers must try to produce crops in such a way that they can maximize benefits and they must also use techniques for crop production that are friendly to the environment. This shows the importance of using modern strategies and techniques for the production of crop plants (Robertson et al., 2014). Accordingly, the importance of agriculture applies not only to the production of yield, but to providing clean and healthy water, ensuring beneficial conservational and recreational aims, and biodiversity and stabilization of the climate.

According to the 2012 data of the United Nations, organic agriculture was practiced by 164 nations of the world and among them 88 nations have legislation for such activities. The market size was 63.8 billion US dollars. However, there are some important questions which must be addressed including: (1) Who consumes the majority of organic products and how much do they spend? (2) What is the reason for the consumption of organic products? (3) Are they really healthier? (4) If they are, from which point of view? (5) In what way are organic products preferable to the products of conventional cropping systems? (6) What is the sustainability of organic products? (7) Are agricultural fields more sustainable than nonagricultural fields? (8) How does organic agriculture affect the climate? (Rahmann and Aksoy, 2014).

Research is able to address such questions. Accordingly, the importance of science lies not only in its merit, but also in addressing the questions that can enhance the life level. For example, how are the properties of soil, such as its biology and fertility, enhanced under dry climates. How might the rate of yield be increased under different conditions including stress by using strong varieties, planting organic, increasing the efficiency of crop production, etc.

Competiveness, sustainability, and diversity are the results of research and innovations that can be spoken about in different parts of the world by researchers. However, to develop modernization in the production of crop by using organic farming, etc., more must be spent, relative to conventional crop production. The congress of IFOAM Organic World has been a place for the discussion of the related subjects for the exchange of knowledge and experience by specialists and practitioners.

Such a congress can bridge between specialists and practitioners, as well as institutions of the world, more and less developed nations, rural and urban, and producers and consumers. Although organic farming is practiced worldwide, less than 1% of fields globally are planted organically. The most important organic markets (94%) (and the related research) are located in the Western world including North America, Europe, and Australia at 2,790,162, 10,637,128, and 12,001,724 ha, respectively. Nevertheless, many challenges must be addressed to make organic farming popular worldwide. Europe devoted less than 1% of its expenditure on the development of organic farming, relative to conventional crop production.

Soybean organic production is a fine method of producing healthy and nutritious food. Although the plant is able to acquire a high amount of N by the process of N fixation, the remaining part of the N requirement and the other nutrients must be supplied by chimerical and/or organic fertilization. The organic production of soybean necessarily requires the use of organic products, which can provide the essential nutrients for the use of plant and keep the plant healthy. Accordingly, healthy food is produced for the use of humans, which is also friendly to the environment.

Biological fertilization

Extensive research has indicated the benefits and importance of organic farming in which chemicals are not used and only organic products are deployed. As a result, the constituents of soil such as microbes and organic manure improve and enable in higher crop production. The benefits of organic farming are not only the production of healthy products but also it can favorably affect the environment and the climate. Although the price of organic products may be higher than that of nonorganic products, due to the benefits of consumption of the former, their production is highly recommendable. Among the advantages of organic farming are its favorable effects on root growth, which hence its interactions with microbes can also be positively affected (Stockdale et al., 2001).

The proper use of ecological knowledge can be a suitable method for the reduction of chemicals which are not friendly to the environment, especially at high levels. The related parameters include the use of organic fertilization, symbiotic plants, and biological methods for the reduction of pathogens, etc. Biological fertilization is among the most recommendable methods in organic farming, which is the use of soil microbes such as AM fungi and PGPR including the N-symbiotic rhizobia, which are able to be in association with their host plant symbiotically or nonsymbiotically (Diskson and Foster, 2011; Miransari, 2011a; Wagg et al., 2011).

AM fungi are soil microbes that are able to develop a symbiotic association with their host plant and increase the uptake of water and nutrients by their host plant through the development of an extensive network of hyphae, under different conditions including stress. Phosphorus is among the most important macronutrients, and is absorbed at a higher rate when the fungi establish a symbiotic association with their host plant. The production of different enzymes such as phosphatases make it possible to increase the solubility of formerly insoluble phosphorus. The fungi are also able to enhance the uptake of micronutrients by the host plant. The marked ability of the fungi to enhance the uptake of water and nutrients by the host plant is the most important feature for enabling the host plant to survive under stress. This may be the most important reason for the use of AM fungi as a biological fertilizer, especially under stress conditions (Miransari, 2010a, b; Wang et al., 2011).

PGPR including rhizobia are soil microbes that are able to increase the uptake of nutrients by the host plant symbiotically or nonsymbiotically. The nonsymbiotic microbes can enhance the solubility of nutrients in the rhizosphere of the host plant by the production of different enzymes. They can also enhance plant growth by the following: (1) alleviation of soil stresses (Miransari, 2011b, c, 2014a, b), (2) availability and solubility of nutrients (Abbas-Zadeh et al., 2010; Miransari 2010a, 2013), (3) production of different enzymes (Berg, 2009; Glick, 2012), (4) plant hormone production (Jalili et al., 2009), (5) controlling the activity of pathogens (Bailly and Weisskopf, 2012; Nielsen et al., 2002) (6) interactions with the other microbes (Bonfante and Anca, 2009; Miransari, 2011b, c, 2014a, b), and (7) rhizoxin production (Miransari, 2014c).

As previously mentioned, the use of plants, such as legumes, which are able to establish a symbiotic association with rhizobia, is also a useful ecological method to decrease the use of chemicals and hence it is favorable to the environment. Such a symbiotic association is specific and usually there is only one strain of rhizobia that is able to establish the symbiosis with their host plant. The use of biological methods, such as PGPR, is also suitable for controlling the pathogens. Such bacteria are able to produce some biochemicals that are not favorable to the pathogens, and hence can reduce their population and activity (Glick, 2012).

Wang et al. (2011) investigated the effects of coinoculation with *B. japonicum* and AM fungi on the growth of two soybean varieties (with different root architecture). They found that mycorrhizal symbiosis affected the root growth of soybean with a deep-root system when the amount of P was low; however, nodulation was higher at high levels of P, relative to the soybean with a shallow root system. Coinoculation with rhizobia and mycorrhiza increased plant growth even at low amounts of N and P; however, when the amount of N and P was high the effects of inoculation on the growth of soybean were not significant. The effects of inoculation with mycorrhiza and rhizobia were more effective on the growth of soybean with deeper roots (Wang et al., 2011).

Use of legumes

Jensen et al. (2012) reviewed the following important details related to the great advantages of legumes. They indicated that legumes can: (1) reduce the rate of greenhouse gas production including CO_2 and N_2O, in relation to the use of chemical N fertilization (Table 13.3), (2) decrease the use of fossil energy for the production of forage and food, (3) increase the rate of soil carbon, and (4) be used as a source of biomaterial for the production of biofuels using biorefinery (Figure 13.1).

They accordingly estimated that legumes are able to fix 33–46 Tg N each year, which can contribute to the production of 350–500 Tg CO_2. However, the production of 100 Tg fertilizer N results in the production of 300 Tg CO_2 per year with the difference that the CO_2 produced by biological N fixation is respired by the root nodules as a result of the photosynthesis process and the CO_2 produced by production of fertilizer N is a result of fossil fuels (Figure 13.2).

The production of N_2O from different fields has been in the range of 1.20–3.22 kg ha^{-1} (Table 13.4), which is a function of different soil, climate, and plant parameters.

Table 13.3 **Production of N$_2$O, normalized by the optical density (o.d.) of different strains of *Bradyrhizobium* spp., used for different forage and grain legumes (Alves et al., unpublished data)**

Strains of *Bradyrhizobium*	Host species of legume	Optical density	Flux of N$_2$O (mmol N$_2$O h^{-1} o.d.$^{-1}$)
BR 2003/2811	*Crotalaria* sp.	0.72	0.002
BR 446	*Stylosanthes* sp.	0.87	1.13
BR 1435	*Arachis* sp.	0.59	0.42
BR 85	*G. max*	0.86	0.02
BR 86	*G. max*	0.98	0.49

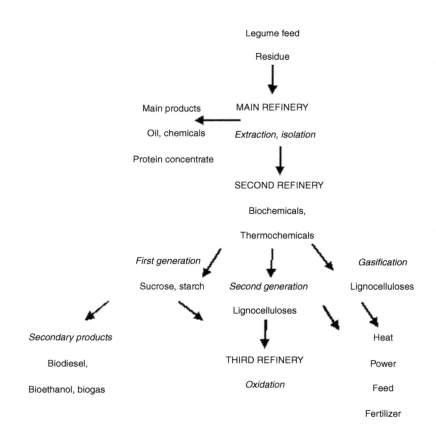

Figure 13.1 The production from a legume biorefinery (Jensen et al., 2012).

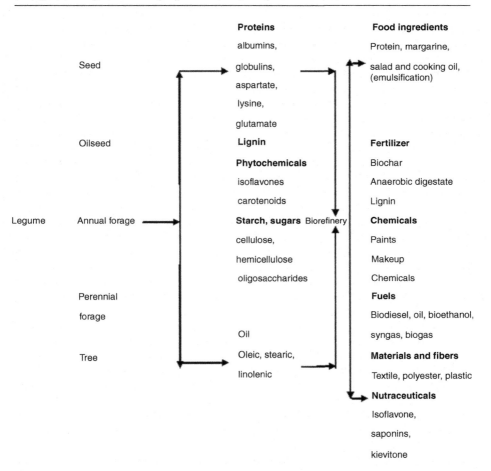

Figure 13.2 Legume ingredients and their use in biorefinery.

The other important role of legumes is that they are the sources of raw industrial materials, which are at present produced from petroleum sources, and can be used to produce food additives, surfactants, and pharmaceuticals. It is interesting that N fertilization, which is essential for the production of other plants, is not essential for the production of legumes (Jensen et al., 2012).

An important topic regarding legumes is their use in biorefinery, which can be a suitable replacement for the present fossil fuels in different parts of industry (Fairley, 2011). Such new uses of biomass can be suitable for: (1) rural development, (2) reduced use of fossil fuels, and (3) reduced production of greenhouse gases, which result in the mitigation of global climate change. The increased concentration of gases such as carbon dioxide (CO_2), ozone (O_3), methane (CH_4), nitrous oxide (N_2O), and chlorofluorocarbons (CFCs) in the atmosphere, results in global climate change. As a result of increased atmospheric concentrations of greenhouse gases, global heat increases and has been predicted to increase by 2°C by the year 2100.

Table 13.4 **The amount of nitrogen per 1000 kg of carbon, and the related C/N ratio, for different crop plants (arial residues), and components of soil**

Plant residues					
Nonlegumes	Maize	Wheat	Rice	Cotton	Canola
Carbon (C)	1000	1000	1000	1000	1000
Nitrogen (N)	21.4	19.6	9.5	38.5	23.8
C/N ratio	47:1	51:1	105:1	26:1	42:1
Legumes	Vetch	Alfalfa	Clover	Faba bean	Field pea
	1000	1000	1000	1000	1000
	100	73.2	83.3	71.4	28.6
	10:1	14:1	12:1	14:1	35:1
Different components of soil	Microbial biomass	Bacteria	Fungi	Organic carbon (soil)	
	1000	1000	1000	1000	
	135	178.5	106	83.3	
	7:1	6:1	9:1	12:1	

From: Jensen et al. (2012).

More than 50% of the concentration of greenhouse gases is related to the production of CO_2, produced by fossil fuels of which 13.5% is related to agriculture including the following activities: (1) the use of fossil fuels for the production of N fertilizer, other chemicals, and agricultural machinery, (2) mineralization of soil organic matter and deforestation, (3) the production of N_2O from the soil, and (4) CH_4 produced by livestock activity, production of manure, and rice production (Table 13.5) (Crews and Peoples, 2004). To decrease the production rate of greenhouse gases, such activities must be adjusted.

Use of leguminous plants can have the following benefits: (1) as a source of protein food, (2) reducing the use of N-chemical fertilization by the process of N fixation and as a source of plant manure, (3) providing better soil structure, (4) directly affecting the soil microbes by enhancing their activities and controlling the activities of unfavorable soil microbes, (5) increasing the tolerance of soil to the chemicals, and hence increased diversity, (6) perennial legumes are able to decrease the rate of groundwater contamination and salinity by absorbing nitrate and salt, and (7) they are useful for the revegetation and reclamation of land (Crews and Peoples, 2004; Jensen et al., 2012; Osborne et al., 2010).

Agricultural systems

Some important issues and challenges related to the use of proper agricultural systems are: (1) the high use of chemical fertilization results in the contamination of the

Table 13.5 Constituents of some cereal and legumes on the basis of dry weight

Crop and component	Starch (%)	Other carbohydrates (C5, C6, a.o.) (%)	Fat (%)	Protein (%)	Lignin (%)
Grain					
Wheat	66	17	3	13	2
Maize	71	14	5	9	2
Soybean	15	14	21	41	6
Faba bean	42	21	1	31	9
Pea	55	18	2	25	6
Lupin	22	23	5	45	16
Above-ground biomass					
Pea residues	0	81	2	7	41
Wheat stubble	0	92	2	3	45
Alfalfa (after flowering)	2	72	3	20	31
Grass clover	2	62	4	22	20

From: Jensen et al. (2012).

environment, (2) the loss of soil as a result of using improper cropping techniques can have unfavorable consequences for the environment, (3) cropping intensification increases the production rate of greenhouse gases, which is not favorable to the atmosphere and the environment, (4) high pressure on the fields for crop production decreases their efficiency and hence adversely affects the environment, and (5) planning efficiently and proper cropping systems is a mandate for a bright future and for the following generations (Drinkwater et al., 1998; Giller, 2001; Timsina and Connor, 2001)

High use of chemical fertilization can adversely affect the environment as the extra amounts of nutrients such as N and P can contaminate different resources such as the ground and surface water. Such nutrients enhance the growth of aquatic plants such as algae and, as a result of decreased soluble oxygen in the water, the growth and activity of other aquatic organisms including microbes decrease. The high rate of nitrate in the water can also make it unsuitable for human use (Cheremisinoff, 2010).

As an important source of production, unsuitable use of soil results in its deficiency and erosion. Such erosion is detrimental to water storage structures such as dams, as a high rate of erosion decreases the efficieny of storage. The surface soil is the most fertile part of the soil and because of a high rate of cropping, it is subjected to erosion and hence the production efficiency of such fields for crop production decreases. Fertilization of soil with organic matter, use of biological fertilization, proper use of agricultural machinery, use of proper cropping systems, etc. can enhance the efficiency of fields for higher and longer-term production, while remaining environmentally robust (Zachar, 2011).

As previously mentioned, the high production of greenhouse gases as a result of agricultural intensification and improper use of cropping techniques, is not conducive to environment friendliness. Hence, it is important to use cropping techniques that are kind to the environment and result in the decreased production of greenhouse gases. High rates of chemical fertilization are also an important factor increasing the production of greenhouse gases, especially under high moisture conditions (Jensen et al., 2012).

There must not be high usage pressure placed on fields as their efficiency will decrease and with time they will be subjected to erosion, decreased levels of organic matter, unsuitable structure, a decreased population of microbes, etc. Hence, it is a good idea not to plant a field after a few seasons of cropping. Fields must be fertilized with organic and biological fertilization regularly and be tilled at an appropriate rate. In an efficient and productive cropping system, the previously mentioned issues must be considered so that the goal of providing food for the world's ever increasing population will be more achievable (Zachar, 2011).

Robertson and Hamilton (2014) investigated three different crop plants (wheat, soybean, and corn) using three different agricultural cropping systems including no-till, reduced input, and biologically based, over a 23-year experimental period from 1989 to 2012. Using long-term global research and experimental databases, they hypothesized (Robertson et al., 2014) that a knowledge of ecology can be used to substitute for the use of most chemicals in highly productive and intensively managed fields.

In a no-till cropping system, no agricultural machinery is used and seeds are planted directly into the field. Under no-till, as a result of less mineralization of organic carbon, the sequestration of carbon in the soil increases and the consumption of fossil fuels decreases; however, use of more herbicide is necessary and the production of the potent greenhouse gas N_2O increases (Van Kessel et al., 2013).

The higher amount of soil organic matter can also cause the soil to retain a higher rate of moisture and hence to be more resistant under stresses such as drought and salinity. Use of no-till is also useful for decreasing the production of sediments as a result of erosion and decreased phosphorus. As previously mentioned, in such situations extra amounts of nitrate in the water can also be unfavorable to the environment. Under no-till, a stronger structure of soil as a result of higher levels of organic matter decreases the rate of erosion and levels of nitrate (Robertson et al., 2014; Six et al., 2004).

The use of winter cover crops, which are eventually cultivated into the soil, is also an effective practice to increase the efficiency of cropping systems, resulting in the following: (1) higher levels of soil organic matter, (2) higher microbial activities, (3) a suitable substitute for chemicals such N fertilizer and herbicide, (4) stronger structure of soil, (5) higher levels of soil moisture, and (6) less production of greenhouse gases such as CO_2 (Robertson et al., 2014; Six et al., 2004; Smith et al., 2008).

In a reduced input cropping system, the use of chemicals is greatly decreased and cover crop is used to provide the important element of N fertilization for the use of crop plant, and weeds are controlled by mechanical methods. Due to the positive

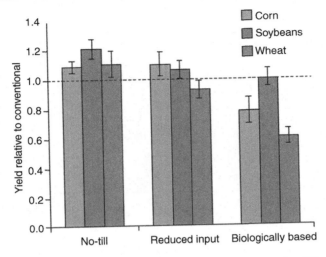

Figure 13.3 Grain yield under three different agricultural systems (no-till, reduced input, biologically based) related to the use of conventional practice (dotted line) in a 23-year-period (1989–2012).
Adopted from Robertson et al. (2014). With kind permission from Oxford University Press, license number: 3539371486324.

effects of a reduced input cropping system on the environment, less N is leached, the production rate of greenhouse gases decreases, and the quality of ground water is better relative to a conventional cropping system. Accordingly, the population and activities of soil microbes and the efficiency and productivity of the field increase (Robertson et al., 2014).

Figure 13.3 illustrates the yield of corn, soybean, and wheat (relative to the conventional cropping) using three different cropping systems including no-till, reduced input, and biologically based, over a 23-year research period (1989–2012) at the Kellogg Biological Station, USA. It is evident that the highest grain yield is associated with the no-till system, followed by reduced input, and biologically based. Under no-till practice the highest yield was associated with soybean and the yield of corn and wheat were almost at the same level. The most important reasons for the higher yield may be that with no-till practice, there is no disturbance to the soil and hence soil microbes including AM fungi and PGPR can be more active and grow at a faster rate.

Under such conditions the levels of soil organic matter are also higher and the mineralization rate is less and, as a result, there is more moisture in the soil. Soybean can develop a symbiotic association with rhizobia; Figure 13.3 shows that no-till is a suitable cropping system for soybean and its symbiotic rhizobia. Similarly, soybean gave the highest grain yield under the biologically based cropping system.

However, the growth of corn and wheat may also be similarly affected by the activities of PGPR and the higher levels of organic matter. N deficiency is the main cause for the decreased level of wheat yield under a reduced input cropping system. Soybean obtained its N by the process of N fixation and corn was planted after a cover crop

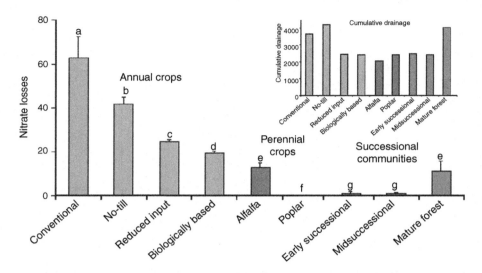

Figure 13.4 The rate of nitrate leaching (kg NO₃/hectare/year) and cumulative drainage (mm) from 1995 to 2006.
Adopted from Robertson et al. (2014). With kind permission from Oxford University Press, license number: 3539371486324.

with the ability to fix N. Under a biologically based cropping system, N deficiency was more evident because N fertilizer was not used. Although grain yield is lower under a biologically based practice, it can have other benefits which compensate for the decreased yield under such conditions including a cleaner environment, lower production of greenhouse gases, and a higher population and activity of soil microbes.

Figure 13.4 shows the effects of different cropping systems and crop plants affecting NO_3 leaching. It is evident that cropping practices can significantly affect the production of NO_3 with the minimum amounts of nitrogen per hectare in the biologically based (19 kg) and reduced input systems (24 kg), and maximum nitrogen per hectare in the no-till (42 kg) and conventional (62 kg) cropping systems (Syswerda et al., 2012). Although the properties of soil affect nitrate leaching, the presence of cover crop in the field is the most important parameter affecting nitrate leaching. During the growth of plants in the field, more N is absorbed by plants and more water is evaporated from the field and hence less N is leached (McSwiney et al., 2010).

The amount of greenhouse gas produced by agriculture is in the range of 10–14%, which is due to the production of N_2O by soil and by manure and the production of methane from crop residue. However, activities such as expansion of agriculture and use of chemicals can also increase the production of greenhouse gases and raise the total production by agriculture to 26–36% (Robertson et al., 2014).

According to United Nations databases, in 1980, and with a population of 4.4 billion, the rural population was 1.53 times greater than the urban population meaning that the number of people who were producing was greater than the number of people who were consuming. However, in 2015, the related rate was 0.85, indicating that the number of people who are producing is lower than the number of people who are consuming. Accordingly, the relative rates for different parts of the world corresponds roughly to the following: in developed countries, 0.27; in less developed countries, 1.05; and in the least developed regions, 2.30 (Rahmann and Aksoy, 2014).

Alleviating challenges

Selecting the appropriate strategy which is both efficient and effective, is one of the most important challenges for the alleviation of stress on and production of soybean in high yields. With respect to parameters including the properties of plant, soil, climate, etc. the optimal strategy must be selected and used. Such a strategy will give the desired results if the associated conditions are evaluated properly and precisely.

The properties of plants indicate their responsiveness to the environment and hence their growth and yield can be determined. Plants must be tested under different conditions and their responses must be evaluated. The most useful evaluations are long-term experiments conducted under field conditions although greenhouse research work is also very useful. According to the obtained results, it is possible to determine how the plant may be responsive under specific conditions and thus the appropriate mitigating strategies can be determined. However, such a course is always challenging, as finding and implementing such strategies may not be easy.

Conclusions and perspectives for future research

In this chapter some of the most important aspects related to strategies and challenges that affect the production of soybean under stress have been reviewed and analyzed. As an N fixer, *B. japonicum* is able to establish symbiotic association with its host plant, soybean. Stress can adversely affect the growth and activities of both symbionts. Hence, it is important that the response of both the bacterium and the host plant be tested under stress and accordingly the suitable alleviating strategies selected and implemented. Among the most applicable strategies and techniques for the alleviation of stress on the growth and yield of soybean are the use of: (1) tolerant varieties and strains of soybean and bacteria, respectively, (2) genetic and molecular techniques, and (3) biochemical and organic products including the use of organic farming. The use of such farming is a suitable method for producing more healthy products; however, relative to the conventional method, it is more expensive. Organic farming is friendly to the environment and can decrease the pressure of crop production on the field. Decreased production of greenhouse gases is also an important aspect, which must be considered under different agricultural systems. Using long-term data is one of the most useful strategies and techniques for evaluating the response of plants to different parameters including stress. The specifics must be pinpointed, analyzed, and put to the test so that the right cropping system can be planned for and used. Using such alleviating strategies is challenging and must be done with precision and care. Among the most important considerations for future research is finding the response of soybean to stress under organic farming systems using different varieties and strains of soybean and *B. japonicum*, respectively.

References

Abbas-Zadeh, P., Saleh-Rastin, N., Asadi-Rahmani, H., Khavazi, K., Soltani, A., Shoary-Nejati, A., Miransari, M., 2010. Plant growth-promoting activities of fluorescent pseudomonads, isolated from the Iranian soils. Acta Physiol. Plant. 32, 281–288.

Abi-Ghanem, R., Carpenter-Boggs, L., Smith, J., 2012. Cultivar effects on nitrogen fixation in peas and lentils. Biol. Fertil. Soils 47, 115–120.

Adesemoye, A., Klopper, J., 2009. Plant–microbes interactions in enhanced fertilizer-use efficiency. Appl. Environ. Microbiol. 85, 1–12.

Alexandre, A., Oliveira, S., 2011. Most heat-tolerant rhizobia show high induction of major chaperone genes upon stress. FEMS Microbiol. Ecol. 75, 28–36.

Aulakh, M., Khera, T., Doran, J., Bronson, K., 2001. Denitrification, N_2O and CO_2 fluxes in rice–wheat cropping system as affected by crop residues, fertilizer N and legume green manure. Biol. Fertil. Soils 34, 375–389.

Bacon, M., 2009. Water Use Efficiency in Plant Biology. Wiley, England, 344 pp.

Bailly, A., Weisskopf, L., 2012. The modulating effect of bacterial volatiles on plant growth. Plant Signal. Behav. 7, 1–7.

Beijerinck, M.W., 1901. On oligonitrophilic microbes. Centralb. Bakteriol. Parasitenkd. Infektionskr. Hyg. Abt. II 7, 561–582.

Berg, G., 2009. Plant–microbe interactions promoting plant growth and health: perspectives for controlled use of microorganisms in agriculture. Appl. Microbiol. Biotechnol. 84, 11–18.

Bergersen, F.J., 1971. Biochemistry of symbiotic nitrogen fixation in legumes. Annu. Rev. Plant Physiol. 22, 121–140.

Blumel, M., Dally, N., Jung, C., 2015. Flowering time regulation in crops – what did we learn from Arabidopsis? Curr. Opin. Biotechnol. 32, 121–129.

Bonfante, P., Anca, I.A., 2009. Plants, mycorrhizal fungi, and bacteria: a network of interactions. Annu. Rev. Microbiol. 63, 363–383.

Chen, S., Huang, Y., Zou, J., 2008. Relationship between nitrous oxide emission and winter wheat production. Biol. Fertil. Soils 44, 985–989.

Cheremisinoff, N.P., 2010. Handbook of Water and Wastewater Treatment Technologies. Elsevier, USA, 576 pp.

Clement, M., Lambert, A., Herouart, D., Boncompagni, E., 2008. Identification of new up-regulated genes under drought stress in soybean nodules. Gene 426, 15–22.

Craine, J., Nippert, J., Elmore, A., Skibbe, A., Hutchinson, S., Brunselld, N., 2012. Timing of climate variability and grassland productivity. Proc. Natl Acad. Sci. USA 109, 3401–3405.

Crews, T.E., Peoples, M.B., 2004. Legume versus fertilizer sources of nitrogen: ecological tradeoffs and human needs. Agric. Ecosyst. Environ. 102, 279–297.

de Ponti, T., Rijk, B., van Ittersum, M., 2012. The crop yield gap between organic and conventional agriculture. Agric. Syst. 108, 1–9.

Denarie, J., Cullimore, J., 1993. Lipo-oligosaccharide nodulation factors: a minireview new class of signaling molecules mediating recognition and morphogenesis. Cell 74, 951–954.

Diskson, T., Foster, B., 2011. Fertilization decreases plant biodiversity even when light is not limiting. Ecol. Lett. 14, 380–388.

Drinkwater, L.E., Wagoner, P., Sarrantonio, M., 1998. Legume-based cropping systems have reduced carbon and nitrogen losses. Nature 396, 262–265.

Du, W., Xu, Y., Liu, D., 2003. Lipase-catalysed transesterification of soya bean oil for biodiesel production during continuous batch operation. Biotechnol. Appl. Biochem. 38, 103–106.

Esfahani, M., Mostajeran, A., 2011. Rhizobial strain involvement in symbiosis efficiency of chickpea–rhizobia under drought stress: plant growth, nitrogen fixation and antioxidant enzyme activities. Acta Physiol. Plant. 33, 1075–1083.

Fairley, P., 2011. Next generation biofuels. Nature 474, S2–S5.

FAO, 2009. FAOSTAT. Food and Agriculture Organization of the United Nations, Rome, Italy. Available from: http://faostat.fao.org.

FAOSTAT, 2014. http://FAOSTAT3.FAO.ORG.

Galili, G., Amir, R., 2013. Fortifying plants with the essential amino acids lysine and methionine to improve nutritional quality. Plant Biotechnol. J. 11, 211–222.

Giller, K., 2001. Nitrogen Fixation in Tropical Cropping Systems. Cabi Publishers, UK, 423 pp.

Gil-Quintana, E., Larrainzar, E., Seminario, A., Diaz-Leal, J., Alamillo, J., Pineda, M., Arrese-Igor, C., Wienkoop, S., Gonzalez, E., 2013. Local inhibition of nitrogen fixation and nodule metabolism in drought-stressed soybean. J. Exp. Bot. 64, 2171–2182.

Glick, B.R., 2012. Plant growth-promoting bacteria: mechanisms and applications. Scientifica 963401.

Graham, P.H., Vance, C.P., 2000. Nitrogen fixation in perspective: an overview of research and extension needs. Field Crops Res. 65, 93–106.

Graham, P.H., Vance, C.P., 2003. Legumes: importance and constraints to greater use. Plant Physiol. 131, 872–877.

Hénault, C., Devis, X., Lucas, J.L., Germon, J.C., 1998. Influence of different agricultural practices (type of crop, form of N-fertilizer) on soil nitrous oxide emissions. Biol. Fertil. Soils 27, 299–306.

Indrasumuna, A., Searle, I., Lin, M.-H., Kereszt, A., Men, A., Carroll, B., Gresshoff, P., 2011. Nodulation factor receptor kinase 1α controls nodule organ number in soybean (*Glycine max* L. Merr.). Plant J. 65, 30–50.

Jalili, F., Khavazi, K., Pazira, F., Nejati, A., Rahmani, H.A., Sadaghiani, H.R., Miransari, M., 2009. Isolation and characterization of ACC deaminase-producing fluorescent pseudomonads, to alleviate salinity stress on canola (*Brassica napus* L.) growth. J. Plant Physiol. 166, 667–674.

James, C., 2009. ISAAA Brief No 41. Global Status of Commercialized Biotech/GM Crops. ISAAA, New York.

Jensen, E., Peoples, M., Boddey, R., Gresshoff, P., Hauggaard-Nielsen, H., J.R. Alves, B., Morrison, M., 2012. Legumes for mitigation of climate change and the provision of feedstock for biofuels and biorefineries. Agron. Sustain. Dev. 32, 329–364.

Jung, G., Matsunami, T., Oki, Y., Kokubun, M., 2008. Effects of water logging on nitrogen fixation and photosynthesis in supernodulating soybean cultivar Kanto 100. Plant Product. Sci. 11, 291–297.

Kaneko, T., Nakamura, Y., Sato, S., Minamisawa, K., Uchiumi, T., Sasamoto, S., Watanabe, A., Idesawa, K., Iriguchi, M., Kawashima, K., Kohara, M., Matsumoto, M., Shimpo, S., Tsuruoka, H., Wada, T., Yamada, M., Tabata, S., 2002a. Complete genomic sequence of nitrogen-fixing symbiotic bacterium *Bradyrhizobium japonicum* USDA110. DNA Res. 9, 189–197.

Kaneko, T., Nakamura, Y., Sato, S., Minamisawa, K., Uchiumi, T., Sasamoto, S., Watanabe, A., Idesawa, K., Iriguchi, M., Kawashima, K., Kohara, M., Matsumoto, M., Shimpo, S., Tsuruoka, H., Wada, T., Yamada, M., Tabata, S., 2002b. Complete genomic sequence of nitrogen-fixing symbiotic bacterium *Bradyrhizobium japonicum* USDA110 (supplement). DNA Res. 9, 225–256.

Kaneko, T., Maita, H., Hirakawa, H., Uchiike, N., Minamisawa, K., Watanabe, A., Sato, S., 2011. Complete genome sequence of the soybean symbiont *Bradyrhizobium japonicum* strain USDA6T. Genes 2, 763–787.

Kohler, J., Antonio Hernández, J., Caravaca, F., Roldán, A., 2008. Plant-growth-promoting rhizobacteria and arbuscular mycorrhizal fungi modify alleviation biochemical mechanisms in water-stressed plants. Funct. Plant Biol. 35, 141–151.

Lambin, E., Meyfroidt, P., 2011. Global land use change, economic globalization, and the looming land scarcity. Proc. Natl. Acad. Sci. USA 108, 3465–3472.

Laranjo, M., Oliveira, S., 2011. Tolerance of *Mesorhizobium* type strains to different environmental stresses. Antonie van Leeuwenhoek 99, 651–662.

Lavania, D., Siddiqui, M., Al-Whaibi, M., Singh, A., Kumar, R., Grover, A., 2015. Genetic approaches for breeding heat stress tolerance in faba bean (*Vicia faba* L.). Acta Physiol. Plant. 37, 1737.

Li, Q., Wang, E., Zhang, Y.Z., Zhang, Y.M., Tian, C., Sui, X., Chen, W.F., Chen, W.X., 2011. Diversity and biogeography of rhizobia isolated from root nodules of *Glycine max* grown in Hebei province China. China Microb. Ecol. 61, 917–931.

Liao, Y., Zhang, Y., Chen, S., Zhang, W., 2008. Role of soybean GmbZIP132 under abscisic acid and salt stresses. J. Integr. Plant Biol. 50, 221–230.

Lindström, K., Murwira, M., Willems, A., Altier, N., 2010. The biodiversity of beneficial microbe-host mutualism: the case of rhizobia. Res. Microbiol. 161, 453–463.

Long, S., 2001. Genes and signals in the *rhizobium*-legume symbiosis. Plant Physiol. 125, 69–72.

Mabood, F., Souleimanov, A., Khan, W., Smith, D.L., 2006. Jasmonates induce Nod factor production by *Bradyrhizobium japonicum*. Plant Physiol. Biochem. 44, 759–765.

Mahmood, T., Ali, R., Malik, K.A., Shamsi, S., 1998. Nitrous oxide emissions from an irrigated sandy-clay loam cropped to maize and wheat. Biol. Fertil. Soils 27, 189–196.

Manavalan, L., Guttikonda, S., Tran, L., Nguyen, H., 2009. Physiological and molecular approaches to improve drought resistance in soybean. Plant Cell Physiol. 50, 1260–1276.

McSwiney, C.P., Snapp, S.S., Gentry, L.E., 2010. Use of N immobilization to tighten the N cycle in conventional agroecosystems. Ecol. Appl. 20, 648–662.

Mir, R., Zaman-Allah, M., Sreenivasulu, N., Trethowan, R., Varshney, R., 2012. Integrated genomics, physiology and breeding approaches for improving drought tolerance in crops. Theor. Appl. Genet. 125, 625–645.

Miransari, M., 2010a. Contribution of arbuscular mycorrhizal symbiosis to plant growth under different types of soil stresses. Plant Biol. 12, 563–569.

Miransari, M., 2010b. Biological fertilization. In: Villas Mendez, A. (Ed.), Current Research, Technology and Education Topics in Applied Microbiology and Microbial Biotechnology, No. 2, Vol 1. Formatex Research Center, Badajoz, Spain, pp. 168–176.

Miransari, M., 2011a. Soil microbes and plant fertilization. Appl. Microbiol. Biotechnol. 92, 875–885.

Miransari, M., 2011b. Soil Microbes and Environmental Health. Nova Publishers, Hauppauge, NY.

Miransari, M., 2011c. Interactions between arbuscular mycorrhizal fungi and soil bacteria. Appl. Microbiol. Biotechnol. 89, 917–930.

Miransari, M., 2013. Soil microbes and the availability of soil nutrients. Acta Physiol. Plant. 35, 3075–3084.

Miransari, M., 2014c. Plant Growth Promoting Rhizobacteria. J. Plant Nutr. 37, 2227–2235.

Miransari, M. (Ed.), 2014a. Use of Microbes for the Alleviation of Soil Stresses, vol. 1. Springer, New York.

Miransari, M., (Ed.), 2014b. Use of Microbes for the Alleviation of Soil Stresses. In: Miransari, M. (Ed.), Alleviation of Soil Stress by PGPR and Mycorrhizal Fungi, vol. 2, Springer, USA.

Miransari, M., Smith, D.L., 2007. Overcoming the stressful effects of salinity and acidity on soybean [Glycine max (L.) Merr.] nodulation and yields using signal molecule genistein under field conditions. J. Plant Nutr. 30, 1967–1992.

Miransari, M., Smith, D.L., 2008. Using signal molecule genistein to alleviate the stress of suboptimal root zone temperature on soybean-Bradyrhizobium symbiosis under different soil textures. J. Plant Interact. 3, 287–295.

Miransari, M., Smith, D., 2009. Alleviating salt stress on soybean (Glycine max (L.) Merr.) – Bradyrhizobium japonicum symbiosis, using signal molecule genistein. Eur. J. Soil Biol. 45, 146–152.

Miransari, M., Riahi, H., Eftekhar, F., Minaie, A., Smith, D.L., 2013. Improving soybean (Glycine max L.) N$_2$-fixation under stress. J. Plant Growth Regul. 32, 909–921.

Müller, J., Boller, T., Wiemken, A., 2001. Trehalose becomes the most abundant non-structural carbohydrate during senescence of soybean nodules. J. Exp. Bot. 52, 943–947.

Munoz, N., Soria-Diaz, M., Manyani, H., Sanchez-Matamoros, R., Gil Serrano, A., Megias, M., Lascano, R., 2014. Structure and biological activities of lipochitooligosaccharide nodulation signals produced by Bradyrhizobium japonicum USDA 138 under saline and osmotic stress. Biol. Fertil. Soils 50, 207–215.

National Research Council, 1994. Biological Nitrogen Fixation: Research Challenges. National Academy Press, Washington, DC.

Naveed, M., Mehboob, I., Hussain, M., Zahir, Z., 2015. Perspectives of rhizobial inoculation for sustainable crop production. In: Kumar, A. (Ed.), Plant Microbes Symbiois: Applied Facets. Springer, New York, pp. 209–239.

Nielsen, T.H., Sørensen, D., Tobiasen, C., Andersen, J.B., Christeophersen, C., Givskov, M., Sørensen, J., 2002. Antibiotic and biosurfactant properties of cyclic lipopeptides produced by fluorescent Pseudomonas spp. from the sugar beet rhizosphere. Appl. Environ. Microbiol. 68, 3416–3423.

Nilsen, E., Orcutt, D., 1996. Physiology of Plants Under Stress. Abiotic Factors. CABI Publication, Wallingford, UK, 689 pp.

Osborne, C.S., Peoples, M.B., Janssen, P.H., 2010. Exposure of soil to a low concentration of hydrogen elicits a reproducible, single member shift in the bacterial community. Appl. Environ. Microbiol. 76, 1471–1479.

Priester, J., Ge, Y., Mielke, R., Horst, A., Moritz, S., Espinosa, K., Gelb, J., Walker, S., Nisbet, R., An, Y., Schimel, J., Palmere, R., Hernandez-Viezcas, J., Zhao, L., Gardea-Torres-dey, J., Holden, P., 2012. Soybean susceptibility to manufactured nanomaterials with evidence for food quality and soil fertility interruption. Proc. Natl Acad. Sci. USA 109, E2451–E2456.

Qiu, L.-J., Chang, R.-Z., 2010. The origin and history of soybean 2010. In: Singh, G. (Ed.), The Soybean: Botany, Production, Uses. CAB International, Wallingford, UK, pp. 1–23.

Rahmann, G., Aksoy, U. (Eds.), 2014. Proceedings of the Fourth ISOFAR Scientific Conference. 'Building Organic Bridges', at the Organic World Congress 2014, October 13–15, Istanbul, Turkey.

Rasmussen, P.E., Goulding, K.W.T., Brown, J.R., Grace, P.R., Janzen, H.H., Körschens, M., 1998. Long-term agroecosystem experiments: assessing agricultural sustainability and global change. Science 282, 893–896.

Rigby, D., Caceres, D., 2001. Organic farming and the sustainability of agricultural systems. Agric. Syst. 68, 21–40.

Robertson, G.P., Hamilton, S.K., 2014. Conceptual and experimental approaches to long-term ecological research at the Kellogg Biological Station. In: Hamilton, S.K., Doll, J.E., Robertson, G.P. (Eds.), The Ecology of Agricultural Ecosystems: Long-Term Research on the Path to Sustainability. Oxford University Press, UK, 448 pp.

Robertson, G.P., Gross, K., Hamilton, S., Landis, D., Schmidt, T., Snapp, S., Swinton, S., 2014. Farming for ecosystem services: as ecological approach to production agriculture. Bioscience 64, 404–415.

Ruz-Jerez, B.E., White, R.E., Ball, P.R., 1994. Long-term measurement of denitrification in three contrasting pastures grazed by sheep. Soil Biol. Biochem. 26, 29–39.

Sachs, J., Russell, J., Hollowell, A., 2011. Evolutionary instability of symbiotic function in *Bradyrhizobium japonicum*. PLoS ONE 6, 26370.

Sajedi, N.A., Ardakani, M.R., Rejali, F., Mohabbati, F., Miransari, M., 2010. Yield and yield components of hybrid corn (*Zea mays* L.) as affected by mycorrhizal symbiosis and zinc sulfate under drought stress. Physiol. Mol. Biol. Plants 16, 343–351.

Sajedi, N.A., Ardakani, M.R., Madani, H., Naderi, A., Miransari, M., 2011. The effects of selenium and other micronutrients on the antioxidant activities and yield of corn (*Zea mays* L.) under drought stress. Physiol. Mol. Biol. Plants 17, 215–222.

Salvagiotti, F., Cassman, K.G., Specht, J.E., Walters, D.T., Weiss, A., Dobermann, A., 2008. Nitrogen uptake, fixation and response to fertilizer N in soybeans: a review. Field Crops Res. 108, 1–13.

Schubert, K., 1986. Products of biological nitrogen fixation in higher plants: synthesis, transport, and metabolism. Annu. Rev. Plant Physiol. 37, 539–574.

Showalter, M., 1993. Structure and function of plant cell wall proteins. Plant Cell 5, 9–23.

Silvente, S., Sobolev, A.P., Lara, M., 2012. Metabolite adjustments in drought tolerant and sensitive soybean genotypes in response to water stress. PLoS ONE 7, e38554.

Sinclair, T., 2011. Challenges in breeding for yield increase for drought. Trends Plant Sci. 16, 289–293.

Sinclair, T., Goudriaan, J., 1981. Physical and morphological constraints on transport in nodules. Plant Physiol. 67, 143–145.

Sinclair, T., Tanner, C., Bennet, J., 1984. Water-use efficiency in crop production. BioScience 34, 36–40.

Six, J., Ogle, S., Breidt, F., Conant, R., Mosier, A., Paustian, K., 2004. The potential to mitigate global warming with no-tillage management is only realized when practised in the long term. Global Change Biol. 10, 155–160.

Smith, R., Gross, K., Robertson, G., 2008. Effects of crop diversity on agroecosystem function: crop yield response. Ecosystems 11, 355–366.

Stockdale, E.A., Lampkin, N.H., Hovi, M., Keatinge, R., Lennartsson, E.K.M., Macdonald, D.W., Padel, S., Tattersall, F.H., Wolfe, M.S., Watson, C.A., 2001. Agronomic and environmental implications of organic farming systems. Adv. Agron. 70, 261–262.

Syswerda, S.P., Basso, B., Hamilton, S.K., Tausig, J.B., Robertson, G.P., 2012. Long-term nitrate loss along an agricultural intensity gradient in the Upper Midwest USA. Agriculture. Ecosyst. Environ. 149, 10–19.

Timsina, J., Connor, D., 2001. Productivity and management of rice–wheat cropping systems: issues and challenges. Field Crops Res. 69, 93–132.

Van Kessel, C., Venterea, R., Six, J., Advento-Borge, M.A., Linquist, B., Van Groenigen, K.J., 2013. Climate, duration, and N placement determine N_2O emissions in reduced tillage systems: a meta-analysis. Global Change Biol. 19, 33–44.

Wagg, C., Jansa, J., Stadler, M., Schmid, B., van der Heijden, M., 2011. Mycorrhizal fungal identity and diversity relaxes plant–plant competition. Ecology 92, 1303–1313.

Wagner, S., 2011. Biological nitrogen fixation. Nat. Educ. Knowledge 3, 15.

Wang, X., Pan, Q., Chen, F., Yan, X., Liao, H., 2011. Effects of co-inoculation with arbuscular mycorrhizal fungi and rhizobia on soybean growth as related to root architecture and availability of N and P. Mycorrhiza 21, 173–181.

Wang, D., Yang, Y., Tang, F., Zhu, H., 2012a. Symbiosis specificity in the legume – rhizobial mutualism. Cell. Microbiol. 14, 334–342.

Wang, N., Khanm, W., Smith, D.L., 2012b. Changes in soybean global gene expression after application of lipo-chitooligosaccharide from *Bradyrhizobium japonicum* under suboptimal temperature. PLoS ONE 7, e31571.

You, J., Zhang, H., Liu, N., Gao, L., Kong, L., Yang, Z., 2011. Transcriptomic responses to aluminum stress in soybean roots. Genome 54, 923–933.

Zachar, D., 2011. Soil Erosion. Elsevier, The Netherlands, 544 pp. (e-publication).

Zhang, H., Jiao, H., Jaing, C., Wang, S., Wei, Z., Luo, J., Jones, R., 2010. Hydrogen sulfide protects soybean seedlings against drought-induced oxidative stress. Acta Physiol. Plant. 32, 849–857.

Zilli, C., Balestrasse, K., Yannarelli, G., Polizio, A., Santa-Cruz, D., Tomaro, M., 2008. Heme oxygenase up-regulation under salt stress protects nitrogen metabolism in nodules of soybean plants. Environ. Exp. Bot. 64, 83–89.

Index

Printed in the United States
By Bookmasters